Modern Techniques in Physiological Sciences

Modern Techniques in Physiological Sciences

Edited by

J. F. GROSS

Department of Chemical Engineering,
University of Arizona, Tuscon,
Arizona, U.S.A. *and*
The Rand Corporation, Santa Monica,
California, U.S.A.

R. KAUFMANN

Institute of Clinical Physiology,
University of Dusseldorf, Germany

E. WETTERER

Institute of Physiology, University
of Erlangen, Germany

1973

ACADEMIC PRESS · London and New York

A Subsidiary of Harcourt Brace Jovanovich, Publishers

ACADEMIC PRESS INC. (LONDON) LTD.
24/28 Oval Road,
London NW1

United States Edition published by
ACADEMIC PRESS INC.
111 Fifth Avenue
New York, New York 10003

Library of Congress Catalog Card Number: 73-9462
ISBN: 0-12-304450-2

Printed in Great Britain by
J. W. Arrowsmith Ltd., Bristol

Contributors

A. ANNÉ, Division of Biochemical Engineering, University of Virginia, Charlottesville, Virginia, U.S.A.

J.O. ARNDT, Bateilung für Experimentelle Anaesthesiologie der Universität Düsseldorf, Germany.

E.O. ATTINGER, Division of Biochemical Engineering, University of Virginia, Charlottesville, Virginia, U.S.A.

F.D. CARLSON, Department of Biophysics, Johns Hopkins University, Baltimore, U.S.A.

A. DÖRGE, Institute of Physiology, University of Munich, Germany.

JEAN-PIERRE DUMERY, Laboratoire de Biophysique, Université Paris, Faculté de Médicine, 94 Creteil, France.

GEORGE EISENMAN, Department of Physiology, UCLA, Los Angeles, California, U.S.A.

WALLACE G. FRASHER Jr., University of Southern California School of Medicine, Los Angeles, California, U.S.A.

PIERRE GALLE, Laboratoire de Biophysique, Université Paris, Faculté de Médicine, 94 Creteil, France.

K.H. GAUKLER, Institute of Applied Physics, University of Tübingen, Germany.

K. GEHRING, Institute of Physiology, University of Munich, Germany.

DAVID GLICK, Histochemistry Division, Pathology Department, Stanford University Medical School, Stanford, California, U.S.A.

LEON GOLDMAN, Laser Laboratory, Medical Center of the University of Cincinnati, U.S.A.

W. GRUNEWALD, Max-Planck-Institut für Arbeitsphysiologie, Dortmund, Germany.

J. GULLASCH, Siemens AG, Bereich Mess- und Prozesstechnik, Karlsruhe, Germany.

T.A. HALL, Cavendish Laboratory, Cambridge, England.

G. HAUCK, Institute of Physiology, University of Würzburg, Germany.

Contributors

F. HILLENKAMP, Gesselschaft für Strahlen- und Umweltforschung mbH. München, Germany.

H.J. HÖHLING, Institut für Medizinische Physik, Münster, Germany.

NILS KAISER, Max-Planck-Institut für Plasmaphysik, München, Germany.

R. KAUFMAN, Physiologisches Institut der Universität Freiburg/ Breisgau, Germany.

A.K. KLEINSCHMIDT, Department of Biochemistry, New York University School of Medicine, New York, U.S.A.

ALEXANDER KOLIN, University of California, School of Medicine, Los Angeles, California, U.S.A.

W. KRIZ, Anatomisches Institut, Münster, Germany.

H. LIEBL, Max-Planck-Institute for Plasma Physics, Munich, Germany.

D.W. LÜBBERS, Max-Planck-Institut für Arbeitsphysiologie, Dortmund, Germany.

R. STUART MACKAY, Boston University and Boston University Medical Center, Boston, Massachusetts, U.S.A.

C.M. MALPUS, Cardiovascular Unit, Department of Physiology, University of Leeds, Leeds, England.

W. NAGEL, Institute of Physiology, University of Munich, Germany.

E. REMY, Infratest-Industria, München, Germany.

V.C. RIDEOUT, University of Wisconsin, Madison, Wisconsin, U.S.A. and Institute of Medical Physics TNO, Utrecht, Holland.

A.P. v. ROSENSTIEL, Metaalinstituut TNO, Delft, The Netherlands.

ROBERT F. RUSHMER, Center for Bioengineering, University of Washington, Seattle, Washington, U.S.A.

T.T. SANDEL, Department of Psychology, Washington University, St. Louis, Missouri, U.S.A.

M. SCHALDACH, Department für Biomedizinische Technik, Friedrich Alexander Universität, Erlangen - Nürnberg, Germany.

Contributors

J. SCHNERMANN, Physiologisches Institut, München, Germany.

H. SCHRÖER, Institute of Physiology, University of Würzburg, Germany.

G. SPITELLER, Institute of Organic Chemistry, Göttingen, Germany.

K. THURAU, Institute of Physiology, University of Munich, Germany.

HAROLD WAYLAND, California Institute of Technology, Pasadena, California, U.S.A.

R. WODICK, Max-Planck-Institut für Arbeitsphysiologie, Dortmund, Germany.

U. ZESSACK, Institut für Medizinische Physik, Münster, Germany.

Preface

The rapid progress in the physical and engineering sciences, particularly in such fields as material testing, microanalysis, computer systems and laser technology, has revealed many new approaches for biomedical research. However, despite the methodological interest of these approaches biomedical scientists often have great difficulty in obtaining reliable, comprehensive information about them. At the same time manufacturers are often unaware of the specific needs of biomedical scientists, although in some cases only minor modifications are necessary to existing products to adapt them successfully.

The primary aim of this book is to bridge this information gap by publishing contributions from researchers who, in most cases, have already successfully applied new techniques to their own branch of biomedical science. The text requires only fundamental knowledge of the physical principles needed by scientists working in the physiological sciences. It is intended that the papers will be comprehensible to readers with no practical experience of the techniques described, nor even any prior information on the particular physical principles involved. This will encourage and help scientists to adapt unfamiliar techniques to their own research.

The editors have sought to provide as broad a coverage as possible in one volume. The book will provide a single and conventional source for workers in both the engineering and physiological sciences through which they may be made aware of their mutual interest in new and existing techniques.

July, 1973

J.F. GROSS
R. KAUFMANN
E. WETTERER

Contents

Contents

Contents

MODERN TECHNOLOGY

IN PHYSIOLOGICAL SCIENCES

ROBERT F. RUSHMER

*Center for Bioengineering, University of Washington,
Seattle, Washington, U.S.A.*

During the past quarter century, the nature and scope of
physiological research has been significantly broadened by increased availability of modern technology.   Traditional physiological research focused on the functional significance of
structure, the mechanisms of organ functions and the principles
of biological controls.   These were largely studied by subjec-
tive observation and mechanical measurements prior to World War
II and are now being subjected to much more comprehensive and
definitive measurements utilizing modern electronic equipment.
The new research tools greatly extend the human senses for collecting data and expand the human brain's capacity for analysis
of multiple simultaneous variables.

Traditional physiological research is exemplified by
studies of the relation between the structure and function of
cells, tissues and organs (Fig.1).   Comparative physiology has
opened wide the doors to the vast store of successful phylo-
genetic experiments by which functional requirements have been
met in the millions of different living species of animals in
the world.   Thus, lower forms of life are recognized as biological models providing insight into function of more complex
forms as exemplified by the squid axone, the guinea pig gut, the

RECENT TRENDS IN PHYSIOLOGICAL RESEARCH

TRADITIONAL PHYSIOLOGY		
FUNCTIONAL ANATOMY	ORGAN SYSTEM ANALYSIS	CONTROL SYSTEMS
1. Cells 2. Tissues 3. Comparative Physiology 4. Biological Models	1. Static Functions 2. Dynamic Responses 3. Physico-Chemical Properties 4. Integrated Functions 5. Systems Analysis 6. Simulation Models	1. Metabolic 2. Hormonal 3. Neural 4. Control System Analysis

QUANTITATIVE MICROSCOPY	DYNAMIC ANALYSIS	APPLIED HUMAN PHYSIOLOGY
1. Structure Electron Microscopy Scanning E—M Electron Microprobe Ion Microprobe Interferometry 2. Cytochemistry Radioautography Fluorescence Microscopy Spectrophotometry	Digital Computers Analogue Computers Hybrid Computers Adaptive Computers	Responses to Stress Environment Disease Processes Structural Functional Control

Fig.1 : Recent trends in physiology are illustrated schematically to indicate increased quantitative sophistication in the traditional aspects of physiology leading to complex analysis utilizing computers, improved technology for quantitative microscopy and development of monitoring techniques applicable to intact animals and man exposed to various types of stress.

frog heart and the muscle of barnacles. Quantitative observations of organ system function have been greatly extended by instrumentation with greatly improved performance characteristics as indicated below. The increased quantity of data has required new modes of analysis stemming from engineering including the use of simulation and mathematical modelling. Recently, studies of the control systems of the body have involved improved studies of metabolic, hormonal and neural interactions employing techniques of control systems analysis, tending toward experiments on unanesthetized cats, dogs, monkeys and man.

Major changes in physiological research have resulted from greatly improved performance characteristics of recording instruments. Virtually all of the attributes of measuring systems have been ungraded in the last twenty years. This change may be more readily appreciated by comparing the techniques available in 1945 and in 1970 for measuring pressure, dimensions, flow and chemical composition of blood with respect to the following criteria of quality and usefulness of recording equipment: (functional definitions for these terms are presented in reference 1).

A. Accuracy
1. directness
2. uniqueness
3. stability
4. reproduceability
5. non-reactance
6. range
7. sampling rate or dynamic response

B. Utility
1. applicability
2. selectivity
3. flexibility
4. versatility
5. convenience
6. portability
7. cost-effectiveness
   a. initial investment
   b. maintenance and operation

Each instrument or recording system has its own spectrum of advantages and disadvantages with reference to characteristics like those listed above. This means that there can be no "ideal" or general purpose instrument for any particular recorded variable. The best possible compromise must be sought to attain the optimum combination of characteristics for the specific experiment under consideration. If the investigator has a variety of instruments available with widely different characteristics, he has a better chance of finding one with the

necessary requirements than if he were limited to one or two.
Thus, a variety of different kinds of research tools and measuring instruments is of great advantage to the physiological investigator developing a quantitative experimental design.   He
also should be aware of the various options available in terms
of both types and models and their positive and negative features in order that he arrives at a proper selection of recording
systems on logical grounds.   Of equal importance is awareness
that the recording equipment for each variable must be considered as a total system.   For example, the most precise and
responsive transducer and amplifier combination might be totally
inadequate for a purpose if the recorder cannot follow rapid
changes in the function under study.   While all these properties have been vastly improved, probably the most dramatic have
been manifest in precision, versatility and sampling rates
(dynamic response).

The dynamic response of modern instruments has made it
possible to record and appreciate the significance of rates of
changes in biological functions.   The importance of phenomena
such as visco-elastic properties, creep, slopes, velocities,
accelerations and other aspects of time rate of change of properties or functions have attained a new appreciation with the
application of engineering concepts of measurement and analysis.
The versatility and convenience of modern instrumentation has
unleashed a flood of information in such quantities that traditional modes of analysis are no longer able to cope with the
multiple interacting variables.   The applications of various
forms of computers will be summarized in later chapters of this
volume.   (See Sandel, "The Laboratory Digital Computer in Biomedical Research Applications", Rideout, "The Use of Analog and
Hybrid Computers for Simulation and Modelling in Physiological
Research" and Attinger and Anné, "The Use of Hybrid Computer
Technology in Physiological Research".)

QUANTITATIVE MICROSCOPY

The scales for both time and space have been greatly extended through technology. We can observe phenomena occurring over brief periods and over very small dimensions. The functional significance of the fine structure of cellular components is now being probed by means of a whole spectrum of methods for obtaining quantitative measurements of the geometry and composition of exceedingly minute particles in cells. Included among these new data sources are the electron microscope, the scanning electron microscope for elucidating fine structure and the electron probe for precisely locating differences in concentration of substances at microscopic dimensions. All of these receive attention during the remainder of this volume. (See Hall and Höhling, "The Electron Microprobe" and Liebl, "The Ion Microprobe".) Interferometry provides an opportunity to assess accurately dimensions of cellular structures; radioautography provides information regarding the location of specific chemicals within cells and fluorescence techniques can identify location and rates of enzymatic reactions. These and associated techniques provide the necessary ingredients for quantitative microscopic analysis of the relationship between the functions of cells and their structural components. (See Wayland and Frasher, "Intravital Microscopy" etc.).

Ever increasing requirements for non-destructive testing of electronic equipment in industry has stimulated the development of many sensitive and reliable instruments for evaluating performance of complex and delicate mechanisms without inducing malfunction or damage. The applicability of this approach to the study of physiological functions with minimal distortion or damage is clear. However, a direct transfer of technology and equipment from industrial testing and quality control laboratories to physiological laboratories has not been possible in most instances because of the stringent conditions imposed by living systems. Living organisms at all levels of organization are exquisitely sensitive to changes in the environment and of stimulation of their tissues. In addition to physiological variability, it is necessary to reckon with a "biological uncertainty principle" that any recording device is apt to induce change in the variable being studied. The insertion into the body of transducers and conduits for picking up signals and transmitting them to suitable recorders must be considered as sources of functional artifacts even under the best of conditions. Microminiaturization (2) and chronic implantation have helped overcome some of these constraints but have imposed new requirements for materials which are biologically, chemically and physically compatible with reactive and delicate living tissues. A major convergence of interest has focused on biomechanics and biomaterials, generated by recognition of a need for new and improved substances from which to fabricate research tools, diagnostic instruments and therapeutic devices. Aspects of biomaterials requirements are considered in general terms by Schaldach and with reference to blood compatible materials by Nosé elsewhere in this volume.

MAMMALIAN PHYSIOLOGY: MONITORING REQUIREMENTS

A wide variety of "non-destructive" data gathering techniques has provided more "relevant" data regarding physiological functions in intact animals and man. Ideally, data should be collected under conditions approximating as closely as possible to the state in which the information is to be applied. A wide spectrum of sensitive and reliable recording devices has been developed for chronic implantation in animals for insertion into animals and man in catheters or needles. Information can be gleaned from internal organs with minimal disturbance to their baselines during spontaneous responses or compensation to imposed stresses. The long standing interest in physiological responses to exertion is being widely supplemented by growing concern about the physiological effects of the environment on the earth's surface, under water or in outer space. In addition, the effects of various disease processes are being increasingly considered in terms of abnormal or clinical physiology (Fig.1).

EXTERNAL MONITORING OF INTERNAL FUNCTION

The traditional concepts of physiological function and control of 30 years ago were based primarily on quantitative measurements performed on samples of excised tissues, perfused organs or on body components exposed in anesthetized animals during perturbations induced by the investigator (3). The advent of recording techniques, by which various critical variables could be continuously recorded from unanesthetized animals by means of implanted transducers, provided new opportunities to study the nature and magnitude of spontaneous changes. This has been repeatedly shown to differ significantly from corresponding observations on excised or exposed tissues and organs. For example, the heart functions at nearly maximal dimensions in intact animals at rest but shrinks to a much smaller size in anesthetized thoracotomized dogs. Neural reflexes induced experimentally (e.g. under anesthesia) need not correspond to the responses observed under more nearly normal conditions. Excised segments of arteries apparently distend much less than has been observed *in situ*. It is now becoming clear that normal spontaneous responses cannot be confidently predicted from controlled experiments which necessarily introduce distortions of function and control.

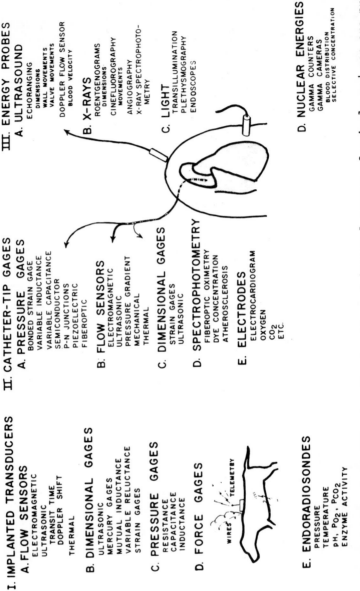

MONITORING FUNCTION OF INTERNAL ORGANS

I. IMPLANTED TRANSDUCERS

A. FLOW SENSORS
ELECTROMAGNETIC
ULTRASONIC
   TRANSIT TIME
   DOPPLER SHIFT
THERMAL

B. DIMENSIONAL GAGES
ULTRASONIC
MERCURY GAGES
MUTUAL INDUCTANCE
VARIABLE RELUCTANCE
STRAIN GAGES

C. PRESSURE GAGES
RESISTANCE
CAPACITANCE
INDUCTANCE

D. FORCE GAGES

E. ENDORADIOSONDES
PRESSURE
TEMPERATURE
pH, PO_2, PCO_2
ENZYME ACTIVITY

II. CATHETER-TIP GAGES

A. PRESSURE GAGES
BONDED STRAIN GAGE
VARIABLE INDUCTANCE
VARIABLE CAPACITANCE
SEMICONDUCTOR
P-N JUNCTIONS
PIEZOELECTRIC
FIBEROPTIC

B. FLOW SENSORS
ELECTROMAGNETIC
ULTRASONIC
PRESSURE GRADIENT
MECHANICAL
THERMAL

C. DIMENSIONAL GAGES
STRAIN GAGES
ULTRASONIC

D. SPECTROPHOTOMETRY
FIBEROPTIC OXIMETRY
DYE CONCENTRATION
ATHEROSCLEROSIS

E. ELECTRODES
ELECTROCARDIOGRAM
OXYGEN
CO_2
ETC.

III. ENERGY PROBES

A. ULTRASOUND
ECHORANGING
DIMENSIONS
WALL MOVEMENTS
VALVE MOVEMENTS
DOPPLER FLOW SENSOR
BLOOD VELOCITY

B. X-RAYS
ROENTGENOGRAMS
   DIMENSIONS
CINEFLUOROGRAPHY
   MOVEMENTS
ANGIOGRAPHY
X-RAY SPECTROPHOTO-
   METRY

C. LIGHT
TRANSILLUMINATION
PLETHYSMOGRAPHY
ENDOSCOPES

D. NUCLEAR ENERGIES
GAMMA COUNTERS
GAMMA CAMERAS
BLOOD DISTRIBUTION
SELECTIVE CONCENTRATION

WIRES   TELEMETRY

Fig.2 : Technologies for monitoring the function of internal organs of animals and man are presented under various categories such as: I - Implanted transducers; II - Implanted transducers, including radio-
sondes; II - Catheter-tip gauges and III - Energy probes for gathering information
from external sources.

IMPLANTED TRANSDUCERS

A wide assortment of variables can now be recorded with precision and with adequate dynamic response from internal organs by means of implanted transducers of various types (Fig.2). Among the most useful and widely accepted flow sensors are electromagnetic flowmeters which have come to serve as a standard for comparison. (See Kolin, "Recent Trends in Approaches to Electromagnetic Determination of Blood Flow".) Ultrasonic flowmeters based on either transit-time measurements or Doppler shift give reliable information about blood flow velocity but are not as definitive as measures of volume flow rates (1). Many applications of ultrasound have been exploited. Thermal flow sensors of numerous types have been developed and used. An assortment of gauges for recording changes in dimensions, pressures and forces are based on different configurations of strain gauges as well as other common principles such as variation in capacitance, inductance and reluctance. The information from these chronically implanted gauges can be recorded either through direct wire connections or by radio transmission to more distant sites (Fig.2). Using such techniques, extensive data regarding changes in cardiovascular function have been recorded from wild animals (4) such as free ranging baboons and giraffes in Africa and aquatic mammals in various situations in and out of water. In addition, tiny capsules incorporating radio transmitters (5) can be swallowed or inserted into various body channels or cavities to transmit information regarding pressure, temperature, pH, dissolved gases and many other important physiological parameters. (See Mackay, "Advances in Miniaturized Electronics for Biomedical Applications".) This expanding array of new techniques is at last providing clearer insight into both the spontaneous or natural responses under "normal" controls and the effects of abnormal environmental or pathological conditions to which the animals can be exposed.

CATHETER-TIP GAUGES

Hollow catheters and needles have been utilized for many
years to extract samples of blood, body fluids and biopsy specimens from within the body.   Recordings of pressure within hollow organs and channels have been acquired through such tubes
using external pressure gauges, but these may have significant
artifacts in either static or dynamic measurements.   In recent
years, an expanding assortment of devices has appeared for recording critical variables directly from the tip of the catheter
(or needle).   For example, a wide variety of pressure sensors
have been utilized to provide dynamic information at the catheter
tip with extremely high frequency response.   Microminiature
bonded strain gauges, semi-conductors and P-N junctions produce
sensing elements of extremely small size and acceptable sensitivity.   A fiberoptic bundle capped with a reflecting membrane
provides a catheter-tip pressure gauge with no internal wires,
where safety is a consideration.   Miniature electromagnetic and
ultrasonic transducers can now be mounted on the tip of small
calibre catheters.   In addition, pressure differences between
two sites along an artery can be analyzed to provide flow information.   Direct mechanical or thermal flow sensors have also
been employed for flow detection.

Techniques for studying the dimensions of hollow chambers,
channels and organs are being developed utilizing ultrasonic
echo ranging techniques and various types of electromechanical
gauges.   Fiberoptic bundles coupled with laser light sources
have combined for the use of spectrophotometry of internal organs and fluids, (additional details are presented by Lübbers,
Goldman and Carlson).   Fiberoptic oximetry and dye concentration indication is now well advanced.   According to Jacobson,
identification of atherosclerotic lesions on the walls of ar-

teries can now be accomplished by spectrophotometry through fiberoptic catheters.   Catheters may carry electrodes for recording electrocardiograms from within the heart and also oxygen electrodes and similar devices for measuring oxygen, CO_2 and other components of the blood.   A most intriguing potential for the future is the combination of various techniques to provide multifunctional catheters, continuously monitoring a select group of critical variables in animals and man over extended periods of time.   Installation of needles and catheters requires mechanical penetration of the skin and therefore imposes limitations on both physiological and clinical applications.

ENERGY PROBES

Extraction of vital information from internal organs without mechanical penetration of the skin can be accomplished by means of energy probes.   Beams of X-rays have been utilized for this purpose since the turn of the century, and now are prominent as sources of information about size, shape, movement and function of internal organs.   In addition, ultrasonic beams can be readily generated on the external surface of the body and are transmitted and backscattered in tissues so that the energy emerging can be analyzed to elicit information regarding dimensions, location, movement, velocity and flow of internal organs and fluids.   Similarly, beams of light can be directed into the body for transillumination (e.g. to detect early hydrocephalus), for photoelectric plethysmography or through tubes and fiber optics as endoscopes.   The expanding role of lasers in biomedical research is discussed by Goldman and by Carlson.   Holography can be employed for visualizing images from light and from ultrasonic waves.

Radio-isotopes have become widely used as sources of energy which can be introduced into the body and localized by external counters or cameras for estimating blood flow or blood distri-

bution in lungs, brain and other organs.   Selective concentration of isotopes may be utilized to evaluate function (e.g.
I^{131} concentrated by the thyroid gland).   The development of
injectable chemicals with discrete affinity for specific tissues or systems could open up whole new vistas for evaluating
size and rate of function of internal organs or enzyme systems.

Additional details and other examples of most of these
various techniques for measuring or monitoring from internal
organs will be discussed elsewhere in this volume.

FUTURE APPLICATIONS OF ENERGY AS PROBES

Rapid technological advances in instrumentation can be
projected to include an increasing number of energy probes (6),
utilizing various bandwidths of the electromagnetic spectrum
and mechanical waves as indicated in Fig.3.   Major divisions of
these spectra in terms of frequency bands are indicated on the
left and the more common subdivisions of radiation are indicated
on the right.   Examples of characteristics of these energies
which might be the basis for recording systems are indicated at
the top of the figure under three major headings; Amplitude,
Frequency and Velocity.   Twenty examples of applications of the
various wave energies are listed at the bottom with their identifying numbers inserted on the chart to suggest characteristics
measured to employ the wave energies as probes in some of the
examples covered in this volume.   There is some justification
for criticism of details and overlap in this kind of matrix presentation but its principal purpose is to stimulate discussion
and innovative thought processes.   The chart illustrates applications of emission of energy from radio-isotopes, from fluorescent materials and from heat sources.   Similarly, spectroscopy
and spectrophotometry have encompassed an expanding bandwidth to
include X-rays, ultra-violet and infra-red portions of the spectrum.   Spectral analysis is employed in many applications to

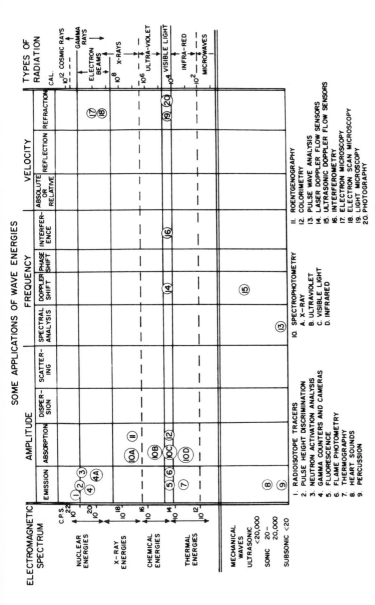

Fig.3 : Applications of various bandwidths of energy are indicated in terms of some recordable characteristics on a matrix which can serve as a framework for projecting some future prospect for utilizing energy to elicit information about biological systems in the future.

elicit information from complex mixtures of frequencies. The Doppler shift principle has been used for velocity detection utilizing laser and ultrasonic beams. Focusing or refraction of electrons and light have added much to our ability to visualize objects through photography, microscopy and more recently electron microscopy and electron scan microscopy. Not all of the wave energies display all of the recordable variations listed but it is clear that many additional opportunities to employ wave energies will be developed in the future. Some of the blank spaces on Fig.3 may be occupied by future developments just as the wider bandwidths have been incorporated under spectroscopy. The wavelengths beyond microwaves were not even included in the chart because there are no obvious biological applications of radiowaves. This might well turn out to be a narrow view in light of future developments. In any case, we may be certain that the development of new research tools will open new vistas for physiological research. Their application to human subjects and patients will be greatly extended if they are truly non-traumatic so that they can be used without discomfort or hazard.

REFERENCES

1. Rushmer, R.F., Baker, D.W. and Stegall, H.F. Transcutaneous Doppler flow detection as a non-destructive technique. *J. Applied Physiol.* 21: 554-6, 1966.

2. Slater, L.E. What's ahead in biomedical measurements? *J. Inst. Soc. Am.* 11: 55-60, 1964.

3. Rushmer, R.F., Van Citters, R.L. and Franklin, D.L. Some axioms, popular notions and misconceptions regarding cardiovascular control. *Circulation XXVII*: 118-141, 1963.

4. Van Citters, R.L. and Franklin, D.L. Cardiovascular research in wild animals: Telemetry techniques and applications. *Acta Zool. et Path. Antverp.* 48: 243-63, 1963.

5. Jacobson, Bertil. Endoradiosonde techniques - a survey. *Med. Electron. Biol. Engng.* 1: 165-180, 1963.

6. Rushmer, R.F. What medical instrumentation in 1964? *Med.-Surg. Rev.* First Quarter, 1969.

MATERIALS

IN BIOMEDICAL ENGINEERING

M. SCHALDACH

Department für Biomedizinische Technik,
Friedrich Alexander Universität, Erlangen – Nürnberg, Germany.

SUMMARY

At present, no polymers meet the required specifications of antithrombogenicity and, therefore, their use is limited to certain areas.

A review of current and past literature indicates that a synthetic surface has not yet been fabricated which will permanently or indefinitely fulfil the biological requirements. Significant and encouraging progress has, however, been made in the development of both synthetic and biological surfaces which tend to retard thrombosis, and there is an indication that surfaces of increased thromboresistivity will appear. A number of interesting and varied approaches have been exploited in an effort to retard surface thrombosis on materials intended for prolonged intravascular implantation. These include such techniques as surface heparinization by either adsorptive or chemical bonding techniques, use of metals with a negative surface charge, low surface energy polymers, polyelectrolytes, electrets, bioelectric polymers, hydrogels and pyrolitic deposited carbons. Other investigators have attempted to circumvent the strong propensity to thrombus formation upon synthetic materials by resorting to the creation of biological coatings or liners for intravascular prosthetic devices.

INTRODUCTION

Interest in implanted materials has grown immensely as a direct consequence of several remarkable developments in surgery and biomedical engineering.   These include heart-lung machines, artificial kidneys, cardiac pacemakers, arterial and heart valve prostheses and orthopedic implants (Fig.1).   Perhaps the most dramatic development is that of artificial implantable hearts, various types of circulatory assist pumps and biological energy sources, as for example, biogalvanic elements and bioautofuel cells.   It is well to recall that many of these developments have their origins in the beginning of the century and in some instances even earlier.   Thus, the heart-lung machines are an outgrowth of the organ perfusion equipment developed by von Frey and Gruber (1) in 1885, Jacoby (2) in 1890, as well as many others.   However, the first clinical application occurred about 1955, an achievement that was the result of extensive research by many workers.   Similarly, the first artificial kidneys were investigated by Abel, Roundtree and Turner (3,4) in 1913, and the first clinically useful machine was introduced by Kolff (5) about 1944.   The question arises why these developments were delayed.   Perhaps two main reasons can be given: the problems arising from blood clotting, and the lack of suitable construc- tion materials.   The first problem has been solved to a great extent by heparin and other anticoagulants while the avail- ability of relatively inert physiological materials such as Teflon, silicones, dacron, and metals like the 316 series stain- less steels, vitallium and titanium have alleviated the second difficulty.   Two important problems associated with implanted materials will be briefly discussed: the deterioration of im- plants in living tissue and the effect of the implant surfaces on blood.   The aim is to describe electrochemical reactions in

the body-solid interface restricted to metals, semiconducting
materials and synthetic polymers.

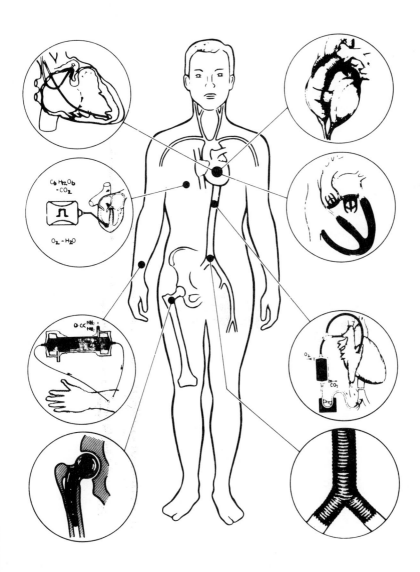

Fig.1 : Implants and artificial organs.

MATERIALS USED IN LIVING TISSUE

Considerable empirical information is available on the use
of many materials in the body and in the related physiological
environment.   At present, however, correlation of this information into a meaningful picture is difficult.   Direct translation of *in vitro* results into *in vivo* applications is not
usually possible.   Furthermore, these solid materials are not
the inert substances they are assumed to be by many investigators, but do react with the biological environment.   Most
polymers are typical commercial materials, in which impurities,
additive imperfections, and structural abnormalities result in
clotting, tissue reactions, mechanical failures, etc.   From
these experiences and from an analysis of the physiological
system in which the foreign material must be used, the following
tentative specifications must be met.   (See Table I in Fig.2.)

1. Not Toxic
2. No Inflammatory Response
3. No Allergic Effect
4. Not Cancerogenic
5. No other Adverse Effects on Cells or Body Fluids

---

1. Can be Reproducibly Obtained
2. Can be Fabicated into the Desired Shape
3. Will have the Needed Chemical, Physical and
   Mechanical Properties
4. Can be Sterilized without Change in Properties
5. Will not Adversely Altered by Biological Environment
6. Will not Dissolve and Migrate to Distant Body Sites

Fig.2 : Effects of materials on recipient, desired characteristics

At present, no solid materials meet these specifications and,
therefore, their use is limited to certain selected areas.   In

METALS

Titanium	Heart Valves, Orthopedic Implants
Platinum	Stimulation Electrodes
Aluminium	Biogalvanic Element
Zinc	" "
Platinum - Iridium	Stimulation Electrodes
Elgiloy	" "
Vitallium	Orthopedic Implants
Stainless Steel	Pacemakers, Orthopedic Implants

SEMICONDUCTORS

Carbon	Heart Valves, Blood Pumps
Bioelectric Polyurethane	Shunts, Heart Assist Devices

INSULATORS

Silicone Rubber	Shunts, Heart Valves, Oxygenators
Teflon	" " "
Polyethylene	Catheters
Polypropylene	Heart Valves
Cellophane	Artif. Kidney
Polyvinyl Fluorine	Heart Assist Devices
Polyurethane	" " "
Epoxy	Pacemakers, Heart Assist Devices
Acrylates	Surgical Adhesive
Dacron	Vascular Implants
Polyvenyl Chloride	Catheters
Polycarbonate	Heart Assist Devices

Fig.3 : Materials for short and long term use in living tissue.

Table II (Fig.3) materials are presented which are now in clinical use or in experimental evaluation.

In a number of applications, metals have been proven tissue acceptable as there are cages of heart valves, cardiac stimulator electrodes (6,7,8) and cardiac pacemaker encapsulations (9), orthopedic implants (10) etc.. The use of pure metals in biogalvanic elements as power sources for pacemakers, however, re-

sulted after 12-18 months of operation in heavy tissue reactions
due to toxic concentrations of the reaction products (11,12,13,
14).

Carbon as well as bioelectric polyurethane promise biocom-
patibility and antithrombogenicity.   All listed polymers, if
treated with heparin or other surface active groups, showed non-
thrombogenic behavior.

ELECTROCHEMISTRY OF IMPLANTS

It is known that the body provides a hostile environment
for metals and plastics.   All implanted materials elicit tissue
reactions which can range from the extremely intense to almost
nil.   Indeed, it would appear that a limited reaction is needed
in order that the prosthetic device be securely fixed within the
surrounding tissue.   In man, only foreign body granuloma reactions have been observed (15).   These reactions have ranged
from the mild response, in which the material may be considered
compatible with the local tissue, to heavy encapsulation and
tissue irritation.

The reaction of materials in a biological system can be
arbitrarily divided into two main types:

reaction with tissue

reaction with blood.

The material classification plays an extremely important
role in the following.   To demonstrate the interface characteristics of metals, semiconductors and insulators in contact with
an electrolyte the potential and charge distribution is depicted
in Fig.4.

The solid-tissue contact can be described for our purpose
in a manner somewhat similar to a pn-junction.   The charge carriers have different natures in the two phases of contact, the
solid part differing only in the charge concentration from metal
to insulator.   When in equilibrium (no current flowing) the

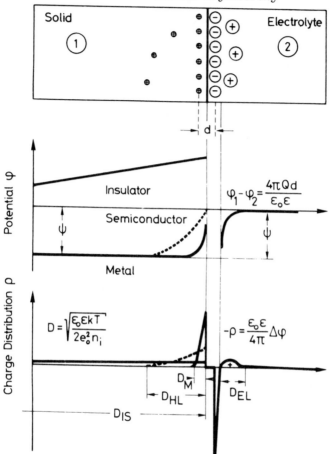

Fig.4 : The influence of specific adsorption on the charge and potential distribution.

electrochemical potential of the electrons in the solid and in the electrolyte has to be equal.   When the solid is immersed in the electrolyte, charge flows from one phase to the other until the electrochemical potentials are equal and the space-charge layer at the surface of the solid is modified.   A space-charge layer will also be present at the electrolyte side of the interface (GOUY-layer).   The schematic diagram of the solid electrolyte system and the variation of the energy with the coordinate normal to the interface shows that the ions cannot get closer to

the surface of the solid than their radius and hence there is a
thin layer near the surface which does not obey the equations of
the electrostatics of continuous media.   This is the so-called
HELMHOLTZ-layer and the potential distribution in it is difficult
to estimate.   The thickness of the space-charge region or GOUY-
layer in this electrolyte depends on the ion concentration.   The
DEBYE-length D_{EL} is in the order of a few microns since the
carrier concentration is relatively high in the living tissue.
In contrast to metals and semiconductors where at body temperature a sufficient charge concentration is always available in
the space-charge region and, therefore, the surface charge distribution is unaffected by the biological environment, the electrochemical behavior of the insulator-tissue contact can be
easily inverted by different electrochemical reactions.   For
example, an adsorbed layer can cause an inversion of the potential profile with totally changed electrical performance of the
solid.   Uncertainty may arise when the surface becomes covered
with layers of adsorbed material such as protein molecules.
Besides adsorption, chemisorption reactions such as corrosion
and redox processes have to be considered because they are able
to influence the equilibrium.   The equilibrium of the implant
should never disturb the biological state due to adsorption,
corrosion and redox processes.

    A comparison of tissue and blood compatibility of materials
is complicated by the fact that the blood is fluid and therefore
the so-called electrokinetic phenomena have to be considered.
This leads to a separation of the problem of the solid electrolyte interface in tissue and blood.   The metal tissue, resp.
polymer-tissue interface should first be described.

METAL IMPLANTS

    As indicated earlier, the body presents a rather hostile
environment to a number of metals including stainless steel.

The combined presence of chloride ions and oxygen in physiological environments results in pitting of the latter metals.
Though this tendency is least apparent among the widely used 316
and 317 austenitic steels, nonetheless after several years of
implantation pits may be developed which will ultimately reduce
the mechanical strength of the implant. In addition, the corrosion products may serve as local irritant. The austenitic
stainless steels are also subject to stress corrosion, cracking
in the presence of oxygen, chloride and applied or internal
stresses. The Co-Cr-Mo-alloy, vitallium, is more resistant to
corrosion in the presence of chloride than 317 steel, and has
been widely used despite the fact that it is difficult to machine.
Unlike the stainless steels and vitallium, titanium is resistant
to pitting and stress corrosion cracking. Consequently, these
metal alloys are commonly used as orthopedic implants (16,17).

In considering the corrosion resistance of implanted metals,
it must be appreciated that a large number of interrelated
physiochemical factors must be taken into account, including the
magnitude and presence of cyclic stresses, the wear occasioned by
the rubbing of the metal against a hard surface, galvanic couples
between dissimilar metals, the presence of active chemical agents
such as halogen ions, as well as nonuniformities in their distribution and stray potentials arising from neuromuscular action and
other sources. However, it is not possible in this brief survey
to enter into a detailed discussion of all factors and their complex interactions.

Since corrosion resistance is considered to be most important and has been placed at the head of the required properties
list, a short description of the electrochemical sequence during
anodic polarization should be given. (See Fig.5.)

In an activated state, i.e. when there is sufficient anodic
polarization, metals generally dissolve. In strongly oxidizing
electrolytes, however, a rapid polarization at the metal surface

M. Schaldach

occurs, which reduces the current flow and results in corrosion
to significantly low values. The polarization reaction elicits
the formation of a thin gelatinous type film which acts as a
diffusion barrier by separating the metal surface from the cor-
rosive environment. The corrosion rate is markedly decreased
due to the formation of this film barrier. This phenomenon is
called passivity. The most important condition for passivation
without current flow is the fact that the oxidation potential of
the metal is less than the redox potential of the electrolyte,
in the case of a Co-Cr-Mo-alloy. There exists a passive region
which shows transpassivity in the presence of chloride ions.
Behind,transpassivity oxygen evolution determines the curve.

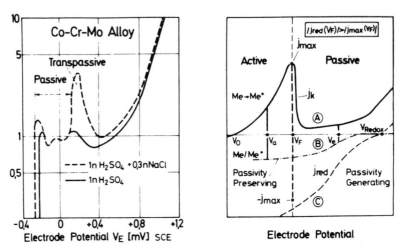

Fig.5 : Schematic anodic dissolution behavior to demonstrate
        active-passive characteristics for metallic implants.

More details about passivation explains the right hand portion of Fig.5. Passivation in electrolytes with low redox
potential occurs only if the value of the corrosion-current density, I_k, reaches the maximum current, I_{max}, at the so-
called FLADE-potential, V_F, the redox-potential of the metal-
oxide electrode (18).

The currentless deposition of the passive-layer in redox electrolytes is influenced by the critical corrosion-current density with the condition:

$$\left| I_{red} \, (V_F) \right| \; > \; \left| I_{max} \, (V_F) \right|$$

and
$$V_{redox} \; > \; V_F$$

If this condition cannot be fulfilled, only an existing passive layer can be preserved. The oxidation of the metal occurs at a so-called mixed potential.

For implants, therefore, only oxide-type material provides sufficient corrosion protection, which possesses FLADE-potentials more negative with respect to the hydrogen electrode. Only the reduction of hydrogen generates the passivity without the need for other oxidizing agents in the tissue electrolyte.

Chromium, titanium, aluminium and manganese reach these requirements, but only the Ti/TiO_2-system forms a tight non-porous passive layer. The negative FLADE-potential and negative mixed potential determine the tissue as well as the blood compatibility supplying the required negative surface charge. The effect of surface potential, corrosion potential, and electrokinetic behavior on thrombogenicity, in the case of metals, was investigated by Boddy, Brattain and Sawyer (19,20,21). Tubes made of various metals were inserted into animal venae cavae and the aorta, and the patency determined in terms of the time required for complete thrombosis. (See Fig.6.)

Clearly, metals which are high in the series and hence tend to ionize readily, have a greater patency than those lower in the series. The explanation offered is that the more positive metals, on ionizing, tend to build up an electrical double layer with the negative charge on the metal surface. The higher metals, on the other hand, do not readily ionize and may absorb positive ions to acquire a positive surface charge.

M. Schaldach

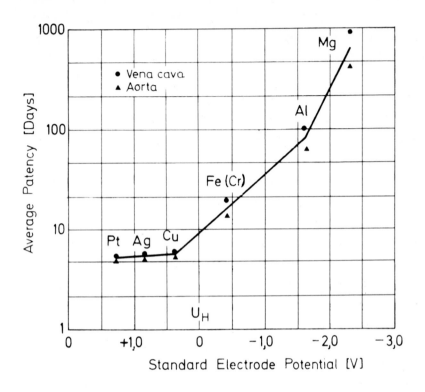

Fig.6 : Duration of patency of metal tubes implanted in canine
        thoracic aorta or vena cava as a function of the stan-
        dard electrode potentials (21).

In conclusion, from the variety of metals and alloys which
have been studied and evaluated experimentally, two basic metal
groups and two unalloyed metals evolve which are in clinical use
at the present time, namely, cobalt-chromium base alloys, fer-
rous base alloys, titanium and tantalum.

The application of metals is not restricted to orthopedic
implants; cages of heart-valves, capsulation of heart-pace-
makers and stimulation electrodes are also made from alloyed or
unalloyed metals.

CARDIAC STIMULATION ELECTRODES

The behavior of metals employed as stimulation electrodes is of considerable interest in experimental medicine and bio-medical engineering, particularly in such applications as cardiac pacemakers. In most commercial pacemaker electrodes the material used so far is either platinum-irridium or ELGILOY, an alloy containing cobalt, nickel, molybdenum, manganese, iron, carbon and beryllium. These electrodes are either sutured into the muscle of the heart or inserted into the right ventricle by means of a transvenous catheter. The energy required to stimulate the heart depends on the electrode and is usually 1-5 μJ when the pacemaker electrode is first implanted (22,23,24,25).

Since a high current density vs. a high electric field strength at a small electrode is required locally to stimulate the cardiac muscle, several polarization voltages are developed with the consequence that most of the delivered energy from the pacemaker circuit is wasted in the electrode-electrolyte inter-face. This is particularly important in the case of implanted pacemakers since their life is determined by the energy stored in their batteries. Considerable extension of the lifetime could be achieved if the stimulation system worked more economically.

The electrochemical reactions in the electrode interface are very complex and involve possible electrode corrosion, gas evolution and oxidation vs. reduction of metabolites with potentially toxic effects on a long-term basis.

It is recognized that ideally the processes taking place at stimulation electrode surfaces should solely involve motion of electric charges without any irreversible chemical reaction taking place. To demonstrate the electrochemical aspects, the potential model of a metal electrode in contact with the endo-cardium is shown in Fig.7. The potential distribution at the

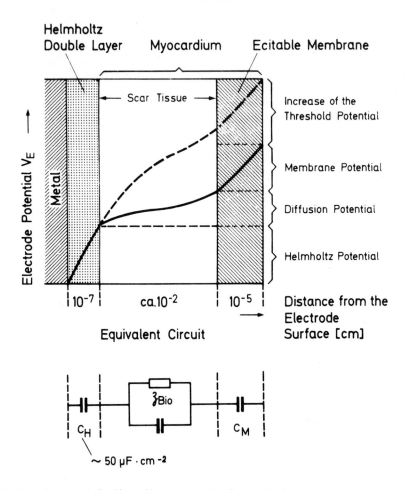

Fig.7 : Potential distribution and electrical properties of the
        metal-myocardium interface.

interface gives a simplified view of the electrochemical behavior under constant conditions. The potential depends on the
distance from the electrode surface and is made up of the drop
in the rigid HELMHOLTZ-double layer, the space charge region and
the membrane potential. The diffusion potential represents the
electrical changes between the excitable membrane and the electrode surface. Thus, the rise in stimulation threshold can be
expressed in the portion of the potential distribution, so that

an increase in the thickness of the connective tissue layer results in a threshold rise. The electrical performance can be
described in the equivalent circuit. Because of the nonlinearity of the electrochemical processes, the value of each component itself is a function not only of the time but also of the
applied electrode voltage, V_E. The electrochemical polarization of metal pacemaker electrodes in the current density vs.
electrode potential curves for platinum and ELGILOY are shown in
Fig.8. During polarization the change of the electrode potential results in the HELMHOLTZ potential as well as in the poten-

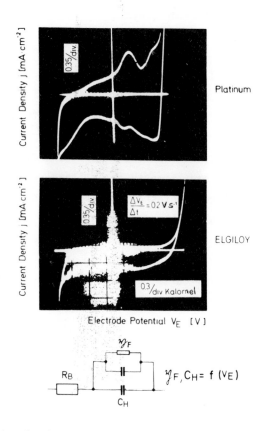

Fig.8 : Electrochemical behavior of Pt- and Elgiloy-stimulation
electrodes in physiologic saline solution.

tial drop of the diffusion layer.   This causes an energy loss
in charging and recharging the HELMHOLTZ capacity.   When the
applied electrode potential rises above a critical value, ions
discharge, chemical reactions take place, current flows across
the interface and the electrical energy is now irreversibly
wasted.   As a result of this current flow, the field strength
necessary for triggering in the next excitable cardiac membrane
breaks down.   The stimulation occurs only as the result of a
change in ion concentration due to a current flow near the excitable membrane.   The FARADAY-impedance, Z_F, now determines
the electrochemical behavior (26,27).

THE DCD-ELECTRODE

One way to avoid the wastage of energy due to polarization
effects has been discussed by Parsonnet (28) using a so-called
"Differential Current Density Electrode".   In this system there
is a low current density at the metal to avoid polarization and
a high current density at the contact to the endocardium to produce a significant field strength to activate responsible cells.
As a result, stimulation thresholds are found whose energy is
1/10 of that achievable with conventional metal electrodes.
Fig.9 shows the dimensional drawing of the DCD-electrode.   The
electrochemical behavior is demonstrated in the current density
vs. electrode potential curve.   The shape is determined by a
current-limiting resistor, R_H, in series as described in the
equivalent circuit below.   The value of the internal resistor
ranges from 2-4 kΩ.   The only disadvantages of this method are
difficulties in stable fixation of the electrode and electrochemical instability, due to the development of fibrotic tissue
within the chamber.

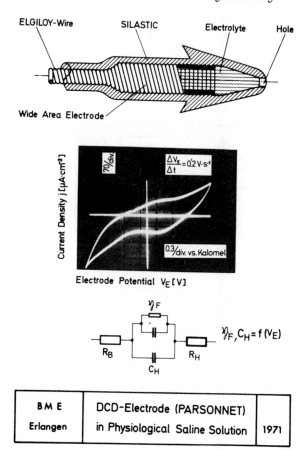

Fig.9

THE DIELECTRIC PACEMAKER ELECTRODE

To prevent electrochemical reactions, a thin dielectric layer is used on the surface of a metallic electrode tip (Fig. 10). The passage of an electrical pulse through the interface of such a coated electrode involves only a change of the charge distribution in the metal and ions in the electrolyte. Charge transport is not possible through the oxide-film. That means no FARADAY reactions can occur and, as a consequence, such a system avoids electrochemical reactions such as electrode cor-

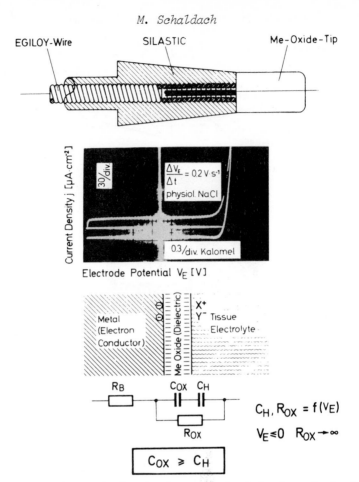

Fig.10 : Electrochemical behavior and electrical equivalent circuit of a dielectric stimulation electrode.

rosion, gas evolution and reactions with metabolites. The potential model of this electrode and its equivalent electrical circuit are shown in Fig.10. The equivalent circuit of this electrode must be completed by an additional in-series capacitor,
C_{ox}, of the same magnitude as the HELMHOLTZ capacitance, C_H. The
current density vs. electrode potential curve demonstrates the
electrochemical behavior of this dielectric pacemaker electrode.
In the electrode potential range below 1 V vs. calomel, the electrode acts as a pure capacitor. This results in the fact that

a part of the delivered pulse energy is also lost due to the fact that the charge stored in the capacitor during a pulse is not utilized for stimulation (26).

The electrodes consist basically of a metal tip which is coated by a thin dielectric layer of electrochemical deposited tantalum oxide.   A helical ELGILOY wire is used for connection to the pacemaker pulse generator.   Silastic medical grade tubing has so far been used as encapsulating material.

BLOOD-MATERIAL REACTIONS

Blood circulating within the body has an infinite clotting time.   However, when blood comes in contact with foreign surfaces, either *in vivo* or *in vitro*, coagulation usually occurs within minutes.   The role of the contacting surface and the coagulation of blood has been investigated for many years and much data are available.   The activation mechanism is described in the clotting sequence of Fig.11.   The surface presented to the blood system by an artificial organ is different from that of the normal endothelium, and various changes in cellular elements and proteins occur as a result of contact with this new surface.

Not only the physical state of the solid surface but its electrochemical behavior has to be considered, as well as the mechanical performance of the implant (Fig.12).

Many adverse results can only be understood at the molecular level with a consideration of electrokinetic effects.   As mentioned before, the solid-blood interface consists of an equal amount of excess opposite charges on both sides of the dipol-layer, thus preserving the electroneutrality of the system.   A streaming electrolyte, e.g. blood in an artificial vessel (Fig.13) results in a charge gradient with a potential drop.   Depending on the conductivity and the surface charge, the repulsion and attraction forces on the charged blood par-

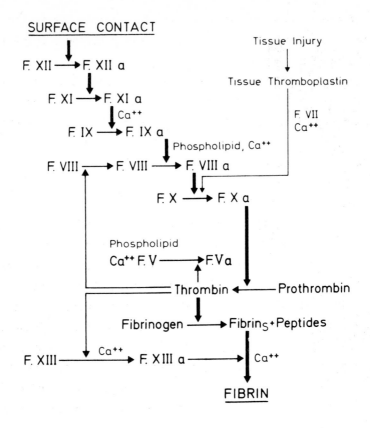

Fig.11 : The clotting mechanism as cascade reaction (Davie and
Ratnoff).

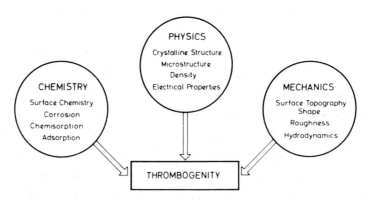

Fig.12 : Effects of thrombogenicity.

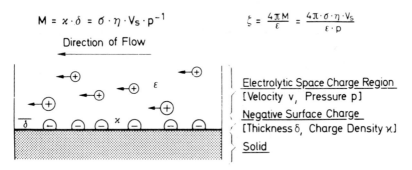

$$M = \varkappa \cdot \delta = \sigma \cdot \eta \cdot V_s \cdot p^{-1} \qquad \xi = \frac{4\pi M}{\varepsilon} = \frac{4\pi \cdot \sigma \cdot \eta \cdot V_s}{\varepsilon \cdot p}$$

Direction of Flow

Electrolytic Space Charge Region
[Velocity v, Pressure p]
Negative Surface Charge
[Thickness δ, Charge Density ϰ]
Solid

Repulsion by Electrostatic Shielding of the Double Layer

Attraction by van der Waals - London Forces

Fig.13 : The electrokinetic effect in the solid-electrolyte
interface.

ticles will be influenced. Or stated medically, the rheologi-
cal property of the implant will be changed. If the surface
of a nonconductive polymer adsorbs protein molecules, the initially negative surface charge will be compensated and the
material becomes thrombogenic.

NATURAL BOUNDARIES

Understanding the interaction of blood and the surrounding
vessel wall is critical to both the design and the construction
of the artificial interface used in the replacement of sections
of the cardiovascular system. Because red blood cells, white
blood cells, fibrinogen and other components of blood are negatively charged at the normal pH of blood, there have been attempts to show that positive electrical charges applied to the
walls of blood vessels induce thrombosis under a positive electrode. Sawyer *et al.* also pioneered this work and have shown
that the artificially precipitated thrombus obtained when using
an applied potential is histologically similar to thrombi produced spontaneously in injured blood vessels. Whether the
charge itself or secondary factors resulting from the influence

M. Schaldach

of the charge on the vessel's intima or its adsorbed monolayer is responsible is not known. The considerations of chemical events and physicochemical forces which act in the interface of a solid, resp. arterial wall or cell boundary are principally the interaction of charged particles and have the same base (29,30). Therefore a survey of general electrical events in biological boundaries is given in Fig.14.

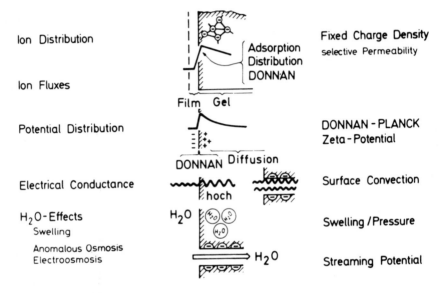

Ion Distribution	Fixed Charge Density selective Permeability
Ion Fluxes	
Potential Distribution	DONNAN-PLANCK Zeta-Potential
Electrical Conductance	Surface Convection
H_2O-Effects Swelling Anomalous Osmosis Electroosmosis	Swelling/Pressure Streaming Potential

Fig.14 : Summarized electrical events on the cell boundary.

The cell membrane, contiguous with both outside and inside solutions, is built of an organized interlinked negatively charged net. The membrane matrix links an ion exchanger and a DONNAN effect exists (31,32). This is comprised of three characteristics.

1. a discontinuity, i.e. the jump of the concentration distribution of free ions
2. a development of a boundary potential (DONNAN-potential)
3. an osmotic pressure effect

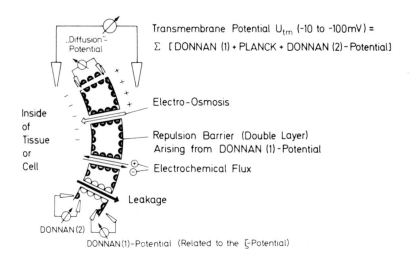

Fig.15 : Electrical behavior of the cell wall or vascular interface.

On this basis the interface should be discussed as a whole,
at least with respect to ion and potential distribution. Later
the presentation will be confined to the external boundary where
the electrokinetic effects take place.

For this case it is assumed that a cell wall is a polymeric
arrangement of positively fixed groups X^+. At the membrane
boundaries these positive groups will attract and accumulate the
salt anions A^- from the surrounding solutions and repel the
cations B^+.

The main effect is the jump of the potential at the outer
boundary where the clotting properties reside. The effects are
summarized in Fig.15.

In respect of the special interest of blood clotting, the
outer boundary potential drop and charge distribution should be
magnified in Fig.17 to describe the equilibrium. The thickness
of the electrolytic space charge region is in the order of about
100 Å.

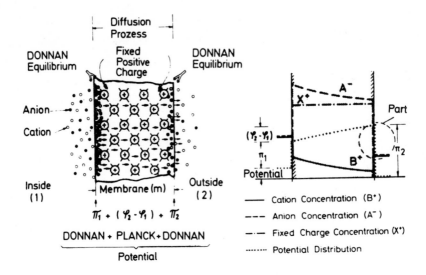

Fig.16 : Charge and potential distribution in the biological
         interface.

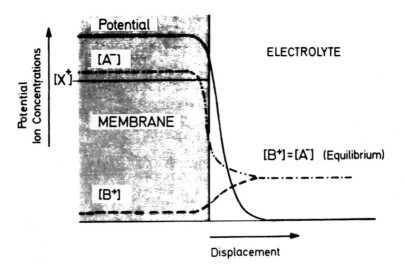

Fig.17 : Magnified picture of charge and potential distribution
         in the outer interface.

The interaction on the outer boundary can be explained only by attraction and repulsion forces. In fact, vaguely defined concepts such as stickiness, wettability and so forth are also merely questions of interplay of forces. At the present time it is not possible to give a satisfactory consistent picture in this respect. The predominant ideas on these problems are very much influenced by the theoretical views of precipitation of colloid solutions. In Fig.18 the repulsive forces are mainly

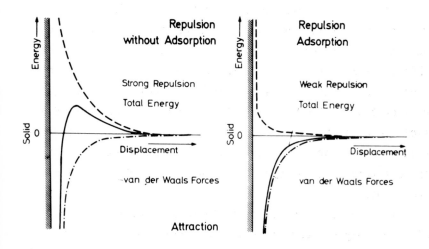

Fig.18 : Weak and strong interaction of charged particles at natural boundaries.

of an electrical nature and are shown as the repulsion curves A and B.   These curves probably have a similar form to the potential profiles discussed above.   The attraction curve is mainly due to long-range forces of the LONDON-VAN DER WAAL types and is represented in the figure with negative values.   The resulting "total energy" will be the sum of repulsion and attrac-

tion.   In the figure these curves are drawn as heavy lines.
Two cases are illustrated, the attraction curve is assumed constant, while two different repulsion curves are assumed.   The
situation in the left hand portion results in a curve, which
has an excess of repulsion over some distance from the boundary.
The situation on the right, i.e., the one with the lower repulsion curve, will be dominated entirely by attraction forces.
The conclusion is that a repulsion screen may be located some
distance out in the ion atmosphere surrounding the particle when
the DONNAN-potential is sufficiently high.

It should be noted that close to the boundary there is a
very narrow attraction zone, then comes the wider repulsion zone,
outside which there is actually a very weak attraction zone.
This could keep the coagulum together and is ascribed to dispersion forces.   As long as the repulsion shell exists around the
cell or tissue it cannot agglutinate with another charged particle, e.g. a platelet.   According to the schematic view the
precipitation, or agglutination, is brought about by bringing
the electrical phase boundary potential towards a certain minimum value.

Therefore, it is now possible to understand that adsorption
of oppositely charged material can diminish or invert the surface
potential of cells, vessels or implanted polymers.   The antithrombogenicity of the arterial wall and of any solids related
to a fixed negative surface charge can now be understood in a
way, that a sufficiently high negative surface charge is the
source of the repulsion barrier, which prevents the charged
blood particles from sticking to implanted materials.   Both the
rheologic effect and the adsorption of some molecular species
can be described in this model.   Crystallographic defects, impurities, crystal faces, edge dislocation or strain points in
the material which may play an important role can also be understood.

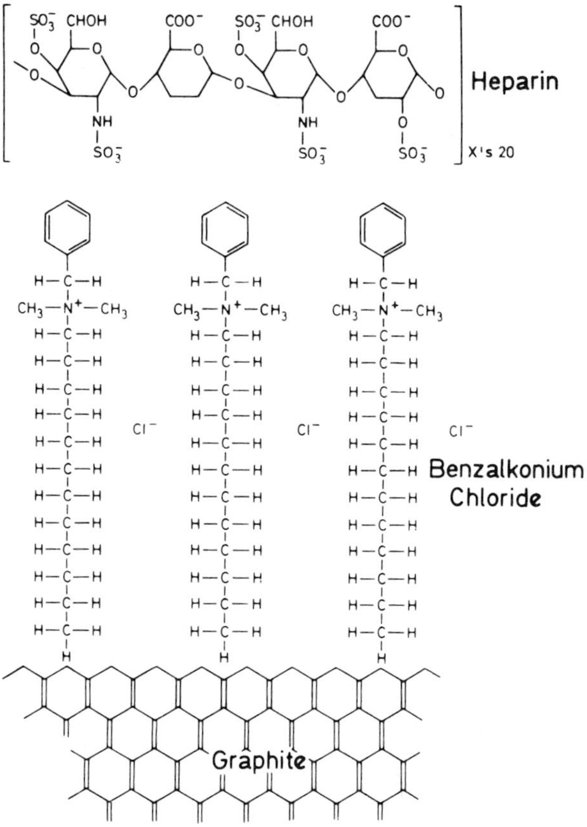

Fig.19 : Surface protection by graphite-benzalkonium-heparin coating.

In the past a large number of investigations have shown that modification of surfaces, such as with coating, decreases the thrombogenic properties of materials used in the vascular system. At first Gott was particularly successful using graphite-benzalkonium-heparin coating to minimize clot formation. (See Fig.19.) Since that time a number of modifications have been tried to prepare plastic surfaces. The long-term nonthrombogenicity of heparinized plastic polymers is at present subject to many research objectives.

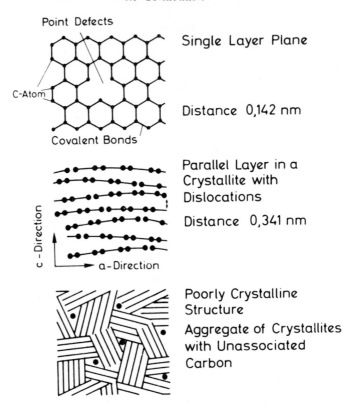

Point Defects

Single Layer Plane

C-Atom

Distance 0,142 nm

Covalent Bonds

Parallel Layer in a
Crystallite with
Dislocations

Distance 0,341 nm

c-Direction

a-Direction

Poorly Crystalline
Structure
Aggregate of Crystallites
with Unassociated
Carbon

Fig.20 : Structure of poorly crystalline carbons.

In addition to heparinized surfaces Bokros *et al.* (33,34)
recently published a paper on the blood compatibility of smooth
isotropic carbons deposited at low temperature (LTI pyrolite carbons). The physical structure is demonstrated in Fig.20. The
most important facts and conclusions are summarized schematically in Fig.21. It was found that clean, smooth LTI pyrolite
carbons have excellent thromboresistance without pretreatment
with benzalkonium chloride or heparin. Heparin is adsorbed
as a monolayer on these surfaces and the amount adsorbed is not
enhanced by pretreatment with benzalkonium chloride. Furthermore, the heparin adsorption did not improve the *in vivo* thromboresistance. In plasma the heparin was elutriated to surface

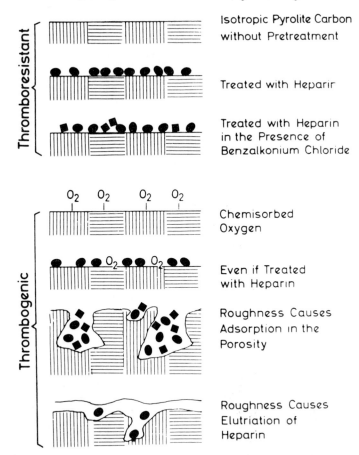

Fig.21 : Influence of chemical and mechanical surface properties.

concentration of less than 1 μg/cm^2 in a very short time.

The study of LTI carbon surfaces treated with oxygen (hydrophilic sited) indicated that the presence of oxygen functionality on the surface markedly reduced the thromboresistance. Although the presence of chemisorbed oxygen did not detectably influence the adsorption of either heparin or the heparin-benzalkonium complex (35,36).

In addition, it was found that increasing the surface roughness of impermeable carbons increased the heparin adsorption by an amount accounted for by the increase in surface area. The

heparin was adsorbed as a monolayer, and the adsorption was unaffected by pretreatments with benzalkonium-chloride.   In all
cases, the thromboresistance of rough surfaces was much less
than for corresponding smooth surfaces.

All of the results taken together indicate that carbon is
a useful material for application in the vascular system. Properly prepared surfaces have a high level of thromboresistance
and do not rely on fragile monomolecular layers bonded to the
surface of a thrombogenic or otherwise chemically active substrate.   The compatibility is a probable consequence of the
ability to maintain an intermediate dynamically adsorbed layer
of proteins at the interface.

A carbon coated ball valve is shown in Fig.22.   The first
clinical experience with many of these valves has demonstrated
promising results.

Fig.22 : Picture of a carbon coated heart valve.

MECHANISM OF CLOTTING

Implantation of materials for observation of thrombus formation at their surfaces has involved a variety of techniques.
It is easy to offer criticism for each of these tests, most of

which predispose the system toward increased, rather than decreased, thrombus formation. One severe test of the blood compatibility of materials is the 2 hour implantation of the standdard prosthetic ring into the inferior vena cava of small dogs (37). Those materials implanted for 2 hours were reanalysed to establish which changes in the interfacial zone might be correlated with their demonstrated thromboresistance.

An extremely sensitive and reliable method to determine surface properties is based on contact angle measurements (38,39). Even small changes in the adsorption layer result in a change of critical surface tension (40), and allow an evaluation of surface constitution and surface cleanliness. Another method is based on the absorption of light in the interface between two materials with a different defraction index during total reflection (41, 42,43). So-called internal reflection spectroscopy allows the measurement of interfacial layers without interfering signals from the bulk. This technique is sensitive enough to identify an optical constant as thin as 100 Å, and follow changes in the configuration and bonding within such films (44). The method is nondestructive and exceeds all other methods. The principle of this technique is shown in Fig.23.

Depending on the incident angle, the penetration length can be varied in the order of 1 - 100 Å. An additional modulation of the electrode potential provides the modulation of the optical constant in the reaction space far beyond the rigid HELMHOLTZ-layer as indicated on the right hand portion of Fig.23. In the application of this new technique further information will be obtained which provides a more informative answer on the activation energy in the clotting enzyme mechanism.

The scheme of measuring equipment for internal reflection spectroscopy, using germanium single crystals as reaction electrodes, is shown in Fig.24. A typical infrared spectrum of internal reflection at germanium-electrodes for serum-albumin

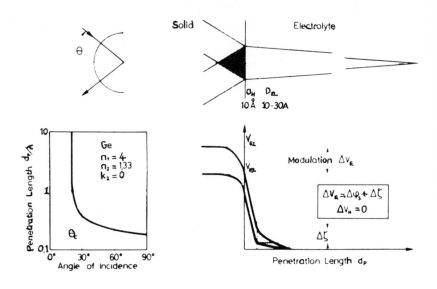

Fig.23 : Electroabsorption in the solid electrolyte interface.

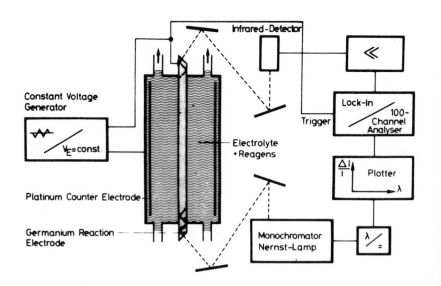

Fig.24 : Scheme of the measuring equipment for internal reflec-
         tion spectroscopy.

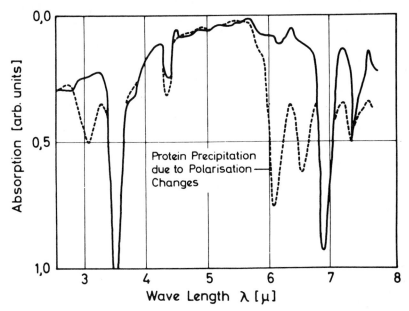

Fig.25 : Internal reflection spectrum of serum-albumin at the germanium electrode. Cathodic and anodic polarization.

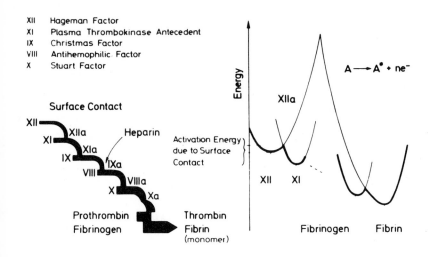

Fig.26 : Clotting mechanism as redox type reaction.

applying anodic polarization is given in Fig.25.   The protein precipitation results in specific absorption peaks at 3, 6 and 6.5 μm.

With the application of this extremely sensitive method a much more favorable understanding of the activation of clotting enzymes may be gained.   The activation energy is defined as the energy necessary for an electron transfer comparable to electro-chemical redox processes.   A number of experimental results in-dicate that the initial thrombus formation occurs due to acti-vation of the HAGEMANN-factor as described in Fig.26.

REFERENCES

1.   von Frey, M. and Gruber, M.   Untersuchungen über den
      Stoffwechsel isolierter Organe.   Ein Respirationsapparat
      für isolierte Organe.   *Arch. Anat. Physiol. 9*: 519 (1885).

2.   Jacoby, C.   Apparat zur Durchblutung isolierter lebender
      Organe.   *Arch. exp. Path. and Pharm. 26*: 388 (1890).

3.   Abel, J.J., Roundtree, L.G. and Turner, B.B.   On the
      Removal of Diffusible Substances from the Circulating Blood
      by Means of Dialysis.   *Trans. Ass. Amer. Physicians 28*:   5
      (1913).

4.   Abel, J.J., Roundtree, L.G. and Turner, B.B.   *J. Pharmacol.
      Exper. Therapie 5*:   275 (1914).

5.   Kolff, W.J. and Berk, H.T.   The Artificial Kidney, a
      Dialyzer with Great Area.   *Acta Med. Scand. 117*:   121
      (1944).

6.   Sauvage, L.R., Robel, S.B., Wood, S.J., Berger, K. and
      Wesolowski, S.A.   Prosthetic Heart Valve Replacement.
      Ann. N.Y. Acad. Sci. 146:   289 (1968).

7.   Sawyer, P.N.   The Effect of Various Metal Interfaces on
      Blood and Other Living Cells.   *Ann. N.Y. Acad. Sci. 146*:
      49 (1968).

8.   Mansfield, P.B.   Myocardial Stimulation.   The Electro-
      chemistry of Electrode-Tissue Coupling.   *Amer. J. of
      Physiol. 212*:   1475 (1967).

9.   Schaldach, M. and Franke, O.   Technischer Stand und
      Tendenzen in der Schrittmachertherapie.   *Acta Medicotech-
      nica 17*:   266 (1969).

10. Debney, D.J.   Cardiac Pacemaker Encapsulation Investigation. *Bio-Medical Engineering 6*: 458 (1971).

11. Roy, O.Z.   Biological Energy Sources. *Bio-Medical Engineering 6*: 250 (1971).

12. Satinski, V.P., Cassel, J. and Salkind, A.   Bioelectric Energy by Reduction of Tissue Oxygen. *JAAMI 5*: 184 (1971).

13. Schaldach, M. and Kirsch, U.   In Vivo Electrochemical Power Generation. *Trans. Amer. Soc. Artif. Int. Organs 16*: 184 (1970).

14. Schaldach, M.   Clinical Experience of a Biogalvanic Pacemaker. *Ann. Cardiol. Angéiol. 20*: 409 (1971).

15. Lyman, D.J.   Biomedical Polymers. *Ann. N.Y. Acad. Sci. 146*: 30 (1968).

16. Weisman, S.   Metals for Implantation in the Human Body. *Ann. N.Y. Acad. Sci. 146*: 80 (1968).

17. Williams, D.F.   Fabrication, Finishing and Selection of Materials. *Bio-Medical Engineering 6*: 300 (1971).

18. Vetter, K.J.   Elektrochemische Kinetik.   Springer-Verlag, Berlin-Göttingen-Heidelberg (1961).

19. Brown, J.C., Lavelle, S.M. and Sawyer, P.N.   Relationship between Electrical and Spontaneous Thrombosis. *Thrombos. Diathes. Haemorrh. 21*: 325 (1969).

20. Sawyer, P.N., Brattain, W.H. and Boddy, P.J.   Electrochemical Criteria in the Choice of Materials used in Vascular Prosthesis. *In:* Biophysical Mechanism in Vascular Homeostasis and Intravascular Thrombosis 337, (1965) N.Y. Appleton-Century Crofts.

21. Sawyer, P.N. and Srinivasan, S.   Metallic and Plastic Prosthetic Devices in Vascular Wall Substitutes: Biophysical Criteria and Methods for Evaluation. *J. Biomed. Mater. Res. 1*: 83 (1967).

22. Irnich, W.   Der Einfluss der Elektrodengrösse auf die Reizschwelle bei der Schrittmacherreizung. *Elektromedizin 14*: 175 (1969).

23. Kirsch, U.   Untersuchungen zur Bestimmung der günstigsten Impulslänge bei Herzschrittmachern. *Elektromedizin 15*, 185 (1970).

24. Furman, S., Parker, B., Escher, D.J. and Solomon, N.   Endocardial Threshold of Cardiac Response as a Function of Electrode Surface Area. *J. Surg. Res. 8*: 161 (1969).

25. Schaldach, M.   Entwicklungsaussichten der Schrittmacher-
    behandlung in technischer Sicht.   *Verh. dtsch. Ges.
    Kreislauf-Forsch. 35*: 127 (1969).

26. Schaldach, M.   New Pacemaker Electrodes.   *Trans. Amer.
    Soc. Artif. Int. Organs 17*: 29 (1971).

27. Schaldach, M., Franke, O. and Blaser, R.   Advances in
    Pacemaker Electrodes.   *Proc. of the 9th ICMBE, Melbourne
    1971.*

28. Parsonnet, V., Zucker, I.R. and Gilbert, I.   Clinical Use
    of a New Transvenous Electrode.   *Ann. N.Y. Acad. Sci. 167*:
    756 (1969).

29. Sawyer, P.N., Burrowes, C.B., Ogoniak, J.C., Smith, A.O.
    and Wesolowski, S.A.   Ionic Structure at the Vascular Wall
    Interface.   *Trans. Amer. Soc. Artif. Int. Organs 10*, 316
    (1964).

30. Scrinivasan, S. and Sawyer, P.N.   Electrochemical Tech-
    niques for Studies on the Intravascular Thrombosis.   *JAAMI
    3*, 116 (1969).

31. Kortüm, G.   Lehrbuch der Elektrochemie.   Verlag Chemie
    Weinheim (1966).

32. Teorell, T.   The Role of Electrical Forces at Cell Bound-
    aries.   (Sawyer, ed.)   *In:* Biophysical Mechanism in
    Vascular Homeostasis, Appleton-Century, N.Y. (1965).

33. Gott, V.L., Whiffen, J.D. and Valiathan, S.M.   Graphite-
    Benzalkonium-Heparin Coatings on Plastics and Metals.
    Ann. N.Y. Acad. Sci. 146: 21 (1968).

34. Bokros, J.C., Gott, V.L., Lagrange, L.D., Fadali, M.M.,
    Vos, K.D. and Ramos, M.D.   Correlations between Blood Com-
    patibility and Heparin Adsorptivity for an Impermeable Iso-
    tropic Pyrolytic Carbon.   *J. Biomed. Mater. Res. 3*: 497
    (1969).   *J. Biomed. Mater. Res. 4*: 145 (1970).

35. Bokros, J.C., Lagrange, L.D. and Schoen, F.J.   Control of
    Structure of Carbon for Use in Bioengineering.   Gulf
    General Atomic-Report GA 10100, June (1970).

36. Epstein, B.D. and Dalle-Molle, E.   Surface Charge Behavior
    of Pyrolytic Carbon in Saline and Blood Plasma.   *Trans.
    Amer. Soc. Artif. Int. Organs 17*: 14 (1971).

37. Gott, V.L., Rames, M.D., Najjar, F.B., Allen, J.L. and
    Becker, K.E.   The In Vivo Screening of Potential Thrombo-
    resistant Materials.   *In:* Proc. Artificial Heart Program
    Conf. (F.W. Hastings, ed.) N.I.H. (1969).

38. Sawyer, P.N. and Srinivasan, S.   New Approaches in the

Selection of Materials Compatible with Blood. *In:* Proc. Artificial Heart Program Conf. (F.W. Hastings, ed.) N.I.H. (1969).

39. Baier, R.E., Gott, V.L. and Feruse, A. Surface Chemical Evaluation of Thromboresistant Materials before and after Venous Implantation. *Trans. Amer. Soc. Artif. Int. Organs 16:* 50 (1970).

40. Zisman, W.A. Relation of Equilibrium Contact Angle to Liquid and Solid Constitution. *Adv. Chem. 43:* 1 (1964).

41. Harrick, N.J. Internal Reflection Spectroscopy. Intersciences Publishers, N.Y. (1967).

42. Schaldach, M. Elektrochemische Kriterien bei der Auswahl von Materialien für den Gefäss- und Klappenersatz. *Verh. dtsch. Ges. Kreislauf-Forsch. 36:* 69 (1970).

43. Lyman, D.J., Brash, J.L., Chaikin, S.W., Klein, K.G. and Carini, M. The Effect of Chemical Structure and Surface Properties of Synthetic Polymers on the Coagulation of Blood. *Trans. Amer. Soc. Artif. Int. Organs 14:* 250 (1968).

44. Symposia of the FARADAY-Society, No.4. Optical Studies of Adsorbed Layers at Interfaces. (Faraday Soc., ed.) Academic Press Inc., London and New York.

BLOOD COMPATIBLE MATERIALS

FOR

CARDIOVASCULAR PROSTHESES*

YUKIHIKO NOSÉ and YOHJI IMAI

*Department of Artificial Organs, Division of Research,
The Cleveland Clinic Foundation, Cleveland, Ohio, U.S.A.*

Blood compatible materials are essential in the composition
of any prosthetic device that is in contact with blood.   The
necessity of such materials is best shown in the records of the
development of an artificial heart.   Death of the animals each
with a complete prosthetic heart was caused by a thromboembolus,
a gradual increase in plasma hemoglobin, a decrease in platelets,
a rapid terminal prolongation in the clotting time, and a decrease in fibrinogen levels (1).   Although errors in design and
function of an artificial heart and in management of animals are
factors in the cause of death, some factors are related to the
materials from which the prosthesis is made.   Ideal materials
for the device should not promote clotting, not damage erythrocytes, leukocytes and platelets; not cause denaturation or
changes in the immunologic properties of plasma proteins; not
interfere with the body's normal defense mechanisms; not modify
blood electrolyte composition nor cause allergic or toxic reactions; and should not cause or promote the development of

* The majority of the text was taken from the manuscript prepared
  for publication in the Journal of Biomedical Engineering.

cancer.   No materials studied to date have met all these requirements as is shown in Table 1 (2).   In many materials some
properties are good, but others are not.

About 10 years ago when initial investigations of cardiac
prostheses were begun, investigators felt that the properties
of the material necessary to make it thromboresistant were
rather easy to understand.   In the past decade many investigators have tried to define "thromboresistant material".   A
great amount of data has been generated, however, and has probably caused more confusion as to what a thromboresistant
material is.   Perhaps to understand these properties a rather
different approach or philosophy is necessary.   Current attempts and progress to make blood-compatible materials will be
discussed here, and in addition some of our current opinions
and hypotheses will be introduced.

SYNTHETIC AND/OR ARTIFICIAL MATERIALS

Thrombus formation is one of the most serious problems
encountered in the development of artificial devices used in
the cardiovascular system, such as heart valves, artificial
hearts, arteries, veins, pump-oxygenators and artificial kidneys.
Therefore most of the efforts have been directed to obtain antithrombogenic materials as a part of blood compatibility.   Most
of the potentially antithrombogenic materials that have been
studied can be categorized as in Table 2.

The first significant advance toward antithrombogenic surfaces came with the development of heparinized surfaces by Gott
et al. (3).   In this procedure, heparin (a naturally occurring
anticoagulant with molecular weight in the range of 4,000 to
18,000) is bonded to benzalkonium chloride which has been absorbed on a graphite coating applied to a plastic surface (GBH
surface).   The GBH treatment is based on the principle that
heparin forms water insoluble complexes with quaternary ammonium

TABLE 1:

MATERIALS WITH SIGNIFICANT BLOOD COMPATIBILITY EVALUATED BY SEVERAL TESTING METHODS[a]

Material	Blood Clotting Tests			Glass Clotting Time	Effect On Plasma Proteins	Effect on Plasma Enzymes	Immuno-genicity
	In Vitro	*In Vivo* 2 Hr	*In Vivo* 2 Weeks				
Polytetrafluoroethylene (Teflon)	X[b]	X	X	0[c]	0	△	-
Polydimethylsiloxane (Silastic)	X	X	X	0	0	0	-
Blend of polydimethyl-siloxane and polytrifluoro-propylmethylsiloxane (25)	0	0	△[d]	-	-	-	-
Polydimethylsiloxane heparinized by TDMAC technique (5)	0	0	0	X	X	△	X
Polydimethylsiloxane heparinized by APTES technique (9)	0	0	△	0	0	0	-
Polydimethylsiloxane heparinized with covalent bond technique (14)	0	△	△	-	-	-	-
Polyurethane (polyester base)	0	-	-	X	0	△	-

Material						
Polyurethane (polyether base)	O	—	X	O		—
Polyurethane heparinized	O	O	—	—	—	—
Carbon	O	△	O	O	△	X
Polyelectrolyte complex (Ioplex) 0.5 meq. excess anion	O	O	X	X	O	△
1.3 meq. excess anion	O	X	△	X	O	
Carboxymethyl cellulose	O	△	△	—	—	—
Polyacrylamide gel	O	O	—	△	O	—
Butylacrylate-methyl methacrylate-methacrylamide terpolymer	O	△	△	—	—	—
Ethylene-vinylacetate copolymer	O	X	O	O	O	—
Epoxy polymer heparinized	O	△	X	X	X	X

[a] For explanation of test procedures see the section of testing method in this article.
[b] Indicates unsatisfactory results.   [c] Indicates satisfactory results.
[d] Indicates that the material shows promise in that 50% or more of the test runs were satisfactory.

TABLE 2:

CLASSIFICATION OF MATERIALS WITH SIGNIFICANT THROMBORESISTANCE

	Characteristic	Basic Materials
Heparinized Surfaces	Three Step Process Graphite-Benzalkonium Chloride – Heparin	Polycarbonate (3)
	Two Step Process Cationic Surfactants – Heparin	Polyethylene (4,5), polypropylene (4,5), silicone rubber (5), poly(vinyl chloride) (5), polyurethane (5), polycarbonate (5), Teflon (5)
	Polymeric quaternary ammonium group – Heparin	Silicone rubber (6,9), polypropylene (6), Hydrin rubber (6), cellophane (6,7), polyurethane(8), etc.
	One Step Process Heparin Blend Cationic Surfactants – Heparin complex blend	Silicone rubber (10), epoxy polymers (11), acrylic polymers (12), polyurethane (12), Hydrin rubber (12), poly(vinyl chloride) (12)

Polyelectrolyte complex with excess sulfonate group	Poly(vinyl benzyltrimethyl ammonium chloride) and poly(sodium styrene sulfonate) (Ioplex) (17)
Polyelectrolyte with carboxyl group	Carboxymethyl cellulose (19), ethylene-acrylic acid copolymers (18)
Negatively Charged Surfaces	
Anionic surfactant	Butylacrylate - methylmethacrylate - methacrylamide terpolymer + dodecylbenzene-sulfonate (20)
Carbon black	Polyurethane containing 10% of carbon black (bioelectric polyurethane) (22)
Electret	Poly(vinylidene chloride) (24), Teflon (24), Hypalon (24), Hypalon - polysulfonate (24)
Hydrophobic Materials	
Fluorinated Silicone Rubber	Polytrifluoropropylmethylsiloxane (25)
Modified Silicone Rubber	Polysiloxane and polyurethane block copolymer (Avcothane-51) (26)
Saturated Hydrocarbon block copolymer	Poly(vinyl cyclohexane) - ethylene/propylene block copolymer (27)
Carbonaceous material	Carbon (LTI Pyrolite carbons) (28), Graphite (Dag 35) (29)
Hydrophilic Materials	
Hydrogel	Poly(hydroxyethyl methacrylate) gel (Hydron) (32), Polyacrylamide gel (33)

salts. This principle has been modified and applied successfully to a wide variety of materials by many investigators, with
elimination of a graphite coating by direct impregnation of cationic surfactants (4,5), or by direct attachment of quaternary
ammonium groups to polymeric surfaces (6-9). Moreover, elimination of the introduction of quaternary ammonium groups has led
to the direct impregnation of heparin (10,11) or heparin derivatives (12). Some attempts to attach heparin to polymeric surfaces by means of covalent bonds was made by Halpern (13) and
are being continued by Grode *et al.* (14), in the expectation of
achieving greater stability of heparinized surfaces because of
the relatively permanent nature of covalent bonds as compared
to ionic bonds that are the basis for the currently predominant
heparinization techniques.

Most of the heparinized materials seem to have initial *in
vivo* thromboresistance because heparin appears to be gradually
leached out from the surface. Not all the heparinized materials
show the same degree of compatibility as can be seen from Table
1. At this time, one of the materials that is most blood compatible may be heparinized silicone rubber by the γ-aminopropyl-
triethoxysilane technique (9).

It is generally accepted that blood cells are negatively
charged at the normal pH of blood, that the normal vascular surface appears to be negative to blood flowing through vessels and
that either a decrease or a reversal in this negative surface
charge would frequently lead to intravascular thrombosis (15).
It seems reasonable to try to prepare a moderately negatively
charged surface as a thromboresistant one. There have been
many unsuccessful attempts to correlate thrombogenicity of materials and surface negative charge as reviewed by Lyman (16).
Introduction of anionic groups such as sulfonate or carboxyl
groups to material surface is a more direct and popular method,
but is technically rather difficult and not reproducible. A

polyelectrolyte complex with 0.5 meq excess of anion or with
cation was thrombogenic (17).   The effect of degree of neutral-
ization or cation species on thrombogenicity of ethylene-acrylic
acid copolymers was observed (18).   Silicone rubbers containing
2-carboxylpropylmethylsiloxane unit were reported to be quite
thrombogenic (25).   In the use of the acrylic terpolymer containing the surfactant, it is not clear whether or not the
anionic surfactant plays an important role in thromboresistance,
because the post-treatment was necessary to obtain good results
(20), and epoxy polymers containing nearly the same amount of
the same surfactant were not thromboresistant (21).   The addition of 10% carbon black to polyurethane increased a negative
electrical potential from 0-100 mV (unmodified plain polyurethane) to 200-300 mV and gave improved thromboresistance (22,
23).   The materials that have electrical asymmetry in the molecules can be polarized in an electric field to give electrets,
which have permanent internal, electric polarization.   Negatively polarized halocarbon polymers showed improved blood compatibility (24).

Hydrophobic material is sometimes considered to be thromboresistant because of its low wettability and its chemical inert-
ness.   Silicone rubber having a dimethylsiloxane unit is one of
the most biocompatible materials available today.   Modification
of the traditional dimethylsiloxane polymer by introduction of a
trifluoropropylmethylsiloxane unit or a polyurethane unit (26)
further improved the blood compatibility.   Moreover, block copolymers synthesized from polysiloxane and polyurethane showed
favorable mechanical properties (26);   for example, Avcothane-51
has a tensile strength of 4,800 psi and breaks at 430% elongatior
as compared with 1,300 psi and 600% respectively for the usual
silicone rubber.

Another hydrophobic elastic material is a saturated hydrocarbon block copolymer with elastomer segments corresponding to

ethylene/propylene copolymer rubber and end segments polyvinyl-
cyclohexane, and was reported to have good blood compatibility
(27). This elastomer has a tensile strength in the range of
3,500 to 5,000 psi and breaks at from 700 to 1,600% elongation.
Studies of the effect of physical microstructure and surface
topography of carbonaceous materials on their thromboresistance
showed that only outgassed and smooth surfaces of isotropic carbons deposited at low temperatures (LTI Pyrolite carbons) had
significant blood compatibility (28), and only Dag 35 (of
Acheson Colloids Company, Alkid resin graphite solution), which
has a greater degree of internal porosity and permeability than
any other type of graphite, showed good thromboresistance different from that of the other graphite products (29). It is
reported (29) that there is mounting evidence that the antithrombogenicity of the GBH coating is not entirely dependent on
the presence of heparin. That there may be an important role
of Dag 35 and a lesser one of heparin is supported by the lack
of significant inhibition by protamine of the GBH surface (30).
Polyurethane and poly(vinyl chloride) blend containing Dag 154
is also claimed to be a thromboresistant coating material (31).
From the effects described above, polyurethane containing carbon
black may fit the category in this section better than in that
of the negatively charged surface, regardless of the authors'
emphasis on negative charge.

Normal body tissue may be characterized as a collagenous
gel system in which water is largely contained within the gel
network of the tissue protein. A highly hydrated surface is
only a partial surface, and also will be less receptive to adsorption of blood elements and therefore more thromboresistant.
These concepts have led to synthetic hydrogels such as cross-
linked poly(hydroxylethyl methacrylate) (Hydron) (32) or polyacrylamide (33). Natural rubber latex and other components
(34) should also be categorized in this group.

BIOLOGICAL SURFACE MATERIAL

After implantation of artificial material in the bloodstream, rapid surface coating by proteins or cellular lining inevitably ensues. Using this response as a weapon, active generation of neointimal lining on the surface of the artificial
material was attempted by many groups using various methods.

So far the most successful approach to developing a blood
compatible material has been the neointimal lining fabrics (35).
The formation of a neointimal layer is closely related to clot
formation. A fibrin layer deposited on the foreign surface is
infiltrated by fibroblasts and gradually becomes organized into
a layer of fibrous tissue that forms a smooth, glistening, white
coat resembling the endocardium. Once this layer has been
formed and entrapped by fabrics such as nylon or Dacron velour
backed with silicone rubber (36), blood is protected from
further contact with a foreign interface. The role of the
fabric is to encourage fibrin deposition and to anchor the
fibrin layer to the surface. Nylon and Dacron velour fabrics
gave the best results in contrast to Teflon, polypropylene and
rayon (35). The problem in regard to the neointimal lining is
how to control its thickness and to attach it firmly to the base
material. A Dacron velour fabric backed with polyurethane containing carbon black (Electrolour) was reported to give notice-
ably thinner neointima (37). Spontaneous organization of the
coagulum is slow and in some locations does not occur. For
this problem, tissue culture techniques have been applied: the
velour fabrics attached with tissue fragments were cultured *in
vivo* (38) and a fibrin coagulum membrane, which was attached
firmly on the velour surface by a steady pressure applied on it
during the preparation of the coagulum, was used as a substrate
for *in vitro* culture of autologous tissue (39).

The use of a natural tissue *per se* is a logical alternative, although its availability is limited. Homograft and heterograft valves have been used clinically with success for replacement of the aortic valve in an attempt to avoid the complications of thromboembolism, anticoagulation, infection and mechanical failure which may accompany the use of prosthetic valves (40-42). Autologous fascia lata grafts and fresh homografts are considered to be better than preserved homografts or heterografts (43).

In regard to the availability of the materials, preserved homologous tissues, and especially heterologous tissues, which have the same good blood compatibility as autologous or fresh homologous tissues, should be considered first. Heterologous tissues cannot be used in their natural state since rejection would immediately ensue. Therefore, they must be altered to preclude rejection. Beta propiolactone and formaldehyde have been used mainly as a preservative or fixing agent, and recently glutaraldehyde and dialdehyde starch have been introduced (44) to eliminate the problem of an antigen-antibody reaction and tissue deterioration which was sometimes observed in the formaldehyde-treated heterografts. The heterografts used are nonviable, sterilized, preserved tissues. Their long-term fate is uncertain, despite the fact that some have been functioning satisfactorily in animals up to 7 years post-operatively.

Excellent antithrombogenicity *in vivo*, as has been observed clinically in aortic valve heterografts, has also been confirmed from *in vitro* clotting tests (34).

The steps involved in thrombus formation on foreign surfaces are: (1) deposition of a layer of noncellular blood components, (2) deposition of platelets on top of the layer, (3) agglutination of platelets to form platelet aggregates containing an abundance of fibrin, (4) formation of fibrin strands that can arise from the noncellular layer and (5) formation of an interaggregate mesh of fibrin strands that trap cells in a manner

similar to that in a blood clot (45).   Therefore, a deposited
proteinaceous layer will determine subsequent reactivity to
blood, and blood compatibility of materials should be closely
related to the nature of the initially adsorbed protein layer.
The treatment of artificial material with proteins will provide
another type of new material and should be useful.   Albumin has
been attached to a polystyrene surface after chloromethylation
of the polymer (46).   Natural rubber, containing albumin or
gelatin, after treatment with formaldehyde or glutaraldehyde
showed good thromboresistance *in vitro* based upon laboratory
study.   Albuminated natural rubber did not give good results
without aldehyde treatment;   this may be similar to a rejection
phenomenon occurring in the untreated heterografts.   (Details
will be discussed in a later chapter.)

METHODS OF TESTING BLOOD COMPATIBILITY AND MATERIALS

A suitable and effective method for experimental evaluation
of blood compatibility is indispensable for the development of
blood compatible material.   Although various techniques have
been proposed, all of which are irrelevant to the actual *in vivo*
situation, the methods currently used will be described.

Thrombus formation
The modified Lee-White method is most popular as an *in vitro*
screening test.   The time required for a firm clot to appear is
determined by adding calcium chloride to ACD blood which is
placed in a tube made from or lined with a test material.   The
Lee-White method is rather qualitative and the end point of clot
formation is not clear, especially in the case of hydrophilic
materials.   Therefore, a more quantitative method, called the
kinetic method, has been developed so that detailed information
in regard to *in vitro* clot formation is acquired.   The point of
this method is to measure gravimetrically the amount of thrombus

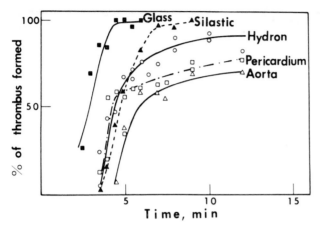

Fig.1 : Thrombus formation curve. This curve was plotted by the
so-called "Kinetic method" (61) developed by the authors.
If the curve is located on the right side and also the
lower side it is considered to be more thromboresistant
to the material located at the left and upper portion.
Pericardium and aorta were both treated with 4% buffered
formaldehyde solution for over 24 hours.

formed at an appropriate interval of time after calcium chloride
is added to ACD blood which is in contact with a test material
(34).   Typical graphs of thrombus formation plotted against
time are shown in Figs.1 and 2.   The percentage of thrombus
formed was calculated on the basis of the equilibrated or final
amount of thrombus formed on glass, which was used as a standard
material.   The curves obtained seem to provide the basis of the
interpretation of the degree of *in vivo* patency of aortic surface or Hydron.

The glass clotting time is a measure of the ability of
plasma to retain its clotting properties when transferred to a
glass container after 48 hours of exposure to finely divided
particles of the material *in vitro* (47).

The Stypven time, a measure of release of platelet factor 3
or other thromboplastic materials, and the partial thromboplastin
time, a measure of the intrinsic coagulation system, is deter-

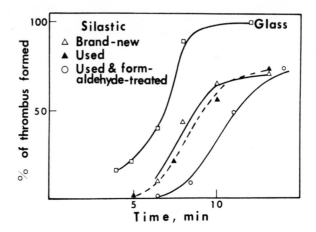

Fig.2 : Thrombus formation curve generated by the kinetic method.
Silastic rubber and Silastic rubber exposed to blood protein showed the same thromboresistant properties. This
will improve after the blood protein coated surface is
treated by 4% buffered formaldehyde solution for over 24
hours.

mined after fresh whole blood in a test tube is stirred with a
rod coated with a test material (48).

The inferior vena cava implant system in dogs provides a
well-established *in vivo* testing method for the detection of
thrombi (49). The standard vena cava ring is 9 mm long, has an
o.d. of 8 mm and an i.d. of 7 mm. Materials that remain free
of thrombus after a 2-hr implantation in the vena cava appear to
have significant thromboresistance, because all the ordinary
materials available form severe thrombus under these conditions.
Those materials that pass the 2-hr test are then subjected to the
2-week test. The inferior vena cava ring method provides consistently reliable results; however, turbulence can greatly
alter the results, and it is imperative for the leading and
trailing edges of the ring to be properly streamlined. The
modified caval test ring with a circumferential ridge or "curb-
stone" in the middle of the ring, which provides turbulent flow,

gave completely different results (50). The GBH-coated turbulent rings with a 1-mm wide and 1-mm high ridge showed significant amounts of thrombus after 30 min of implantation, whereas the nonridged GBH rings were free of thrombus after 2 hr or 2 weeks.

The vena cava ring method seems to provide the most consistent information with regard to thromboresistance of materials when compared to the atrial "flags" or "swords" method (49) that is sometimes employed to evaluate thrombogenicity of materials, and which gives inconsistent results due to fibrinolytic activity in the animal (51).

The caval ring method described above is useful to detect thrombotic material actually present on the surface of the implant, but is somewhat less suited to demonstrate the prior occurrence of minute thromboembolism that may be lethal or seriously injurious within the systemic arterial system. The renal vascular bed and parenchyma have been used as a biological indicator of thromboembolic phenomena (52). Ligation of the abdominal aorta below the origin of the renal arteries was accompanied by insertion of a ring of test material immediately above the origin of these vessels; thus, blood flowing through the test ring must course through the kidneys (organs that are highly susceptible to infarction). The finding that a GBH-coated poly(vinyl chloride) ring was completely free of thrombus yet there were a number of renal infarcts suggests that the absence of thrombotic material on the surface of intravascular prostheses in no way excludes the possibility of antecedent embolization (with possible infarction) from such surfaces.

Attempts to fill the gap between the static *in vitro* tests, and the somewhat complicated, invisible *in vivo* tests have included the use of the reflected-light microscopy technique (53, 54). The direct microscopic observation of thrombus formation on artificial surfaces has been undertaken by use of a stag-

nation flow chamber in which blood flow is known precisely, so
that shear rates at the surface and diffusion of platelets and
proteins to the surface are well controlled.   A sequence of reactions in thrombus formation was observed on artificial
materials after the blood was directly introduced from a dog's
carotid artery into the stagnation-flow chamber.   The test is
so designed that when experimental observations were limited to
a small region in the immediate vicinity of the stagnation
points, blood was being observed which had no prior surface contact, and the blood withdrawn was not returned to the animal, to
avoid modification of the animal's circulating blood.   Platelets
were always the first-formed element to attach to a surface, but
did not necessarily lead to thrombus formation.   In all regions
where thrombi formed, leukocytes also became attached to the surface, and the least possible tendency for leukocytes to adhere
seems to provide one criterion for judging thromboresistance (55).

Hemolysis

Accelerated destruction of erythrocytes by extracorporeal
circulation (hemodialyzers, pump oxygenators) and artificial heart
devices has been observed.   Hemolysis appears to be due primarily
to physical and mechanical factors such as pumping, flow rate or
circuit design which affect turbulence, shear forces, or frequency
of collision between erythrocytes and foreign surfaces in the vascular system.   The relationship between materials and hemolysis
is not clear, and has not been studied extensively.

Four different *in vitro* testing methods have been reported
(6,34,48,56).   A test tube containing ACD blood was incubated at
37°C in a shaker-incubator that was reciprocating at the rate of
180 cpm for 4 hr (6).   In a container made of testing materials,
blood was stirred at 60 rpm for 4 min with a paddle made of the
same materials (48).   ACD blood and powdered, sieved material in
a glass flask at room temperature, were stirred gently for various

periods up to 24 hr with a small magnetic Teflon-coated stirring
rod (56).   The test tube filled just with ACD blood and without
air was rotated at 50 rpm at 30°C for 24 hr (34).   After these
procedures all the blood underwent assay of plasma hemoglobin.
In general, polymers having a reactive functional group are more
hemolytic, and low molecular weight organic compounds contained
as additives or impurities are the basis of hemolytic activity
of materials.

In clinical trials, specimens of plasma from chronically
dialyzed patients showed no measurable free hemoglobin in plasma
collected before, during and after hemodialysis; thus indicating
that either only small amounts were being produced or that reti-
culoendothelial uptake of free hemoglobin was so rapid as to
prevent detection.   *In vitro* evaluation of clinical dialyzers
has shown that they produced only slight hemolysis (57).   Considering these facts and the fact that cellulosic materials used
for hemodialysis membrane are the most hemolytic material (6,48),
the chance that materials *per se* cause hemolysis in the vascular
system seems to be slight.

Effect on other cellular elements

The amount of adenine nucleotides released into plasma as a
measure of cellular damage, especially to platelets, has been
determined as a part of blood-compatibility tests (48).   A decrease in leukocyte phagocytic capability and oxygen utilization
and morphologic alterations of leukocytes has been observed in
in vitro closed-loop pumping of blood at room temperatures (58).
The effect of materials on leukocytes in regard to blood compatibility has not been studied.

Denaturing effects on plasma proteins

The potentially direct denaturing effects of various polymeric materials on the proteins in human plasma *in vitro* have

been evaluated in a specially devised experimental system (47, 56). The plasma sample was exposed to a relatively large surface of granular test material in a silicone rubber chamber, to enhance any alterations that might occur, and was gently rotated periodically through 180 degrees at 37°C for as long as 48 hr. Seventeen different plasma proteins were quantified by radial immunodiffusion assay; eight plasma enzyme activities were measured; immunoelectrophoresis was performed with antiserum to whole human serum; disk electrophoresis patterns on acrylamide gel were examined. Silicone rubber, ethylene-vinyl acetate copolymer, and polyacrylamide gel were found to be inert to the plasma proteins, while several other materials revealed only minor alterations.

Immunogenic effects on plasma proteins

A possible characteristic of materials which may have long-term toxic potential is the promotion of immunogenicity in adsorbed autologous plasma proteins. Test materials and rabbit plasma were sealed in poly(vinyl chloride) bags and agitated gently at 37°C for 24 hr in the absence of gas interfaces. A sample of the exposed plasma was retained for analysis and later used as antigen, and the remainder was injected into the original donor rabbit in an attempt to stimulate an immune response. The formation of antibodies was determined by interfacial ring, gel diffusion, passive hemagglutination and passive cutaneous anaphylaxis technique (59). Some of the materials became immunogenic for the host, but no comparison or assessment of materials is yet possible.

"Biolization" - Hypothetical process to generate a biocompatible material

It has been well documented that any artificial material, whether or not it is hydrophilic, hydrophobic, inert, electroconductive or heparinized is covered with protein almost instantaneously after implantation in the vascular system. Regardless of this protein coating, some of the material still showed better thromboresistance over other materials. Many theories have been proposed to explain these thromboresistant properties of the material, but it is certainly difficult at this moment to interpret what properties make the material thromboresistant. Coating the surface of a cardiac prosthesis with monomolecular layered undenaturated albumin has been proposed by Lyman *et al.* (60), but its effect was rather difficult to reproduce. Through our efforts we found that not every protein surface shows thromboresistance; however, when it is treated by aldehyde then it shows reasonable and reproducible thromboresistance (61,62,63).

This protein can be either mixed in the synthetic material such as in albumin mixed natural rubber, coated over plastic material, or it can be from a natural origin such as in natural tissue (61,62,63) (Fig.1). Other biological components, such as polysaccharides (e.g. heparin), probably have a similar effect (61).

It is obvious that protein alone on a surface will not enhance its thromboresistance; however, when this protein is treated either by aldehyde, heat, or some other method, it becomes quite thromboresistant (Fig.2). This treatment makes the protein insoluble, cross-linked or denaturated. What kind of process is actually taking the main role is not completely understood at this time so it was difficult to select an appropriate term to describe it. Thus, the term "biolization" was made to describe this process. Together with this term, hypothetical processes for the production of biocompatible and particularly blood compatible materials were proposed (Fig.3).

Material to be used for this process will be either of natural or artificial origin, but it should contain protein, polysaccharides or other biological components which are not cer-

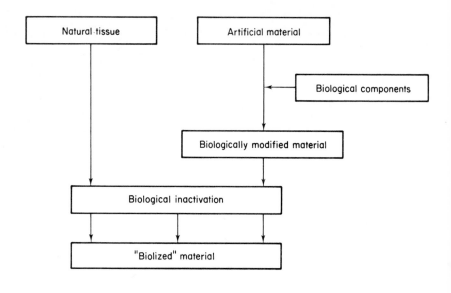

(Thromboresistance) (Biocompatible)

Fig.3 : "Biolization" - Hypothetical process to generate bio-
        compatible material.

tain at this moment.   Artificial synthetic materials can be
subjected to this condition either by the mixture of biological
components or coating with biological components.   When the
surface is treated it becomes "biologically inactivated".   "Bio·
logical inactivation" means insolubilization of the biological
components on the surface or cross-linkage of the biological com·
ponents.
        Natural tissue of a tissue typing group different or remote
can be treated in the same way and inactivated in the same way
so that this tissue can be used for implantation out of the target for the intensive immunorejection process.   Heterologous or
homologous preserved aortic valves have already been successfully used clinically.   A low incidence of thromboembolic complications was unanimously reported.

In the past attempts have been made to implant biological tissue, as in transplantation, and to keep the biological activity as normal as possible.   However, when the species is different this biological activity stimulates the defense mechanisms of the body.   By making the surface protein nonspecific heterologous tissue can be implanted.   Some of the tissue used is actually serving, not as a biologically functioning organ, but as part of the mechanical structure, e.g., vascular grafts or heart valves.   For this type of application this nonviable graft will serve the purpose of implantation.   If nonviable tissue material can be used as parts for cardiovascular prostheses combined with synthetic material it will open up a new field for the construction of thromboresistant and hemodynamically ideal devices.

In summary, for biolization to occur, the surface must have: 1) a protein or biological component on the surface; 2) this protein or biological component must become nonspecific, crosslinked, or insoluble by aldehyde, heat or some other method; 3) the surface must be hydrophilic.   Probably from the viewpoint of the recipient of the device, it is impossible to make a differentiation between transplantation of the natural tissue and implantation of the prosthetic material.   The reaction of the recipient's system to the foreign system must be the same whether the foreign system is of natural or synthetic origin.

CONCLUSION

Blood compatible materials are essential for the development of any prosthetic devices that are in contact with blood, such as prosthetic heart valves, vascular prosthesis, pump oxygenator, artificial kidneys and artificial hearts.   None of the synthetic materials available today are satisfactory, and efforts to solve the problem of blood incompatibility are con-

tinuing. A hopeful development is the evidence that the biological materials derived from natural origin demonstrate excellent blood compatibility. Current approaches to the problem
of blood compatibility are reviewed, and the various materials
and tests are discussed.

In order to give some clear understanding of this problem
to the authors one hypothetical explanation was attempted utilizing the term "biolization". However, it must be proved by
further experimentation.

REFERENCES

1. DeVries, W.C., Kwan-Gott, C.S. and Kolff, W.J.   Consumptive
    coagulopathy, shock, and the artificial heart.   *Trans.*
    Amer. Soc. Artif. Int. Organs 16: 29 (1970).

2. Cardiac Replacement - A Report by Ad Hoc Task Force on Car-
    diac Replacement.   National Heart Institute, page 76,
    October 1969.

3. Gott, V.L., Whiffen, J.D. and Dutton, R.C.   Heparin bond-
    ing on colloidal graphite surfaces.   *Science 142*: 1297
    (1963).

4. Eriksson, J.C., Gillberg, G. and Lagergren, H.   A new
    method for preparing nonthrombogenic plastic surfaces.   *J.*
    Biomed. Mater. Res. 1: 301 (1967).

5. Grode, G.A., Anderson, S.J., Grotta, H.M. and Falb, R.D.
    Nonthrombogenic materials via a simple coating process.
    Trans. Amer. Soc. Artif. Int. Organs 15: 1 (1969).

6. Leininger, R.I., Falb, R.D. and Grode, G.A.   Blood-com-
    patible plastics.   *Ann. N.Y. Acad. Sci. 146*: 11 (1968).

7. Britton, R.A., Merrill, E.W., Gilliand, E.R., Salzman, E.W.,
    Austen, W.G. and Kemp, P.S.   Antithrombogenic cellulose
    film.   *J. Biomed. Mater. Res. 2*: 429 (1968).

8. Gardner, D.L., Sharp, W.V., Ewing, K.L. and Finelli, A.F.
    Stability of heparin S^{35} attached to a modified polyurethane
    vascular prosthetic.   *Trans. Amer. Soc. Artif. Int. Organs*
    15: 7 (1969).

9. Merker, R.L., Elyash, L.J., Mayhew, S.H. and Wang, J.Y.C.
    The heparinization of silicone rubber using aminoorgano-
    silane coupling agent.   *Proc. Art. Heart Program Conf.*,
    page 29 (1969).

10. Hufnagel, C.A., Conrad, P.W., Gillespie, J.F., Pifarre, R., Ilano, A. and Yokoyama, T. Characteristics of materials for intravascular application. *Ann. N.Y. Acad. Sci. 146*: 262 (1968).

11. Salyer, I.O. and Weesner, W.E. Materials and components for circulatory assist devices. *Proc. Art Heart Program Conf.*, page 59 (1969).

12. Imai, Y. and Masuhara, E. Preparation of non-thrombogenic materials. Reports of the Institute for Medical and Dental Engineering, Tokyo Medical and Dental University, 3: 72 (1970).

13. Halpern, B.D. and Shibakawa, R. Heparin covalently bonded to polymer surface. Interaction of liquids at solid substrate. Advances in Chemistry Series 87, American Chemical Society, page 197 (1968).

14. Grode, R., Falb, R. and Anderson, S. Development of materials for circulatory assist devices. *Proc. Art. Heart Program Conf.*, page 19 (1969).

15. Sawyer, P.N., Srinivasan, S., Chopra, P.S., Martin, J.G., Lucas, T., Burrowes, C.B. and Sauvage, L. Electrochemistry of thrombosis - An aid in the selection of prosthetic materials. *J. Biomed. Mater. Res. 4*: 43 (1970).

16. Lyman, D.L. Biomedical polymers. *Ann. N.Y. Acad. Sci. 146*: 30 (1968).

17. Bixler, H.J., Cross, R.A. and Marshall, D.W. Polyelectrolyte complexes as antithrombogenic materials. *Proc. Art. Heart Program Conf.*, page 79 (1969).

18. Costello, M., Stanczewski, B., Vriesman, P., Lucas, T., Srinivasan, S. and Sawyer, P.N. Correlation between electrochemical and antithrombogenic characteristics of polyelectrolyte materials. *Trans. Amer. Soc. Artif. Int. Organs 16*: 1 (1970).

19. Nemchin, R.G., Patel, A.R., Able, H.I., Sims, L. and Speaker, D.M. Anionic cellulose: Potential blood compatible material. *Proc. Art. Heart Program Conf.*, page 71 (1969).

20. Leonard, F., Nielson, C.A., Fadali, A.M. and Gott, V.L. Thromboresistant polymers by emulsion polymerization with anionic surfactants I. *J. Biomed. Mater. Res. 3*: 455 (1969).

21. Salyer, I.O. and Weesner, W.E. Materials and components for circulatory assist devices. *Proc. Art. Heart Program Conf.*, page 59 (1969).

22. Sharp, W.V., Gardner, D.L. and Anderson, G.L. A bioelectric polyurethane elastomer for intravascular replacement. *Trans. Amer. Soc. Artif. Int. Organs 12*: 179 (1966).

23. Taylor, B.C., Sharp, W.V., Wright, J.I. and Ewing, K.L. Factors that affect success of small diameter polyurethane vascular prosthetics. *Proc. 23rd Ann. Conf. on Eng. in Med. and Biology 12*: 146 (1970).

24. Murphy, P., Lacroix, A., Merchant, S. and Bernhard, W. Studies relative to materials suitable for use in artificial heart. *Proc. Art. Heart Program Conf.*, page 99 (1969).

25. Musolf, M.C., Hulce, V.D., Bennett, D.R. and Ramos, M. Development of blood compatible silicone elastomers. *Trans. Amer. Soc. Artif. Int. Organs 15*: 18 (1969).

26. Nyilas, E.   Development of blood-compatible elastomers: Theory, practice and in vivo performance. *Proc. 23rd Ann. Conf. Eng. Med. Biology 12*: 147 (1970).

27. Bishop, E.T. and O'Neil, W.P.   Block copolymers for use in blood pumps and oxygenators: Preparation and characterization.   *Proc. Art. Heart Program Conf.*, page 133 (1969).

28. LaGrange, L.D., Gott, V.L., Bokros, J.C. and Ramos, M.D. Compatibility of carbon and blood. *Proc. Art. Heart Program Conf.*, page 47 (1969).

29. Milligan, H.L., Davis, J.W. and Edmark, K.W.   The search for the nonthrombogenic property of colloidal graphite. *J. Biomed. Mater. Res. 4*: 121 (1970).

30. Whiffen, J.D., Young, W.P. and Gott, V.L.   Stability of the thrombus resistant graphite-benzalkonium-heparin surface in an anti-heparin environment. *J. Thoracic Cardiovasc. Surg. 48*: 317 (1964).

31. Wakabayashi, A.   Discussion for the paper entitled "compatibility of carbon and blood". *Proc. Art. Heart Program Conf.*, page 57 (1969).

32. Levowitz, B.S., La Guerre, J.N., Calem, W.S., Gould, F.E., Scherrer, J. and Shoenfeld, H.   Biologic compatibility and applications of Hydron. *Trans. Amer. Soc. Artif. Int. Organs 14*: 82 (1968).

33. Halpern, B.D., Cheng, H., Kuo, S. and Greenberg, H. Hydrogels as non-thrombogenic surfaces. *Proc. Art. Heart Program Conf.*, page 87 (1969).

34. Imai, Y., von Bally, K. and Nosé, Y.   New elastic materials for the artificial heart. *Trans. Amer. Soc. Artif. Int. Organs 16*: 17 (1970).

35. Hall, C.W., Liotta, D., O'Neal, R.M., Adams, J.G. and DeBakey, M.E.   Medical application of the velour fabrics. *Ann. N.Y. Acad. Sci. 146*: 314 (1968).

36. Liotta, D., Hall, C.W., Akers, W.W., Villanueva, A., O'Neal, R.M. and DeBakey, M.E.   A pseudoendocardium for implantable blood pumps.   *Trans. Amer. Soc. Artif. Int. Organs 12*: 129 (1966).

37. Sharp, W.V., Gardner, D.L., Anderson, G.J. and Wright, J. Electrolour: A new vascular interface.   *Trans. Amer. Soc. Artif. Int. Organs 14*: 73 (1968).

38. Ghidoni, J.J., Liotta, D., Hall, C.W., O'Neal, R.M. and DeBakey, M.E.   In vivo culture of tissue fragments which produce viable cellular linings covered by endothelium (neointimas) in impermeable, velour-lined arterial prostheses, bypass pumps, and valvular prostheses.   *Trans. Amer. Soc. Artif. Int. Organs 14*: 69 (1968).

39. Adachi, M., Suzuki, M. and Kennedy, J.H.   Preparative coating of velour-lined circulatory assist devices with a fibrin coagulum membrane (FCM).   *Trans. Amer. Soc. Artif. Int. Organs 16*: 7 (1970).

40. Ross, D.N.   Homograft replacement of the aortic valve. *Lancet 2*: 487 (1962).

41. Barratt-Boyes, B.G.   Homograft aortic valve replacement in aortic incompetence and stenosis.   *Thorax 19*: 131 (1964).

42. Binet, J.P., Duran, C.G., Carpentier, A. and Langlois, J. Heterologous aortic valve transplantation.   *Lancet 2*: 1275 (1965).

43. McGoon, D.C.   Proceedings of the first international workshop on tissue valves.   *Ann. Surg. 172*, No.1, page 21, (1970).

44. Keshishian, J. and Carpentier, A.   Discussion for the paper "Deterioration of formalin-treated aortic valve heterografts".   *J. Thoracic Cardiov. Surg. 60*: 678 (1970).

45. Dutton, R.C., Webber, A.J., Johnson, S.A. and Baier, R.E. Microstructure of initial thrombus formation on foreign materials.   *J. Biomed. Mater. Res. 3*: 13 (1969).

46. Denver Research Institute: Described in the paper "The in vivo screening of potential thromboresistant materials". *Proc. Art. Heart Program Conf.*, page 185 (1969).

47. Halbert, S.P., Ushakoff, A.E. and Anken, M.   Compatibility of various plastics with human plasma protein.   *J. Biomed. Mater. Res. 4*: 549 (1970).

48. Mason, R.G., Scarborough, D.E., Saba, S.R., Brinkhous, K.M., Ikenberry, L.D., Kearney, J.J. and Clark, H.G. Thrombogenicity of some biomedical materials: Platelet-interface reactions. *J. Biomed. Mater. Res. 3:* 615 (1969).

49. Gott, V.L., Whiffen, J.D. and Valiathan, S.M. Graphite-benzalkonium-heparin coatings on plastics and metals. *Ann. N.Y. Acad. Sci. 146:* 21 (1968).

50. Gott, V.L., Ramos, M.D., Najjar, F.B., Allen, J.L. and Becker, K.E. The in vivo screening of potential thrombo-resistant materials. *Proc. Art. Heart Program Conf.*, page 181 (1969).

51. Jacobs, L.A., Klopp, E. and Gott, V.L. Studies on the fibrinolytic removal of thrombus from prosthetic surfaces. *Trans. Amer. Soc. Artif. Int. Organs 14:* 63 (1968).

52. Kusserow, B., Larrow, R. and Nichols, J. Observations concerning prosthesis-induced thromboembolic phenomena made with an in vivo embolus test system. *Trans. Amer. Soc. Artif. Int. Organs 16:* 58 (1970).

53. Dutton, R.C., Baier, R.E., Dedrick, R.L. and Bovman, R.L. Initial thrombus formation on foreign surfaces. *Trans. Amer. Soc. Artif. Int. Organs 14:* 57 (1968).

54. Petschek, H., Adamis, D. and Kantrowitz, A. Stagnation flow thrombus formation. *Trans. Amer. Soc. Artif. Int. Organs 14:* 256 (1968).

55. Petschek, H. and Madras, P.N. Thrombus formation on artificial surfaces. *Proc. Art. Heart Program Conf.*, page 271 (1969).

56. Cordis Corporation: Compatibility of blood with materials useful in the fabrication of artificial organs. PB-176, 200 (1967).

57. Hyde III, S.E. and Sadler, J.H. Red blood cell destruction in hemodialysis. *Trans. Amer. Soc. Artif. Int. Organs 15:* 50 (1969).

58. Kusserow, B., Larrow, R. and Nichols, J. Metabolic and morphological alterations in leukocytes following prolonged blood pumping. *Trans. Amer. Soc. Artif. Int. Organs 15:* 40 (1969).

59. Stern, I.J., Kapsalis, A.A. and Neil, B.L. Immunogenic effects of materials on plasma proteins. *Proc. Art. Heart Program Conf.*, page 259 (1969).

60. Lyman, D.J., Brash, J.L. and Klein, K.G. The effect of chemical structure and surface properties of synthetic polymers on the coagulation of blood. *Proc. Art. Heart*

Program Conf., page 113 (1969) and personal communication from Dr. Lyman to the authors.

61. Imai, Y., Tajima, K. and Nosé, Y. Biolized materials for cardiovascular prosthesis. *Trans. Amer. Soc. Artif. Int. Organs 17*: 6 (1971).

62. Nosé, Y., Tajima, K., Imai, Y., Klain, M., Mrava, G., Schriber, K., Urbanek, K. and Ogawa, H. Artificial heart constructed with biological material. *Trans. Amer. Soc. Artif. Int. Organs 17*: 482 (1971).

63. Nosé, Y., Imai, Y., Tajima, K., Ogawa, H., Klain, M. and von Bally, K. Cardiac prosthesis utilizing biological material. *J. Thoracic Cardiov. Surg.* (in press).

RECENT TRENDS IN APPROACHES

TO ELECTROMAGNETIC

DETERMINATION OF BLOOD FLOW

AND OUTLOOKS FOR THE FUTURE*

ALEXANDER KOLIN

*University of California, School of Medicine,
Los Angeles, California, U.S.A.*

INTRODUCTION

The development of the electromagnetic flow meter has been stimulated a third of a century ago by the need for a method to measure blood flow in unopened blood vessels (1,2). Injury to blood vessels was to be avoided to obviate the necessity of heparinizing the animal in order to avoid blood clotting. The solution of this problem by the electromagnetic flow meter resulted in a device which proved useful not only in animal experimentation but also in diagnostic applications to human subjects. The original approach required, however, extensive surgery to

* The author's work on external magnetic field catheter flow meters has been supported by Medical Testing Systems, Inc. and all other work on electromagnetic catheter flow meters by the Office of Naval Research. The recent animal experiments have been carried out in the Leo Rigler Center for Radiological Sciences where the able assistance of Miss L.M. Jaco, Miss N. Ross and Mr. M. Tsuno proved very helpful.

expose and dissect from the surrounding tissues the blood vessel to which the flow transducer is applied externally.   The
need for extensive surgery was a distinct drawback in applications to research animals as well as to human subjects.   A
trend developed, therefore, during the past decade to search
for an alternative mode of application of the electromagnetic
principle which would obviate the need for extensive surgery.

Such a simplification proved indeed possible.   It was
stimulated by the development of angiographic techniques in
which catheters made of materials which do not tend to induce
blood clot formation (such as Teflon, polyethylene or polyurethane) are introduced into blood vessels for a sufficient
length of time to complete diagnostic procedures.   Such
catheters have been used to withdraw blood samples, to inject
fluids or to link devices, such as pressure transducers, to the
bloodstream and the idea to introduce a flow transducer via a
catheter into the bloodstream seemed to be a logical step.
Even a cutdown on a branch artery or vein to introduce a catheter which could be moved into the aorta, vena cava, or pulmonary
artery is a minor procedure as compared to surgical exposure of
these deep-seated vessels and their separation from surrounding
tissues necessary for the application of a conventional perivascular flow meter.

The physical principles and conventional approaches to
electromagnetic measurement of blood flow have been extensively
reviewed in the literature (3,4,5) and will not be considered
here.   Our attention will be focused in this paper on the exploration of various recent ideas on the design of intravascular
blood flow sensors which appear to reflect the main trend in current developments in the field of electromagnetic determination
of blood flow.

In the original scheme of the electromagnetic blood flow
meter (1,2,3), a magnet external to an intact blood vessel gener-

ates a magnetic field which permeates its lumen.   The e.m.f. induced in the moving blood gives rise to a potential difference across the blood vessel wall which can be optimally sensed by two electrodes contacting the external blood vessel surface along a diameter perpendicular to the magnetic field which is preferably oriented at right angles to the direction of blood flow.

In the first successful scheme of an intravascular flow sensor this configuration was inverted (6).   A miniature magnet was introduced into the bloodstream.   The electrodes were likewise inside the blood vessel lumen disposed in such a fashion as to maximize the flow signal conveyed from them to the amplifier.

The physical background of the currently used catheter blood flow meters go back to 1944-45 (7,8).   We shall distinguish between two main types of flow sensors:

(a)   VELOMETERS, which measure the local velocity of flow and yield the rate of volume flow through the blood vessel lumen under assumption of a uniform velocity throughout the vessel cross-section in pulsating arterial flow by multiplication of the velocity by the blood vessel cross-section area which must be determined (usually neglecting the pulsatile variations in diameter).

(b)   VOLUME RATE OF FLOW METERS, which yield a flow signal which is directly proportional to the rate of volume flow through the blood vessel.   This is the normal mode of action of the standard electromagnetic flow meter based on a circular conduit in a uniform transverse magnetic field.

We shall now review different configurations of magnetic fields and sensing electrodes which can be used for intravascular flow measurements.

THE INSIDE-OUT MAGNETIC SENSOR SCHEME

The early flat disk velometer (7) proved far less effective
than the other investigated schemes and was not pursued beyond
exploratory experiments.   Using more sophisticated electronics
and far superior probe design, Mills (6,9) succeeded in developing a very effective intravascular catheter velometer.   Fig.1
shows the scheme of Mills' catheter tip velocity probe.
Grooves are provided in a cylindrical epoxy body, so that a 30-
turn coil, to be wound as shown, and twisted electrode leads to
be guided toward two diametrically opposite signal electrodes
(so as to minimize the induced quadrature voltage) can be incorporated in the transducer.   The epoxy body is enclosed in a
nylon tube of 3 mm o.d.   A round platinum tip serves as a
grounding electrode.   A central lumen permits measurement of

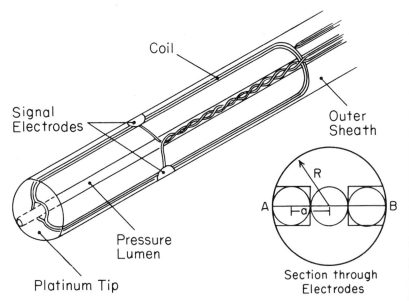

Fig.1 : Catheter tip velometer with provision for concurrent
        blood pressure recording.   From C.J. Mills in New Find-
        ings in Blood Flowmetry; Chr. Cappelen, Ed. (Universi-
        tetsforlaget, Oslo, 1968).

blood pressure at the catheter tip. (More recently the flow
sensor has been reduced in diameter so as to be incorporated in
a nylon tube equivalent to a No.8 cardiac catheter (10).)

This catheter probe is sensitive to flow parallel to the
cylinder axis. Nevertheless, angulation of the probe against
the direction of flow by as much as 30° causes less than 3% drop
in sensitivity and by 60° less than 11%. Hence, no steps are
taken to center and align the probe within the blood vessel.
The lack of centering of the probe entails the danger of an
effect of electrode proximity to the vessel wall upon the probe
sensitivity. In a plastic tube, the sensitivity falls off by
25% when an electrode contacts the tube wall. To avoid this,
the catheter is deliberately bent near the tip so as to angulate
the flow transducer.

This flow probe is a velometer. It integrates the flow
signal over an annular cylindrical space ranging about a probe
radius beyond its surface. To determine the volume rate of
flow in a blood vessel, its diameter must be determined radio-
graphically and the volume rate of flow is computed under the
assumption of a uniform velocity throughout the blood vessel
cross-section (taking the obstruction of the blood vessel lumen
by the catheter into account). The sensitivity of the probe is
13% higher in the laminar regime than in the turbulent range.

Mills' catheter tip velometer has been extensively used
(9,10) on human subjects for recording of blood flow in the
aorta, pulmonary artery and vena cava. It appears at the moment,
in spite of some imperfections, to be the most reliable and
most thoroughly explored means of studying blood flow in major
blood vessels of human subjects by means of catheter techniques.

One of the main advantages of this probe is the ease with
which the base line can be established without stopping the
flow. This can be done by immersing the transducer in a
beaker of saline. This catheter transducer has been under

thorough study for a longer period of time than any of the
others and has been extensively used in clinical investigations
(10,11).

Almost simultaneously with the development of the probe
described above, a catheter tip flow meter utilizing an iron
core transducer was described by Bond and Barefoot (12). Like
the previous probe, it measures about 3 mm in diameter and no
attempt is made to center or align the catheter within the
blood vessel. Commercial probes are available corresponding
to No.8, 9, 10 French cardiac catheters. Errors as large as
20% can be made in flow measurements due to transposition of
the probe within the lumen of the vessel. The zero flow reference is established by "withdrawing the catheter to its point
of insertion into the vessel and occluding the flow distal to
that point". The original paper (12) did not describe the
probe design which was provided in a later publication (13).

A physiological evaluation of the iron core velometer probe
was made in 1969 (14) by experiments on anesthetized dogs. The
authors reported their satisfaction with the correlation between
measurements made with the catheter tip velometer and perivascular electromagnetic flow probes. The catheter probes corresponded to No.8 French cardiac catheters. Among limitations of
this type of probe, these authors cite "... high frequency electrical interference ('noise') ..." and they note that "...
Alterations in the electrical signal of the catheter probe could
occur with changes in position of the catheter probe within the
cross-sectional area of the blood vessel."

A report (15) referring to observations on animals and
patients describes the use of a Bond and Barefoot velometer to
evaluate the reliability of the dye-dilution method of cardiac o
put determination. The authors find the latter to be inaccurate
"... especially in high flow states ... and in hypovolemic state
...". "If the cardiac output falls below 50% of normal the

racings are frequently impossible to interpret." The zero-
low base line has been obtained by momentary interruption of
lood flow. At the aortic arch the base line has been pro-
ided during the diastolic interval.

A third type of cylindrical flow probe which responds to
ongitudinal flow around the transducer has been described re-
ently but has not as yet been studied thoroughly. It will be
escribed in the section on unconventional configurations (16).

The common limitation of all velometers is that they
easure the local fluid velocity at a "point" within the blood
essel cross-section, whereas one is normally interested in the
olume rate of flow through the blood vessel cross-section.
he latter can be obtained strictly and easily only in cases
here the vessel cross-section does not vary significantly
hroughout the cardiac cycle and where the velocity profile can
e assumed to be flat. The latter assumption appears to be an
cceptable approximation in the pulsating flow not far from the
oot of the aorta (17,18). Other limitations are due to poss-
ble falsification of flow velocity records due to whipping
otion of the flow-sensor tip throughout the cardiac cycle and
o possible lateral translocations within the blood vessel
umen.

UNCONVENTIONAL SCHEMES BASED ON CONVENTIONAL CONFIGURATIONS

The schemes described above were velometers designed for
elocity measurements in the major blood vessels. To solve
he problem of measuring the volume rate of flow to various or-
ans through branches of the aorta by means of a catheter
evice, the following scheme was used (19,20). Fig.2a shows a
atheter inserted into the aorta via a femoral artery. Near
he tip of the catheter is a flow transducer with a lumen L
hose axis is perpendicular to the catheter axis. A flexible
ail T made of silicone-rubber-covered spring steel forms the

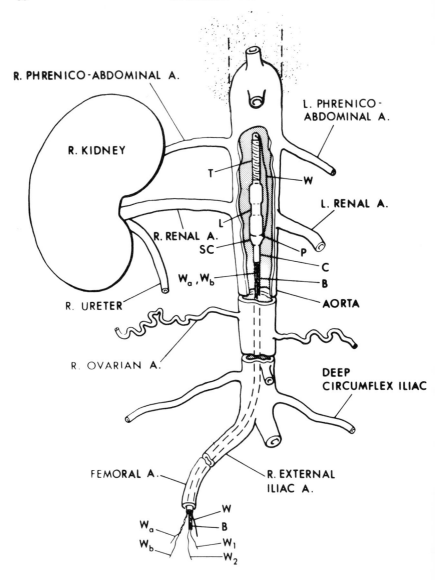

Fig. 2a

Fig.2a : Catheter flow sensor introduced into the aorta via the
femoral artery. L: lumen; T: flexible "tail";
W: pull wire emerging at P; C: catheter tubing;
W_1, W_2: electrode lead wires; W_a, W_b: magnet coil
lead wires.

end of the catheter.   A pull wire W attached to the tip of
the tail and entering the catheter below the transducer body at
point P can now accomplish the following.   As W is pulled,
the tail T is flexed as shown in Fig.2b and the lumen of the
transducer is pressed against the ostium of the branch artery
(the renal in this case).   All of the blood flowing through
the renal artery must now pass through the transducer.   Thus,

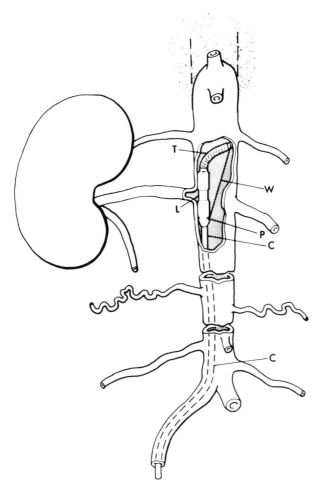

Fig.2b : Deformation of the tail T by pull on wire W presses
flow meter lumen against ostium of right renal artery.

Alexander Kolin

the volume rate of flow through the kidney can be measured according to the basic principle of the electromagnetic flow meter. To
ensure that no blood will leak into the artery around the transducer and thus escape detection, a flexible washer (not shown) is
cemented around the opening of lumen L. Fig.2c shows the skeleton of the flow transducer. The magnetic field between two
cylindrical iron-core electromagnets traverses an epoxy body in
which a lumen L permits blood to flow at right angles to the
coil axis. Electrodes E_1, E_2 convey the flow signal to the
amplifier. The coils consist of 75 turns of No.32 enamelled
copper wire per coil and two iron cores of 1.25 mm diameter,
6 mm long. The best way of establishing a zero base line is to
pull the catheter so that the lumen L shifts away from the
branch artery ostium and is obstructed by the wall of the aorta.
This operation takes a fraction of a second and provides a highly
reliable base line. It has been shown that a blood flow tracing
thus obtained in a dog's renal artery is identical with a record
taken by means of a perivascular flow meter applied to the same
artery (20). To demonstrate that the intravascular flow meter

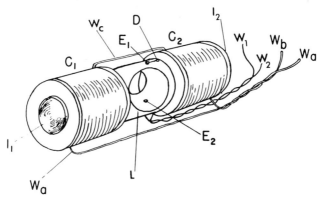

Fig.2c : Skeleton of the electromagnetic flow sensor. I_1, I_2:
cylindrical iron cores; C_1, C_2: coils surrounding the
iron cores; E_1, E_2: electrodes; W_1, W_2: electrode
lead wires; W_a, W_b: coil lead wires; L: flow transducer lumen through which the bloodstream passes.
From: A. Kolin: *Proc. Nat. Ac. Sc. 57*, 1331, 1967.

does not obstruct the flow through the artery, it was repeatedly withdrawn and reapplied while the renal flow was monitored by the perivascular flow meter. There was no trace of variation in flow. The dimensions of this transducer are comparable (being somewhat larger) to those of the preceding devices. The main weakness of the transducer is the thinness of the side walls of the lumen through which the electrodes are inserted. Another drawback of this device is the difficulty of alignment of the probe lumen against the branch artery ostium.

A slight variation of the preceding design resulted in a centerable velometer capable of measuring blood velocity in a major blood vessel, such as the aorta (21). The transformation of the above catheter flow meter into the centerable catheter velometer is accomplished by eliminating the tail T of Fig.2 and replacing it by a flexible neck just under the transducer at points C of Figs.2a and 2b. A pull on the pull-wire PW then bends the transducer at right angles to the artery axis and aligns the axis of the lumen L parallel to the flow. The transducer whose length should be such as to span the artery diameter is thus roughly centered and well stabilized within the artery. It can be smaller than the artery diameter since moderate lateral movements are not disturbing. Accurate orientation of the transducer at right angles to the artery axis is essential because of the angular dependence of the probe sensitivity. The main disadvantages of this type of catheter velometer are the complexity of the transducer construction and the need for a pull wire.

A particularly versatile device which, like the preceding ones, utilizes a lumen through which the fluid must pass in order to be exposed to a transverse magnetic field and to make contact with pick-up electrodes, has been described by Stein and Schuette (22). It can serve both as a velometer for flow measurements in large vessels and as a volume rate of flow meter

Alexander Kolin

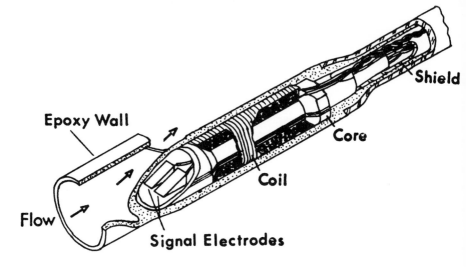

Fig.3 : Flow transducer with an epoxy cylinder, through which
the flow passes around the electrodes.   From:
P.D. Stein and W.H. Schuette.   *J. Appl. physiol. 26*,
851, 1969.

in smaller branch vessels.   Fig.3 shows the transducer which
is placed at the tip of a catheter.   A solenoid coil surrounding a hollow iron core generates a magnetic field which is
roughly perpendicular to the signal electrodes near their surface and to the line joining them.   This field possesses a
transverse component relative to the direction of flow which
enters a short epoxy-walled cylinder which is attached like a
"balcony" to the transducer tip making the device somewhat
eccentric.   The two electrodes are disposed inside this cylinder so as to pick up a signal voltage induced in the fluid as it
moves through the cylinder across the magnetic field.   The
electrode leads pass through the interior of the hollow iron
core.   The maximum outer diameter of the epoxy cylinder is 3.5
mm.   The gauge of the attached catheter is French 7.

The main advantage of this scheme is the ease with which a
reliable zero-flow base line can be obtained.   This is accomplished by the same method as used in the previously described

flow sensor (19,20) by pushing the open tip of the entry cylinder of the transducer against the wall of an artery or a heart
chamber.   The probe records forward and backward flow although
with unequal sensitivity.   The sensitivity also changes with
angulation of the probe in the bloodstream.   A 20^0 deviation
may cause a 20% fall-off in sensitivity.   Nevertheless no
measures are taken normally to center, align and stabilize the
catheter in a major blood vessel.   Due to shielding of the
electrodes from ambient currents, recordings in the vicinity of
the heart are not perturbed by an E.C.G. artifact.   Good fid-
elity has been obtained in recording the flow wave form in the
aorta as well as in branch vessels, including the left circum-
flex coronary artery.   The measurement of branch flow is made
possible by adding a short flexible tube to the end of the
cylinder.   This tubular projection can be "plugged" in into an
ostium of an artery so that blood entering it must first pass
through the transducer.

The considerable virtues of this instrument are balanced
by a few drawbacks.   It is necessary to administer heparin to
prevent blood clotting, the epoxy cylinder forms a hazardous
projection potentially injurious to the artery wall and the
outer diameter of the transducer is larger than for other intravascular flow sensors, which is undesirable, especially for
applications to human subjects.   A rather pessimistic report
has been published recently based on evaluation of commercial
catheter flow meters based on Stein and Schuette's design *in
vitro* as well as in animals (23).   They were found to deviate,
when used in the velometer mode very markedly from anticipated
velocity sensitivity values in tubes of different diameters.
The sensitivity to forward and reversed flow was found to vary
by a factor of 9 and angulation of the transducer at 45^0 to the
flow axis caused an 80% drop in sensitivity.   The results when
using the instrument in the volume rate flow meter mode were

reported as likewise discouraging.   The flow meter introduced
a hydrodynamic resistance which caused the flow value to drop
by about a factor of 3 as compared to unobstructed flow.
Similar obstruction effects were experienced when using the
commercial instrument *in vivo*.   A noncommercial catheter obtained from one of the originators of this design performed,
however, very much better, introducing only about a 7% change
in flow resistance.

UNCONVENTIONAL CONFIGURATIONS AND FURTHER MINIATURIZATION

The radial field velometer

We shall now return to the "inside-out" flow sensor scheme
the treatment of which was deferred to this section.   It uses
a magnetic field configuration which is a drastic example of an
ineffective scheme (24).   Fig.4 shows a radial magnetic field
symbolized by vectors B.   If we disregard the radial wall S
for the moment, a fluid flow directed at right angles to the
page will induce circulating currents indicated by counter-
clockwise bent arrows in the conductive field which flows around
the cylindrical source M of the radial magnetic field.   Electrodes placed anywhere along the periphery of M will not
record a potential difference.   The picture is changed, however,
if we stop the flow of the circulating eddy currents by a radial
septum S (25).   The induced e.m.f. is no longer short-circuited
and we can record a maximal potential difference between the
electrodes E_1, E_2 placed next to each other across the septum
S.   Even a very short septum projecting a millimeter or two
beyond the surface of a 5 mm wide cylinder is sufficient to produce a usable flow signal.

If we represent the radial magnetic field B by

$$B = \frac{B_o}{r} \qquad\qquad (1)$$

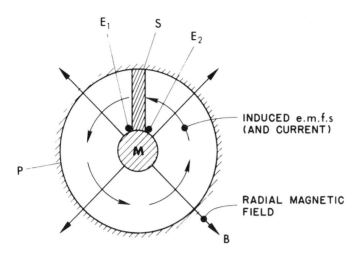

Fig.4 : Radial magnetic field velometer. The field B is generated by transducer magnet M which is centered in a pipe P. The induced currents (curved arrows) flow only in the absence of the dielectric septum S. E_1, E_2: electrodes. From: A. Kolin. *Trans. Biomed. Engr.* BME-16, 220, 1969.

we can show (25) that the induced flow signal between the electrodes will be in a uniform velocity field:

$$V = 10^{-8}. 2\pi B_0 v \qquad (2)$$

(V is measured in volts, B_0 in gauss, and the velocity v in cm/sec.) The flow signal V thus turns out to be independent of the radius of the magnet M. This implies that the transducer diameter could be miniaturized without loss in sensitivity. This type of transducer has not been further developed beyond the stage of demonstration of its principle.

The variable gauge collapsible flow sensor

Flow sensors exceeding about 3 mm in diameter require an incision in the branch blood vessel for their introduction. The following idea was stimulated by a search for a probe design

which would permit further miniaturization of the flow transducer without making it so small that fabrication would become
difficult while making it small enough for insertion through a
narrow intravascular catheter into a blood vessel by a percutaneous procedure which does not require a cut-down but merely
a needle puncture in the skin and the underlying tissues (26).
Imagine a probe which is large in its initial state but can be
collapsed so as to pass through a narrow opening as illustrated
in Fig.5. A basic part of the probe is a lens-shaped resili-
ent stainless steel frame (not shown in this diagram). This
frame can collapse so as to pass through a narrow branch tube
B and will expand when it enters the wider main tube so as to
make contact with its walls across a diameter. Imagine now a
fine flexible wire wound along this frame so as to form a flexible lens-shaped coil. A current passed through this coil
through the lead wires W_1, W_2 will give rise to a magnetic
field indicated by circular flux lines H in Fig.5. The magnetic field can be approximated roughly by the field between two
long parallel wires carrying equal currents in opposite directions. In the vicinity of each leg of the coil (C_1 and C_2)
the Lorentz forces exerted upon the ions in the moving liquid
by the current in the given leg will be substantially radial.
A potential difference due to flow can be detected by placing
two electrodes E_1 and E_2, as shown, in a central location.

The transducer sensitivity S = V/v depends on the diameter of the artery (where V is the flow signal voltage and v,
the velocity of flow in the artery, is assumed to be uniform).
The logarithmic relation

$$S = c \, ln \frac{D}{2d} \tag{3}$$

is a good approximation (26) (where d is the diameter of the
wire bundles forming the legs of the lenticular coil, D the

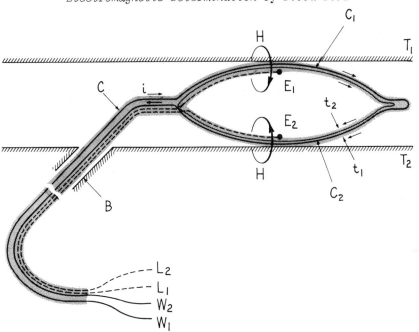

Fig.5 : Collapsible electromagnetic flow sensor is introduced
through a branch vessel B into the main vessel where
it expands so as to make contact with the vessel wall
on opposite sides T_1, T_2. The catheter stem C har-
bors the coil lead wires W_1, W_2 and the electrode
lead wires L_1, L_2; C_1, C_2: legs of the coil generating the magnetic field H. Arrows indicate the direction of the current i in the lead wires W_1, W_2 as
well as the turns t_1, t_2 of the coil which are shown.
E_1, E_2: electrodes. From: A. Kolin. *Proc. Nat.*
Ac. Sc. 63, 357, 1969.

artery diameter and c is a constant which depends on the
choice of units and intensity of the field-generating current).
It is noteworthy that the sensitivity of the transducer can be
increased according to equation (3) by diminishing the value of
d.

An important and interesting property of this catheter
flow transducer is the minimal flow of induced eddy currents
generated by fluid flow. The flow sensitivity shows so little
dependence on the conductivity of the pipe wall that it is only

moderately reduced (about 20%) in an aluminium pipe as compared
to a lucite pipe or to the performance of the probe without dielectric enclosure. The calibration is thus practically independent of the conductivity of the artery wall.

   In view of the logarithmic relation between tube diameter
and sensitivity, one calibration for a given pipe diameter is
sufficient to provide a calibration for other conduit dimensions. This probe is to be used as a velometer assuming a
uniform velocity profile in the blood vessel. To obtain the
rate of volume flow with a velometer, one must know the diameter of the artery. Fortunately, an x-ray radiogram of the
probe at right angles to its plane in Fig.5 provides a measure
of the artery diameter (26). It is also possible to measure
this diameter purely electromagnetically by a slight modification of this probe (26,27).

   The main experimental difficulty to be overcome to make
this transducer effective is the large Quadrature e.m.f. induced in the loop formed by the electrode leads L_1, L_2 of Fig.
5 bridged by the electrolyte resistance between the electrodes
E_1, E_2. One way of reducing the e.m.f. induced in this loop
to zero is by crossing the lead wires. Another way is described in the next section in connection with the modification
of this approach.

The external field electromagnetic catheter flow meter
   It is obvious that the greatest extent of miniaturization
could be achieved only by eliminating the means of generating
the magnetic field from the catheter flow sensor.   In this
case, the magnetic field would have to be established by a large
external magnet (28,29,30). Fig.6 shows a patient with a large
circular coil below him to generate the required magnetic field.
The frame of the sensor S has then merely to carry tiny electrodes in the same position as in the flexible probe described

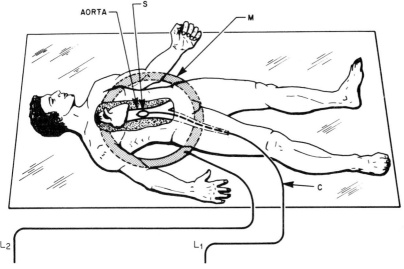

Fig.6 : Scheme of the external field electromagnetic catheter
flow meter.   M: coil above or below the patient
generating a magnetic field.   S: loop of the flow
sensor at the end of catheter   C   (actually the stem of
the probe is passed through a catheter tube which is
not shown). L_1:   electrode lead wires; L_2:   lead
wires conveying current to coil   C.

in the preceding section.

Actually the lateral dimensions can be greatly diminished
by using the frame itself (Teflon - insulated beryllium - copper
wire 0.2 mm in diameter) as one of the electrode leads (connected to E_1) as shown in Fig.7a.   The second lead to electrode
2 is a Teflon-insulated copper wire 0.075 mm in diameter.
(Both wires form a bifilar lead encased in a Teflon jacket.)*
Fig.7b shows how thin the resulting loop probe is in comparison
to a millimeter.   Fig.7c shows the loop emerging from a trans-
lucent No.6 French catheter tube.   It is easy to make probes
passing through even thinner tubes.

Actually, the central wire leading to E_2 does not have to
terminate at this electrode.   In the bifilar wire, the thin

---

* At present a bifilar, polyurethane-insulated beryllium-copper
  wire pair is used.

175 73 2

Alexander Kolin

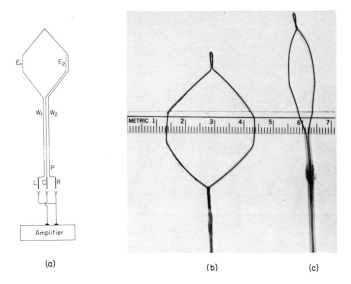

(a)

(b)   (c)

Fig.7a : Loop probe for external magnetic field catheter flow
meter.
Wiring scheme. Electrodes E_1, E_2 are formed by partial
removal of insulation from 0.2 mm beryllium copper
wire (E_1) and 0.075 mm copper wire (E_2). The connections shown at the plug P are used when the probe
measures flow. When the product of magnetic field
times artery diameter is to be ascertained for calibration, center pin C is disconnected and pins L
and R are connected to amplifier. W_1, W_2: loop lead
wires.

7b : Loop probe for external magnetic field catheter flow
meter.
Photograph of loop probe in expanded state. The scale
is in millimeters.

7c : Loop probe emerging from a transparent No.6 French
catheter tube.

copper wire leading to E_2 continues beyond it descending past
the electrode E_1 along the lead wire W_1 until it terminates
without being connected to anything. The two electrodes are
simply produced by scraping the Teflon insulation off the beryllium-copper and thin copper wire along the dashed line sections

labelled E_1 and E_2.   These electrodes are platinized prior
to use.   Fig.7a shows the connections to the amplifier when the
probe is used for measurement of flow.   An alternative connection used for calibration is described below.

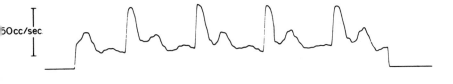

Fig.8 : Blood flow in a dog's abdominal aorta just above the
        renal arteries.   The base line is established by
        short-circuiting the amplifier output.

Fig.8 shows a record of blood flow obtained in a dog's
abdominal aorta just above the renal arteries.   Since the
electrodes are disposed across an artery diameter in a substantially homogeneous magnetic field, the measurement is governed
by the well-known flow meter equation:

$$S = V / \overline{v} = 10^{-8}\, B\, d, \qquad (4)$$

where S is the instrument sensitivity as determined by the
ratio of the flow signal V and the average flow velocity $\overline{v}$.
If an alternating magnetic field is used, the values for the
induced voltage V, and for the magnetic field can stand for
the peak values, d is the artery diameter and B is the component of the magnetic field which is effective in producing
the flow signal, i.e., it is perpendicular to the direction of
the flow and to the line joining the electrodes.   In the case
of our loop probe it is perpendicular to the plane of the loop.
We shall refer to B as the "effective field component".   In

order to obtain information about the rate of volume flow, we
must determine the diameter of the blood vessel since we are
measuring the average velocity. Thus, whereas the velometers
considered above determined the local velocity of flow within a
tube cross-section, the present flow meter integrates the signal over the entire cross-section and thus provides a measure
of the average velocity without the necessity of assuming a
flat velocity profile. This is of great importance since this
assumption cannot be expected to be valid except in a limited
number of cases.

It was pointed out that the flow signal derived from an
artery embedded in an extended volume conductor of conductivity
equal to that of the artery wall should be 20% lower than that
obtained in a dielectric pipe (31) and that this reduction in
signal should be quite insensitive to variations in the ratio
of the conductivities of blood and surrounding tissues (30).
This factor must be taken into account in the calibration.

The greatest experimental difficulty seems at first glance
to be the following: when the loop probe is in the animal, it
is difficult to ascertain the magnitude of the magnetic field
at the location of the probe and its direction relative to the
probe.   This is however the information which determines the
flow meter calibration.   The following idea solves this problem.   The loop can be used not only as part of a flow probe,
but also as a magnetometer in an alternating magnetic field.
If we leave the pin C of Fig.7a disconnected and connect the
terminals L and R of the loop to the amplifier, we measure
the e.m.f. induced in this transformer secondary which is in
phase quadrature relative to the flow signal and magnetic field.
This induced voltage measures the amplitude of the magnetic
field component which is perpendicular to the area of the loop
and which also happens to be the "effective field component"
determining the magnitude of the induced flow signal and also

proportional to the area A of the loop. If we thus vary the orientation and distance of the loop from the magnet coil we obtain a varying transformer signal which provides information about the variations in the flow meter sensitivity for a tube of constant diameter.

Actually we cannot assume the artery diameter to remain constant as we change the position of the probe within a blood vessel such as the aorta. We have thus two variables which affect the sensitivity: the artery diameter d and the "effective field component" B. Fortunately we do not have to know these values separately since it is the product $Bd = M$ in the flow meter equation (4) which determines the flow meter sensitivity.

The signal from the transformer secondary loop is proportional to the product BA where B is the amplitude of the magnetic field component normal to the loop area and A is the area of the loop. Let us now assume a shape of the loop for which the area A is proportional to the diameter d of the confining artery to a good degree of approximation. Then the signal induced in the loop will be approximately proportional to $Bd = Mc$ which determines the flow sensitivity (c representing the proportionality constant). Our calibration procedure is then as follows: we pass a known average flow through a pipe *in vitro* while recording the flow meter signal per unit flow velocity S_1 as well as the induced transformer signal cM_1 derived from the loop. We measure subsequently in the animal the value of $cM_2 = cB_2d_2$ which depends on the unknown values of the effective magnetic field and artery diameter. The flow velocity sensitivity S_2 in the animal follows then from:

$$\frac{S_2}{S_1} = \frac{cM_2}{cM_1} \tag{5}$$

In other words, the sensitivity ratio is equal to the ratio of

Alexander Kolin

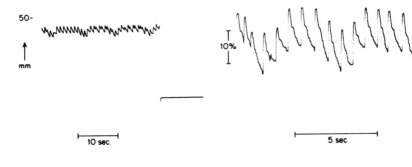

Fig. 9a   Fig. 9b

Fig.9 : Electromagnetic artery diameter gauge and calibration
signal (the probe is in a dog's abdominal aorta).
(a) Transformer signal induced in the probe loop.
This signal is proportional to product of magnetic
field component perpendicular to loop area and area of
the loop.   It is also proportional (with good approximation) to the artery diameter.   This signal (the
average deflection) is used for flow sensitivity calibration.   The pulsations of the signal measure pulsatile variations in the artery diameter as a fraction of
the diameter.   (b) Pulsatile variations in artery
diameter on an expanded scale (at higher amplification).

the transformer signals.   Fig.9a shows the signal derived from
a transformer loop placed in a dog's abdominal aorta.   If the
magnetic field were determined by a magnetometer, this record
could be used to measure the artery diameter.   For determination of relative changes in the artery diameter due to pulsations in blood pressure, the knowledge of B is not necessary.   Fig.9b shows the diameter changes on an expanded scale
obtained at higher amplification.   We have thus a very simple
and effective electromagnetic artery gauge (28).

The signal derived from the loop shown in Fig.9a serves two
purposes: (a) It can be used to adjust the phase setting of
the phase-sensitive detector until this transformer signal disappears so as to optimize the phase-setting for the flow
measurement; (b) It can be used to measure the ratio M_1/M_2 in
equation (5).   (In this case the average value can be measured

at a high time constant setting of the instrument.)

Fig.10 shows to what extent the assumption of approximate proportionality between the area and diameter of the hexagonal loop holds true for a wide range of compression of the loop (dashed line). The same diagram shows the correlation between the loop transformer signal (measuring the M_1/M_2 ratio) and the flow sensitivity S (solid line). The various points have been obtained by changing the "effective magnetic field component" in different ways (see legend).

This method is still in an early stage of its development. The main artifact is a flow-independent signal which is in a large measure due to eddy currents induced in the heterogeneous volume conductor surrounding the probe. The ideal of being able to establish a reliable zero flow base line under all circumstances by turning off the magnetic field or short-circuiting the amplifier output has not been reached yet. Nevertheless, there are some locations at which this does not seem to present a problem (32). At the root of the aorta and in the pulmonary artery a brief diastolic interval of zero flow provides a base line, and in limbs a base line can be obtained by temporary occlusions. The interest in the further development of this method is mainly based on the anticipation that it may make it possible to measure flow in main blood vessels and their branches by percutaneous introduction of miniature probes through very narrow catheters making such measurements a safe and easy procedure. Its value in clinical applications would be based on the possibility of assessing the effects of a drug on a therapeutic procedure on individual patients without having to place reliance on clinical statistics.

It is not difficult to foresee which direction innovative efforts will take in the field of electromagnetic blood flow meter development. It will be most likely the search for intravascular transducers which could be safely and easily introduced

Alexander Kolin

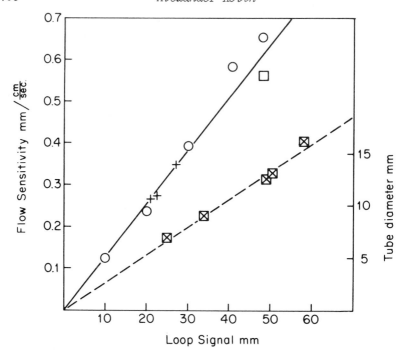

Fig.10 : Performance of the external field loop probe in its
flow meter and calibrator functions.
(a) *Solid line* shows the proportionality between the
flow sensitivity (vertical scale on the left) and the
calibrating transformer e.m.f. signal derived from the
loop.   The *circles* represent cases where the transformer signal was varied by rotating the loop orientation about the tube axis relative to the magnetic
field.   The *square* illustrates a case where the magnetic field axis and the (co-planar) tube axis form an
angle of 45⁰ and the *crosses* represent cases of arbitrary variations in distance and orientation of the
magnet coil from the probe loop.
(b) *Dashed line* shows the proportionality between the
tube diameter (vertical scale on the right) and the
transformer e.m.f. loop signal.   The probe was placed
into lucite tubes of different diameters varying over
a wide range.   The location and orientation of the
probe within the magnetic field remained constant.
The transformer loop signal provides the information
necessary for calibration also in cases where the distance and orientation of the probe relative to the magnetic field as well as the artery diameter are varied
simultaneously.

percutaneously into the vascular tree through a catheter of minimal dimensions and would allow reliable continuous measurement of branch flow as well as of the cardiac output without uncertainties about the base line or the calibration.

REFERENCES

1. Kolin, A.   *Proc. Soc. Exp. Biol. Med. 35*, 53, 1936.

2. Wetterer, E.   *Zschr. Biol. 98*, 26, 1937.

3. Kolin, A.   *UCLA Forum. Med. Sci. 10*, 383, 1970.   (Univ. of Cal. Press.)

4. Wyatt, D.G.   *Medical Electronic Monographs* (B.W. Watson, Ed.), Inst. of El. Eng. 1971.

5. Cappelen, C.   New Findings in Blood Flowmetry.   (Universitetsforlaget, Oslo, 1968.)

6. Mills, C.J.   *Phys. in Med. & Biol. 11*, 323, 1966.

7. Kolin, A.   *J. Appl. Physics 15*, 150, 1944.

8. Kolin, A.   *Rev. Sc. Instr. 16*, 109, 1945.

9. Mills, C.J. and Shillingford, J.P.   *Cardiovasc. Res. 1*, 263, 1967.

10. Gabe, I.T., Gault, J.T., Ross, J.R., Mason, D.T., Mills, C.J., Shillingford, J.P. and Braunwald, E.   *Circulation 40*, 603, 1969.

11. Wexler, L., Bergel, D.H., Gabe, I.T., Makin, G.S. and Mills, C.J.   *Circ. Res. 23*, 349, 1968.

12. Bond, R.F. and Barefoot, C.A.   *J. Appl. Physiol. 23*, 403, 1967.

13. Spencer, M.P. and Barefoot, C.A.   *In:* New Findings in Blood Flowmetry.   (Chr. Cappelen, Ed.)   Universitetsforlaget, 1968.

14. Warbasse, J.R., Hellman, B.H., Gillian, R.E., Howley, R.R. and Rabitt, H.I.   *Am. J. Cardiol. 23*, 424, 1969.

15. Jacobs, R.R., Williams, B.T., Andersen, M.N. and Schenk, W.G.   *J. of Trauma 86*, 178, 1971.

16. Kolin, A.   *IEEE Trans. on Biomed. Engr. BME-16*, 220, 1969.

17. McDonald, D.A.   Blood Flow in Arteries (London, Arnold Publ. 1960).

18. Kolin, A., Ross, G., Grollman, J.H. and Archer, J.   *Proc. Nat. Ac. Sc. 59*, 808, 1968.

19. Kolin, A.   *Proc. Nat. Ac. Sc. 57*, 1331, 1967.

20. Kolin, A., Archer, D.J. and Ross, G.   *Circ. Res. 21*, 889, 1967.

21. Kolin, A., Ross, G., Grollman, J.H. and Archer, J.   *Proc. Nat. Ac. Sc. 59*, 808, 1968.

22. Stein, P.D. and Schuette, H.   *J. Appl. Physiol. 26*, 851, 1969.

23. Meyer, P., Moore, G., Brobman, G.F. and Jacobson, E.D. *Am. Heart J. 80*, 846, 1970.

24. Shercliff, J.A.   Theory of Electromagnetic Flow Measurement.   Cambridge Univ. Press (New York 1962).

25. Kolin, A.   *IEEE Trans. Biomed. Eng. BME-16*, 220, 1969.

26. Kolin, A.   *Proc. Nat. Ac. Sc. 63*, 357, 1969.

27. Kolin, A. and Culp, G.W.   *LIEEE Trans. Bio. Med. Eng., BME-18*, 110, 1971.

28. Kolin, A.   *Proc. Nat. Ac. Sc. 65*, 521, 1970.

29. Kolin, A., Grollman, J.H., Steckel, R.J. and Snow, H.D. *Proc. Nat. Ac. Sc. 67*, 1769, 1970.

30. Kolin, A., Grollman, J.H., Steckel, R.J. and Snow, H.D. *Proc. Nat. Ac. Sc. 68*, 29, 1971.

31. Gessner, U.   *Biophys. J. 1*, 627, 1961.

32. Kolin, A., Grollman, J.H., Steckel, R.J. and Snow, H.D. To be published.

THE USE OF ULTRASOUND

FOR RECORDING

BLOOD VESSEL DIAMETER IN MAN

J.O. ARNDT

*Abteilung für Experimentelle Anaesthesiologie der
Universität Düsseldorf, Germany.*

A general approach for analyzing a biological system such as the cardiovascular system is to study the effect of a certain disturbance on a certain parameter. The results will depend on the extent to which the parameter under investigation is affected by the method and, obviously, the most suited method is that which does not interfere with the parameter to be measured. The ultrasound reflection method for measuring arterial diameter transcutaneously fulfils the above requirements and is, therefore, the method of choice for analyzing arterial mechanics *in vivo*.

However, before discussing the details and limitations of this method it seems appropriate to stress the physiological significance of arterial mechanics; it will then become clear why the mechanical behavior of arteries depends so heavily on the experimental conditions and, therefore, how necessary it is to study arteries nondestructively in order to come up with realistic results.

Arterial mechanics are intimately connected with two major aspects of the cardiovascular system:

1. Pulse transmission, i.e. our understanding of arterial
    hemodynamics (12,17), and

2. The control of arterial blood pressure, because the baro-
    receptor discharge depends predominantly on the disten-
    sibility of the receptor zones (11).

The simplest approach to studying the distensibility characteristics of arteries that is relevant to both these aspects is to
derive the relationship between distending pressure and arterial
diameter (8). From this relationship some coefficients of
elasticity can be derived:

1. Distensibility (W)

    $W = \frac{\Delta D}{\Delta P}$ where D stands for diameter, P for pressure,
    Δ for a change of these parameters.

2a. The volume elastic modulus (K)

    $K = \frac{\Delta P}{\Delta V} \cdot V$ (V stands for volume)

2b. This modulus can be approximated from the pressure diameter
    relationship when assuming constant length of a vessel
    from:

    $K \approx \frac{\Delta P}{2\Delta D} \cdot D$.

The volume elastic modulus is preferred to the pressure
elastic modulus $(E_p = \frac{\Delta P}{\Delta D} \cdot D)$ because the former is closely
related with pulse wave velocity according to:

3. $c^2 = \frac{K}{\rho}$ where ρ is the density of blood.

Thus, pulse wave velocity, which can be measured nondestructively in man, is an indicator of the elastic properties
of arteries and is suited for comparison with the data derived
from the pressure-diameter relationship.

For the following discussion it is important to realize
that the pressure-diameter relationship of an artery does not
depend solely on the properties of the nonreactive wall com-

ponents like elastin and collagen but also on the activity of vascular smooth muscle.   It is common experience that smooth muscle activity is sensitive to mechanical irritations; for example, a simple arterial puncture causes arterial constriction, reducing the diameter by more than half, as shown with X-ray angiography in the carotid vasculature in cats.   Accordingly, the pressure-diameter relationship of a traumatized artery can deviate appreciably from what is physiological. This might well be the reason why, in a certain pressure range, exposed arteries appear to be so much stiffer than the unexposed.

A case in point is the pressure-diameter data of the exposed common carotid artery of man studied by Greenfield and co-workers (10).   From these data the calculated pulse-wave velocity amounts to 17.5 $m.sec^{-1}$ which is out of range of the accepted figures.   Even in the relatively stiff peripheral conduit arteries wave velocities are rarely more than 10 to 12 $m.sec^{-1}$ in man (18).   In fact, it will be shown that the pulse-wave velocity along the intact carotid artery in supine man is of the order of 4 to 5 $m.sec^{-1}$ as measured directly or derived from the pressure-diameter relationship, based on transcutaneously measured diameters.

ULTRASOUND REFLECTION METHOD FOR MEASURING ARTERIAL DIAMETER IN MAN TRANSCUTANEOUSLY

The method is based on the fact that echoes occur whenever sound waves meet with interfaces of changing acoustic impedances. The distance between the interfaces or between the echoes, respectively, is proportional to the time interval between them and can be calculated when the speed of sound in the material is known.

For measurement of arterial diameter this principle has been applied previously by Buschmann (5) and the method was modified and used by the author (1,2).   The experimental set-up

J.O. Arndt

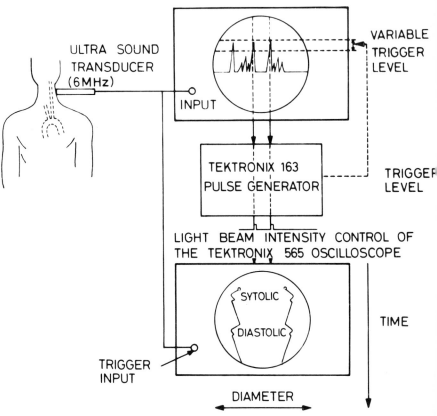

Fig.1 : Experimental set-up for diameter recording.

is presented in Fig.1.

Bursts of 6 megacycle ultrasound were transmitted from the skin surface through superficial arteries in time intervals of 1.3 m.sec. During the time between consecutive bursts the transducer receives the echoes which are displayed on an oscilloscope as indicated. The arterial echoes are identifiable because they are pulsating, i.e., they move away from and towards each other with systole and diastole, respectively.

The leading edges of the echoes triggered a defined pulse (pulse width 0.7 μsec, 25 volts) for either echo from the Tektronix Pulse Generator 163 which also supplied a variable trigger level for eliminating low amplitude echoes from unspecific structures.

These pulses were then used for controlling the light intensity of the electron beam of the Tektronix 565 oscilloscope. Thus the information "wall echo" was converted into two light spots. The vessel diameter can then be calculated from the speed of sound in blood (assumed to be 1500 m.sec^{-1}) and the time interval between the light spots as indicated by the time sweep of the oscilloscope in the x-direction.

Finally, to visualize the change of diameter with time the light spots were deflected in the y-direction from a second sweep generator which was triggered when the shutter of a camera viewing the face of the oscilloscope was opened. Thus the diameter tracings were photographed. Typical recordings from the common carotid artery are shown in Fig.2. The contour of these tracings closely resembles the typical pressure curve of this vasculature with a well developed dicrotic notch. Note, also, the relatively large pulsatile diameter changes which are of the order of 12%.

The following data were derived from these recordings:

1. The diastolic diameter (D) from the smallest distance between the two tracings, and

2. The pulsatile diameter changes (ΔD) by subtracting the diastolic diameter from the systolic, i.e. from the largest distance.

Critique of the method

Since the conclusions rely upon the validity of the diameter measurements it is appropriate to discuss the limitations and problems of the method first, and to analyse the physiologi-

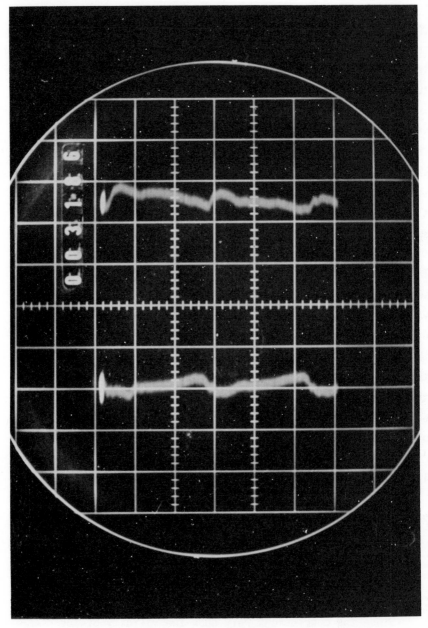

Fig.2 : Diameter recording from an intact carotid artery.
Smallest distance between the tracings is a measure of
diastolic, the largest of systolic diameter.

cal significance of the data thereafter.   The diameter measurements are based upon the difference in the transit time of the ultrasound between the echoes from opposite vessel walls.   Four main sources of potential errors must be considered: 1. the origin of the echoes; 2. the angulation of the transducer in relation to the vessel; 3. the transducer contact with the skin; and 4. the resolution of the method (2).

1.   Studies on excised arteries showed that thick-walled vessels (wall thickness more than 0.5 mm) give rise to two echoes - one from the outer, the other from the inner, wall surface.   With a wall thickness 0.5 mm and less the two echoes usually merged, which was the case with the carotid artery in man.   From its radius-to-wall-thickness ratio of about 7 the wall thickness is approximately 0.5 mm (7,16).

Since the wall echoes were converted into electrical pulses which were triggered by the leading edges of the original echoes from opposite walls, it can reasonably be assumed that the internal diameter plus one wall thickness was actually measured with the technique used.

2.   The wall echoes disappeared when the axis of the transducer deviated by more than 10 degrees from the length axis of the vessel.   However, for this worse-case condition the vessel diameter would be overestimated by not more than 2%. The echoes also vanished when the transducer did not exactly face the center of the vessel.   Thus the presence of the high amplitude, stable echoes assured that the vessel diameter was measured with sufficient accuracy.

The correct adjustment of the transducer is the main handicap of the method because it is an extremely tedious procedure; in fact, the transducer is most easily adjusted with a micromanipulator.

3. Compression of the artery by forceful application of the transducer is another potential source of error. Reduced pulsatile excursions of the wall next to the transducer and distortions of the pulse contour indicate this.

    Considerable care was taken to apply the transducer as loosely as possible to the skin. In any case, the presence of the dicrotic notch (carotid artery) or of the dicrotic wave (femoral artery) in both wall tracings was used as a criterion for adequate recordings. Approximately 70% of all films had to be discarded because of unreliable recordings.

4. Theoretically the resolution cannot be better than 0.25 mm when using 6 megacycle ultrasound according to $\lambda = c.f$ (λ wave length, c speed of sound, and f frequency of the sound). Nevertheless, this resolution is sufficient. In man the intact carotid artery measures on the average 0.76 ± 0.06 cm, and the intact femoral artery 0.74 ± 0.05 cm, and even the pulsatile diameter excursions are larger than 0.25 mm. Pulsatile diameter changes are 0.11 ± 0.01 cm or 14.6% of diastolic diameter in the carotid artery and 0.06 ± 0.01 cm or 6.73% of diastolic diameter in the femoral artery.

    The resolution of the ultrasound reflection method was also studied in model experiments using the excised aorta of cats. The diameter changes were measured simultaneously with a differential transformer-type dimension gauge whose good resolution is well known. The measurements agreed perfectly.

    From the above discussion it therefore appears justifiable to conclude that the diameter of superficial arteries in man can be measured with sufficient accuracy with the ultrasound reflection technique.

PRESSURE-DIAMETER RELATIONSHIP OF THE INTACT CAROTID
ARTERY IN MAN

As was pointed out in the introduction, arterial mechanics
can most easily be evaluated from the pressure-diameter relationship.   Therefore, the diameter data were related to the
arterial pressure, which was measured indirectly in the brachial
artery.   To measure the carotid arterial pressure directly was
not justifiable because of the hazards involved for the subjects,
particularly since the conclusions are not limited by this compromise (1).

The measurements from one subject who was studied repeatedly
in the supine position over several months are plotted in Fig.3.

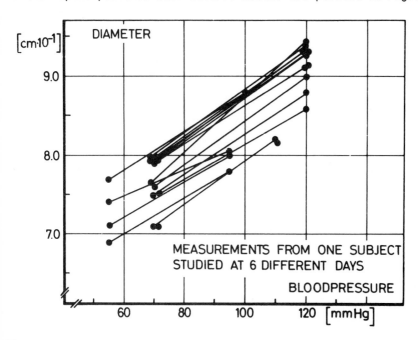

Fig.3 : Pressure-diameter relationship of the carotid artery
        from one subject studied in the supine position
        repeatedly over several months.   Note the relatively
        small fluctuations of the diastolic diameter and of
        the slope of these curves.

Obviously, the diameter fluctuations are relatively small for a
certain pressure range: they amount to about 8%.   Also, the
slope of the curves ($\Delta D/\Delta P$) which by definition is a measure
of distensibility, varies but slightly.   It seemed, therefore,
justified to average the diameter data for each subject for a
certain pressure range.

The results from 16 subjects are summarized in Fig.4.
Note that the diastolic diameter as well as the distensibility
of the intact common carotid artery vary appreciably from subject to subject.   Note, furthermore, that the pulsatile diameter excursions are large relative to diastolic diameter;   they
are on the average 14.6 ± 2.1%.

The relevance of these measurements for pulse transmission
in the intact circulation was pursued further by calculating the
pulse wave velocity.   The parameters to calculate it are:
1. the diastolic diameter (D); 2. the pulsatile diameter
change (ΔD), and 3. the corresponding pressure change (ΔP),
which is the pulse pressure and can be taken directly from the
pressure-diameter relationship.   These data are summarized in
Table I.   Attention is directed to the calculated pulse wave
velocities in the last column.   They range between 3.8 and
5.2 m.sec^{-1} with an average of 4.6 m.sec^{-1}, and moreover, they
compare well with those measured directly in the same subjects,
as indicated by the numbers added in parentheses.

This result is of interest for several reasons.   Firstly,
the close agreement between the calculated and the actual
measured pulse wave velocities gives appreciable weight to the
reliability of the diameter measurements; and second, the dis-
agreement between these low wave velocities in the intact arteries and those in the exposed arteries which are much higher
(10,15) suggests that the latter data have to be taken with
caution.   Several potential sources for the reduced distensibility during vessel exposure are possible, though none is

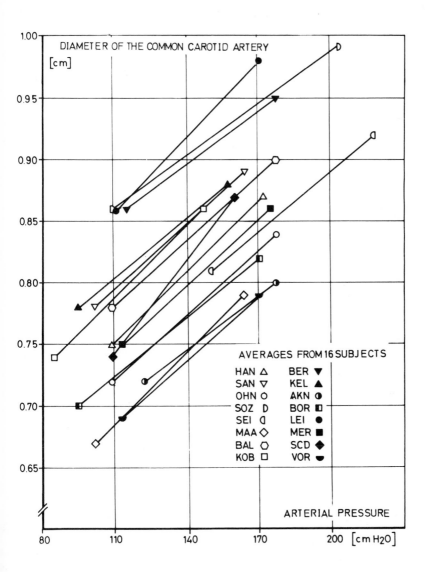

Fig.4 : Pressure-diameter relationships of the common carotid
artery from 16 subjects studied in the supine position.
Note the variations in diastolic diameter and in the
slope of the curves.

TABLE I:

SUMMARY OF THE RESULTS FROM 16 SUBJECTS STUDIED IN THE SUPINE POSITION

SUBJECT	AGE YEARS	BODY WEIGHT (kg)	HEIGHT (cm)	DISTANCE H.-C.A. (cm)	PRESSURE syst.diast. (cm H₂O)		ΔP	DIAMETER syst. diast. (cm)		ΔD	$\frac{\Delta d}{d}$·100 %	$\frac{\Delta d}{\Delta p}$ (10^{-6} dyn^{-1} cm^3)	VOLUME ELASTIC MODULUS ($\varkappa$) (dyn·cm^{-2}·10^6)		PULSE WAVE VELOCITY (cm·sec^{-1})	
Han	22	67	177	25	177	109	68	0,87	0,75	0,12	16,0	1,8	0,43	0,21	4,5	–
San	23	81	190	25	164	102	62	0,89	0,78	0,11	14,1	1,8	0,40	0,20	4,4	4,2
Ohn	29	52	164	23	177	109	68	0,84	0,72	0,12	16,7	1,8	0,41	0,20	4,4	3,2
Soz	23	115	186	26	204	109	95	0,99	0,86	0,13	15,1	1,4	0,63	0,31	5,4	–
Sei	23	81	192	27	218	150	68	0,92	0,81	0,11	13,6	1,6	0,50	0,25	4,9	–
Maa	24	71	178	25	163	102	61	0,79	0,67	0,12	17,9	2,0	0,34	0,17	4,0	3,9
Bal	24	80	193	29	177	109	68	0,90	0,78	0,12	15,4	1,8	0,44	0,22	4,6	–
Kob	28	85	192	27	146	85	61	0,86	0,74	0,12	16,2	2,0	0,38	0,19	4,2	4,8
Ber	22	65	170	24	177	115	62	0,95	0,86	0,09	10,5	1,5	0,59	0,30	5,3	–
Kel	30	76	178	23	157	95	62	0,88	0,78	0,10	12,8	1,6	0,48	0,24	4,7	–
Arn	34	82	178	26	177	122	55	0,80	0,72	0,08	11,1	1,5	0,50	0,25	4,8	5,0
Bor	24	70	182	26	177	95	82	0,82	0,70	0,12	17,2	1,5	0,47	0,24	4,8	4,2
Lei	28	86	189	27	170	109	61	0,98	0,86	0,12	14,0	2,0	0,44	0,22	4,5	–
Mer	28	80	188	27	174	113	61	0,86	0,75	0,11	14,7	1,8	0,42	0,21	4,4	5,0
Scö	27	48	160	22	160	109	51	0,86	0,74	0,12	16,2	2,4	0,32	0,16	3,8	5,0
Vor	28	53	175	26	170	113	57	0,79	0,69	0,10	14,5	1,8	0,39	0,20	4,3	4,0
X̄	26	75	181	26	174	109	65	0,88	0,76	0,11	14,8	1,8	0,49	0,22	4,6	4,4
M	16	16	16	16	16	16	16	16	16	16	16	16	16	16	16	9
SD	±3,4	±16,2	±10,2	±1,8	±17	±14	±11	±0,06	±0,06	±0,01	±2,1	±0,25	±0,08	±0,04	±0,4	±0,6

clearly responsible for the problem at present. There is evidence that the spring-loaded dimension gauges used by some authors in the past (10,15) restrained the vessel wall (9). Yet experimental problems might also be relevant to the problem. Vasoconstriction due to mechanical wall irritation, inadvertent drying, or cold must be considered, since reduced distensibility will, under certain conditions, go along with vasoconstriction (6). From this viewpoint it was challenging to know whether and to what extent vasoconstrictory stimuli might effect arterial mechanics.

ARTERIAL MECHANICS IN RESPONSE TO NOR-EPINEPHRINE INFUSION IN MAN

The effect of nor-epinephrine infusions (10 and 20 µg/min) on distensibility ($\Delta D/\Delta P$), on diastolic diameter (D), and on calculated wave velocity of the intact carotid artery was investigated in 14 subjects. The data are shown in Fig.5. Compared with the controls (hatched columns), infusions of 10 and 20 µg/min nor-epinephrine (open columns) tend to increase the distensibility of the carotid artery while the diastolic diameter remains constant. Thus there is also a tendency for the wave velocity to decrease. However, these results are only typical for the carotid artery inasmuch as the intact femoral artery in man, for example, responded to the same experiment in a qualitatively different manner. As is obvious from Fig.5, distensibility and also diastolic diameter decreased significantly, whereas wave velocity increased. These are data from 12 subjects.

That the two arteries would respond differently to nor-epinephrine was to be expected from their morphological differences, the carotid artery being "elastic", the femoral "muscular" (4,13). It must, therefore, be concluded that attempts to generalize data relevant for arterial mechanics must also take into

J.O. Arndt

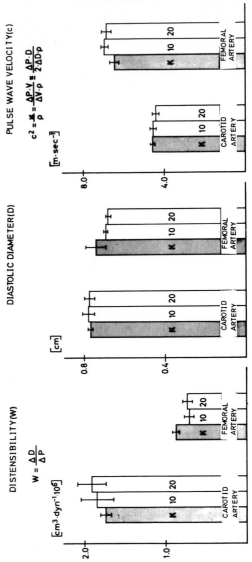

Fig.5 : The effect of continuous nor-epinephrine infusions (10
       and 20 µg/min) on arterial mechanics.   A comparison of
       the "elastic" common carotid artery with the "muscular"
       femoral artery.   Averages with the deviations of the
       mean from 14 and 12 experiments, respectively, in man.
       Note that the two arteries respond in a different
       pattern to nor-epinephrine.

account the morphology of arteries, which changes characteristically along the arterial tree and also from species to species (13,14). Moreover, vasoconstrictory stimuli can obviously reduce distensibility in some arteries. This might be another reason why exposed arteries are less distensible than the intact. At present, the state of vascular smooth muscle activity cannot be estimated. This by itself calls for noninvasive techniques for analysing arterial mechanics, and the ultrasound reflection method proved to be a reliable noninvasive technique. The tedious procedure of adjusting the transducer properly in order to receive stable echoes still hinders routine application of the ultrasound reflection method in clinical research, but a better design of the transducer may help to overcome this particular handicap.

REFERENCES

1. Arndt, J.O., Klauske, J. and Mersch, F.   The diameter of the intact carotid artery in man and its change with pulse pressure. *Pflügers Arch. 301*, 230 (1968).

2. Arndt, J.O.   Über die Mechanik der intakten A. carotis communis des Menschen unter verschiedenen Kreislaufbedingungen. *Arch. Kreisl.-Forsch. 59*, 153 (1969).

3. Arndt, J.O. and Kober, G.   Pressure diameter relationship of the intact femoral artery in conscious man. *Pflügers Arch. 318*, 130 (1970).

4. Benninghoff, A.   Blutgefässe und Herz. *In:* Handbuch der mikroskopischen Anatomie, Band VI: 1 (1930).

5. Buschmann, W.   Zur Ultraschalldiagnostik an oberflächlichen Arterien. Wiss. Ztschr. Humboldt Universität Berlin, *Math.-Naturwiss. Reihe, 14*, 223 (1965).

6. Dobrin, P.B. and Rovrick, A.A.   Influence of vascular smooth muscle on contractile mechanics and elasticity of arteries. *Am. J. Physiol. 217*, 1644 (1969).

7. Evans, R.L., Pelley, J.W. and Quenemoen, L.   Some simple geometric and mechanical characteristics of mammalian blood vessels. *Am. J. Physiol. 199*, 1150 (1960).

8. Frank, O.   Die Elastizität der Blutgefässe. *Z. Biol. 71*, 255 (1920).

9.   Gow, B.S. and Taylor, M.G.   Measurement of viscoelastic properties of arteries in the living dog.   *Circ. Research* *23*, 111 (1968).

10.   Greenfield, J.C., Tindall, G.T., Dilion, M.L. and Mahalley, M.S.   Mechanics of the human common carotid artery in vivo.   *Circ. Research 15*, 240 (1964).

11.   Heymans, C. and Neil, E.   Reflexogenic Areas of the Cardiovascular System.   Boston (1958).

12.   McDonald, D.A.   Blood Flow in Arteries.   London (1960).

13.   Muratori, G.   Contribution of the study of the microscopical structure of the carotid sinus in man and some animals.   *Anat. Anz. 119*, 466 (1966).

14.   Rees, P.M.   Electromicroscopical observation on the architecture of carotid arterial walls with special reference to the sinus portion.   *J. Anat. 103*, 35 (1968).

15.   Peterson, L.H., Jensen, R.E. and Parnell, J.   Mechanical properties of arteries in vivo.   *Circ. Research 8*, 622 (1960).

16.   Peterson, L.H.   Properties and behavior of living vascular wall.   *Physiol. Rev. 42*, 309 (1962).

17.   Wetterer, E. and Kenner, Th.   Grundlagen der Dynamik des Arterienpulses.   Springer-Verlag Berlin-Heidelberg-New York (1968).

18.   Wezler, K. and Böger, A.   Die Dynamik des arteriellen Systems.   *Erg. Physiol. 41*, 359 (1939).

INTRAVITAL MICROSCOPY ON THE BASIS OF TELESCOPIC PRINCIPLES: DESIGN AND APPLICATION OF AN INTRAVITAL MICROSCOPE FOR MICROVASCULAR AND NEUROPHYSIOLOGICAL STUDIES*

HAROLD WAYLAND and WALLACE G. FRASHER, Jr.

California Inst. of Technology, Pasadena, California, U.S.A.
University of Southern California School of Medicine,
Los Angeles, California, U.S.A.

An intravital microscope system suitable for studies of
animals as large as dogs and miniature pigs has been designed

* We are deeply grateful to the administration of the California
Institute of Technology and to the Alfred P. Sloan Foundation
for making the funds available for this development. Although
this instrument was built primarily for use in our own research
program, it was conceived as a prototype, and we hope that some
or all of its features will be useful to our colleagues
throughout the world. Design details will be furnished on re-
quest. If any significant amount of engineering or drafting
is required, it will be furnished at cost.
     The detail drawings and the fabrication of this system
were made by the Central Engineering Facilities of the California Institute of Technology. We greatly appreciate their
skill and cooperation.

and constructed at the California Institute of Technology with the following attributes: (1) a large, stable animal platform capable of both relatively large and micrometer motion in the horizontal plane;   (2) an optical system giving a stable magnified image, but in which the height of the object plane can be changed over considerable limits (± 10 cm) without loss of resolution, change of magnification, or repositioning of viewing and recording systems;   (3) a flexible means of changing from one type of viewing system to another without requiring refocusing;   (4) a flexible system for rapid change between different light sources;   and (5) vibration isolation of cine cameras from the optical train.

The optical and mechanical design principles are discussed and illustration given of some types of physiological studies for which the system has been used.

INTRODUCTION

The physiologist is no longer content to use a microscope merely for observation in his study of living systems.   Microscopic optical devices are an essential part of many quantitative measurement systems used, for example, in measurements of the size of microvessels and rate of blood flow through them;   in the use of microspectrophotometry to follow local chemical changes;   and the use of fluorescent tracers to study transport and diffusion across blood vessel walls or through tissue. Since a single high resolution optical field may be inadequate to give a statistically valid sample of the processes being studied, it is important to be able to move from one field to another with a minimum of difficulty.   If, along with the optical information, microelectrodes are being used to measure action potentials or pressures in individual microvessels, for example, it is important that these systems remain fixed relative to the animal while the optical field is moved from one region to another.

The basic criteria which we have laid down for the design and construction of an intravital microscope system are the following:

(1) Stage

The microscope stage must be large enough to carry an animal as big as a dog or miniature pig, along with the auxiliary equipment for anesthesia and life support, as well as any micromanipulators required, for example, for microelectrode work. This stage must have a high degree of mechanical stability, and be capable of precise horizontal motion in two orthogonal directions.

(2) Recording Equipment

It must be possible to fix the positions of the optical recording equipment such as film cameras, TV cameras and photometric sensors with precision so that they can be stably mounted in such a way as not to transmit vibration from camera mechanisms to the optical train. Rapid interchange between several such devices should be possible so that, for example, still photographs can be interspersed between cinematographic sequences or photometric measurements.

(3) Illumination System

For transillumination, it should be possible to interchange light sources quickly and easily -- e.g. between an incandescent source and a pulsed source. Capability should also be provided for light pipe, fiber optic and epi-illumination systems.

THE BASIC OPTICAL DESIGN

After exploring a number of alternatives, we came to the conclusion that our requirements could most easily be met by using telescopic optics, utilizing, wherever possible, commercially available infinity corrected objectives. Such optical

systems have long been used on metallographs, and form the
basis for the Microstar line of microscopes made by the American
Optical Company, but we have seen no evidence of any manufacturer
taking full advantage of this optical approach to produce an
effective and flexible intravital microscope system.

The basic principle of the system is illustrated in Fig.1.
If the object to be viewed is at the front focal plane of the
objective, the light emanating from any point on the object
which is intercepted by the objective will emerge from it as a
parallel bundle of rays.   If such a bundle is intercepted by

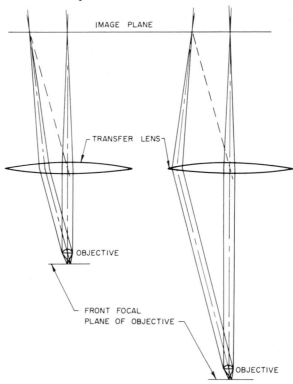

OPTICAL COMPONENTS ARE SHOWN AS SIMPLE LENSES FOR CLARITY
IN ILLUSTRATING THE PRINCIPLES INVOLVED

Fig.1 : Basic principles of telescopic optical system for
microscopy.   The image plane remains fixed regardless
of distance of object-objective combination from the
transfer lens.

another lens, this light will be focused in the back focal plane of that lens. This second system can be thought of as a telescopic or transfer lens. The magnification will depend on the ratio of the focal length of the transfer lens to that of the objective.

The position, magnification and resolution of the final image will be independent of the distance of the objective-object combination from the transfer lens. If this distance is too great, however, there will be a loss of field. Each parallel bundle of rays from the objective strikes only a relatively small area of the transfer lens, so that it is being used at a very low aperture, hence this system does not make severe optical demands on that lens. If one is in a position to design one's own lenses instead of being dependent on commercial suppliers, it is possible to utilize the design of the transfer lens to improve the off axis correction of the system. The size of the image formed by the transfer lens can be further enlarged or reduced by the use of additional auxiliary lenses. These additional lenses can be chosen to improve the flatness of field in the image plane. Another inherent advantage of the telescopic system is that, if the lens can be servocontrolled to follow a moving object, such as an animal's diaphragm, the image plane remains fixed, so that only the objective has to be moved.

A similar approach can be used for the condensing system so the distance of the source from the condenser is relatively unimportant. We have found, however, that for our system we can use a standard Köhler illumination system as shown in Fig.2. For our pulsed lamp source, which has a luminous area about 1 mm in diameter, we use a Nikon long working distance condensing lens identical to that used in our condenser system, but turned in reverse, so that the diaphragm is on the side away from the light source. An enlarged image of the source is formed by this lens on the diaphragm of the condenser, and the condenser images

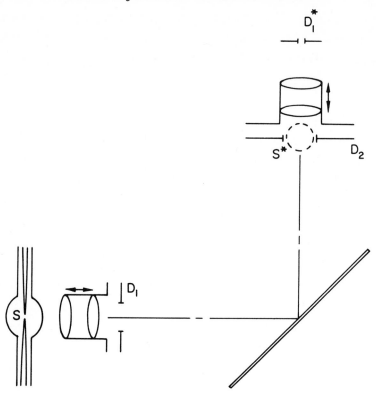

Fig.2 : Köhler illumination system for pulsed flash lamp.

the source diaphragm onto the object plane.   For most transillumination applications the object plane is well defined in
advance of putting the animal preparation in place, so that the
necessary adjustments of light source or sources and condenser
can be made.   This gives an adequate field and aperture of
illumination for use with the highest aperture lenses we have
normally found suitable for living preparations.   For larger
fields with low aperture objectives, an auxiliary lens can be
used with the condenser; or if the resolution demands are not
severe, the condenser can merely be defocused.

THE BASIC MECHANICAL ARRANGEMENT

Once the philosophy of the optical system was determined, we were able to design the mechanical features of our intravital microscope to meet our other criteria. Since this was a prototype model to test out a variety of ideas in intravital microscope design, we introduced a great deal more flexibility into our system than would be needed by most laboratories.

Our instrument, of which a sketch is shown in Fig.3 and a photo in Fig.4, has four working levels, all supported inboard of tripod supports. The lowest level, the source table, carries the light sources for normal transillumination; the second level carries the condenser system and the animal table; the third level, the optical table, carries the objective and transfer lens systems and the TV monitor; and the uppermost level carries the cameras. The three lower levels are all supported on the same set of three heavy columns, so that the optical components and the specimen under observation are firmly linked together mechanically.

The camera platform is carried on a separate set of columns, each concentric with, and interior to, one of the other supports, but in contact only at the base. Vibration due to cinecameras is thus isolated from the remainder of the optical system. Since all imaging lenses are normally carried on the optical table, a micron of motion by the camera results in only a micron of motion of the image on the film, and is not magnified by the optical system. Provision has been made to introduce vibration damping onto the camera table, but to date has not been found to be necessary.

The source table is integral with the base of the microscope and is permanently fixed in position. The animal table and optical table can be raised by means of pneumatic cylinders, shown in Fig.3, from 86 cm to 121 cm from the floor. Changes

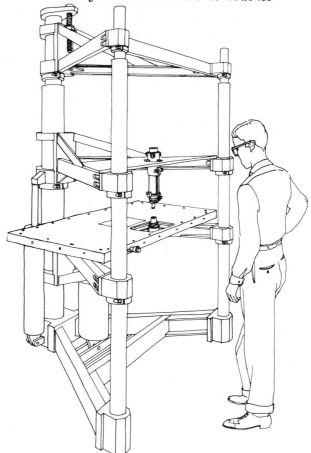

Fig.3 : Perspective drawing of intravital microscope assembly
showing the mechanical support system and pneumatic
lifts.

in vertical position are not made during an experiment; the
great demands on mechanical stability are met by firmly anchor-
ing each level to each of its three supporting columns by fric-
tion clamps.   Once the height for the animal table is attained
the clamp on the rear column is fixed, the table is carefully
levelled with the assistance of hydraulic jacks, and the other
two clamps are fixed.   The optical table is similarly levelled
and clamped in its desired position.   The height of the camera

Fig.4 : Photograph of completed intravital microscope.   The
rack to the left carries camera controls, power sources
for illumination systems, and TV monitor.   The rack to
the right carries recording equipment for physiological
monitors.

platform is then adjusted manually by means of a jack screw
between the rear supporting column and the table.   It is then
levelled and clamped to its three supporting columns.   Alignment and adjustment of the optical components must then be made.
Since this is an operation requiring considerable time, we find
that we change the relative vertical positions of the tables
very seldom.   Even with fixed vertical positions of the tables

the instrument has proved to be sufficiently versatile that for
many laboratories we would not recommend the additional expense
of providing the vertical adjustments.   A major factor in this
versatility is the fact that the objective carrier has a useful
vertical travel of 20 cm.

The animal table is essentially a massive microscope stage
with a clear working surface 75 × 120 cm, capable of precise
horizontal motion in two horizontal directions of ±5 cm.   The
table is designed to carry a load of 110 kg for animal, cradle,
life support systems, micromanipulators, etc.   Horizontal
positioning can be accomplished either manually, or electrically
with a joystick control which permits a wide range of speed of
movement as well as control of direction.   The horizontal
screws give 1/2 mm of motion per revolution.   Revolution counters on each screw, plus graduations on the dials attached to
the positioning screws, permit recording of the stage position,
and return to a prescribed location within a few micrometers.

DESIGN DETAILS

Source table

The source table (lowest level, Fig.4) carries a variety of
light sources for transillumination of transparent specimens.
It is designed to carry five triangular optical benches, each of
which can carry a permanently mounted source with lamp diaphragm
and a holder for various filters.   Since the sources are normally mounted on standard carriers for the optical benches,
sources can easily be changed on any particular mount.   To date
we have employed three sources in routine operation: a Chadwick-
Helmuth 100 watt pulsed xenon arc for both still and cinemato-
graphy; a quartz-halogen incandescent source for visual observation; and a 200 watt high pressure mercury arc, supplied from
an extremely well-filtered full wave rectifier for photometric
studies and streak photography.   A laser source is used for

basic instrument alignment.

A mirror at 45° with the horizontal can be rotated about a vertical axis to select the desired light source. This rotating mirror assembly is provided with adjustable detents which, when once adjusted, permit rapid movement from one source to another so that we can use the incandescent source for focusing and then rapidly move to the pulsed source for photography. A photograph of the rotating mirror assembly on the optical table, which is identical with that on the source table, is shown in Fig.5, and a detail in Fig.6.

Fig.5 : Rotating mirror assembly with partially reflecting mirror in place. Ball holder for detent mechanism on left front of mirror holder. Screen to the left carries phototransistor bank for erythrocyte velocity measurements.

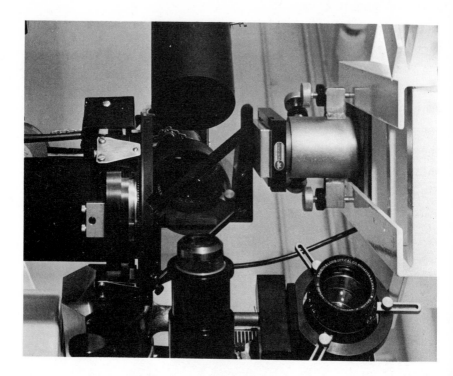

Fig.6 : Details of mirror assemblies.   The lower mirror system
rotates about a vertical axis, and is positioned by the
detent mechanism at lower right.   Note reentrant light
trap between the rotating mirror and the fixed upper
mirror system which deflects the light into the TV
camera.

Optical condenser and animal table

The second level consists of a steel table, clamped to the
main tripod support, which carries the condenser assembly and
the traverse mechanism for the animal table proper. The system
was designed so that the condenser assembly could be located
either at the forward edge of the table, as indicated in Fig.3,
or in the geometric center of the table.   To date we have found
the forward position satisfactory for all of our preparations,
so that the source and mirror systems on the lower table have
not been installed in the central position.   It would require

only the drilling and tapping of mounting holes in the lower table to permit the mirror mount and the required optical benches to be fixed in position to permit the use of this central location.

The condenser holder (Fig.4) consists of two concentric cylinders, the outer of which is supported from the base platform on the second level. The inner cylinder can be moved vertically over a distance of 13 cm by means of a rack and pinion. Only the upper 5 cm of this motion is useful when the rotating mirror system is being used since the moving cylinder begins to obscure the mirror as it is lowered. The moving cylinder is centered in the fixed cylinder by adjustable roller bearings, and counterbalanced by a constant force spring. The axis of the vertical motion can be adjusted by a set of three pairs of push-pull screws attaching the external cylinder to the table. The condenser and its diaphragm are carried on top of the movable cylinder on an adaptor designed for the particular condensing system being used (Fig.7). We have normally employed the condenser from a Nikon inverted phase microscope. This gives a clear working distance in air of 18 mm with a maximum N.A. of 0.70, which we have found to be adequate for all of our work to date. This lens has been mounted on an AO Microstar condenser mount, which gives a suitable condenser diaphragm with a centering capability. The holder for this condenser mount has a three point, spring loaded ball and screw system to allow adjusting the optical axis to be parallel to its axis of vertical motion. Space is provided in this condenser mount for insertion of optical filters, since this location is more accessible to the observer than the filter holders in front of the light sources. A series of extender rings is available to give flexibility in locating the height of the condenser above the animal table.

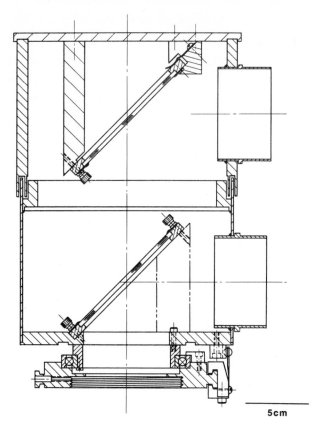

5cm

Fig.7 : Section of condenser adapter and levelling mechanism.

The animal table itself (Fig.4, second level from bottom) has been machined out of a slab of No.316 stainless steel 75 × 120 cm by 2.5 cm thick, weighing approximately 180 kg. A 5 cm deep skirt of the same material adds additional stiffness as well as protecting the traverse mechanism from physiological solutions. A stainless steel gutter is attached to this skirt to catch solutions dripping over the edge and to carry them to a waste reservoir.

A series of threaded holes around the periphery of the table, both in the upper surface and in the reinforcing skirt, permits great flexibility in attaching supports to hold various pieces of equipment in convenient locations. Threaded Teflon plugs are provided to protect the holes when not in use. Greater flexibility could be achieved if the table were made of a magnetic material to permit the use of magnetic clamps for holding auxiliary equipment. The corrosion resistance of the stainless steel has, however, proved to be a great advantage for maintenance.

Two square holes are milled into the table, corresponding to the alternate positions for the optical train. Covers of the same type of steel are provided for both holes, with "O" ring seals to prevent leakage of fluids onto the condenser and traverse mechanisms. When light pipe or epi-illumination systems are used, both covers are sealed in place. For operation with the standard condenser system for transillumination the required hole is uncovered. We originally planned to use a lucite shield to protect the condenser system, but we have found it adequate to introduce a drained catch basin below the open hole, at the level of the supporting table.

The animal table is supported at each of three points on two levels of hardened steel balls running in V-grooves. The two orthogonal directions are determined by a cross slide. Details of this design are available from the authors. A post carrying two pairs of ball bearings transmits the motion from the two elements of the cross slide which are, in turn, driven by the lead screws. The directions of motion of the table are kinematically determined by the V-races. Overdetermination is avoided by the use of hardened flat plates in two of the locations. Each motion is controlled by a moving nut, driven by a ½ mm pitch thread, actuated either manually from knobs situated at the front of the table (Fig.4) or by variable speed electric motors, controlled by a joystick controller, permitting control of both speed and direction of motion of the table. The dials on the knobs for manual control are graduated in 250 divisions, so that each division corresponds to 2 μm of horizontal motion. The drive shafts for the horizontal motion are attached to revolution counters so that the table can be positioned within a few micrometers of a given location by returning to the appropriate readings of the counters and the dials.   Mechanical stops and limit switches are provided to prevent damage to the traverse mechanism or to the condenser system by driving the table too far in any direction.

Optical and camera platforms

The objective mount is suspended from the optical platform (Fig.4, third level) with a system similar to that used for the condenser mount.   It can be located in either the forward or central position, corresponding to the two positions for the condenser.   The central position is considered to be potentially useful for work on the brain or in the thoracic cavity where either epi-illumination or light pipe or fiber optic illumination would be used.   To date none of our programs has required the use of the central position.

The supporting cylinder for the objective mount is suspended from above with three pairs of push-pull screws to permit adjustment of the axis of vertical motion.   An inner cylinder, on the bottom of which the objective holder is carried, is translated vertically over a range of 20 cm by means of a rack and pinion. Fine focusing is obtained by means of a $\frac{1}{4}$ mm pitch screw, driving a sliding objective mount, so that the objective does not turn during focusing.   The objective suspension system is counterbalanced by means of a constant force spring.   The axis of the objective can be made parallel to the axis of motion by a spring loaded three point suspension similar to that used on the condenser mount.

The transfer lens consists of an achromatic doublet, 61 mm diameter, with a focal length of 365 mm.   This is mounted in a tubular holder suspended from the top of the objective assembly, and remains fixed in position as the objective is moved up and down. The vignetting due to the size of the transfer lens will be worst at the lowest position of the objective. Even in the lowest position, the system gives a usable field 500 μm in diameter with a 50/.060 water immersion lens we have built by adapting a Leitz UMK 50/0.60 lens. Used high and dry this same lens gives a maximum usable field of 800 μm in the lowest position, but it should be pointed out that used without the hemispheric front element, this lens only gives a basic magnification of 32 with a numerical aperture of 0.40.

Space is provided above the objective mount to permit the insertion of a K-mirror, which permits the image to be rotated through more than 180⁰ without the chromatic aberrations which would be introduced with a dove prism. The ability to rotate the image is essential, for example, when the image of a vessel must be aligned with the elements of a velocity measuring system. The fact that the K-mirror assembly must be kept within the dimensional limits of the objective traverse tube introduces additional vignetting, but so far this has not proved to be serious. The design of the K-mirror assembly is given in Fig.8.

A rotating mirror system is mounted on top of the objective assembly (Figs.4 and 6) to deflect the image into the various optical recording devices mounted on top of the optical platform or suspended from the camera platform. Detents are positioned for each desired location, so that the mirror can be quickly swung from one position to another.

A microswitch has been incorporated into the circuit for the 16 mm Milliken camera to the right of Fig.9 so that it cannot be operated unless the light beam is pointed towards that camera. We normally monitor the light level at the viewfinder on the Hasselblad camera, shown at the left of Fig.9. Eventually we would like to incorporate film-plane light meters at each camera

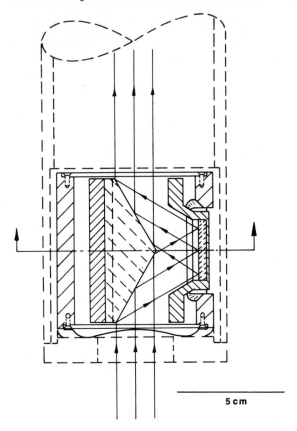

Fig.8 : K-mirror details.   This permits rotation of the optical
        image without moving the animal.

location, with capability for monitoring both continuous and
flash illumination.

We have found it extremely difficult to determine the position of sharpest focus with either the viewfinder supplied with
cameras such as the Hasselblad, or with beam splitters such as
are usually employed with conventional microscopes due to the
wide range of accommodation of the eye.   For many situations we
have found it to be satisfactory to use the TV camera for determining the point of best focus.   For this purpose, the mirror
in the rotating head reflects 70% of the light horizontally and

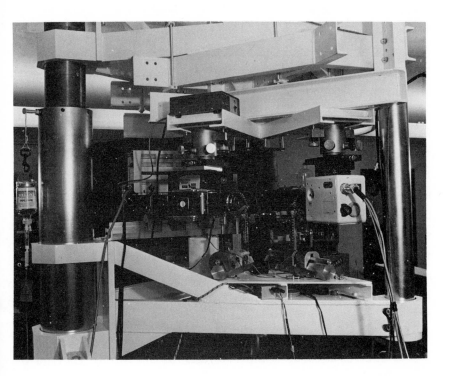

Fig.9 : Optical and camera platforms from left rear.   From
        left to right we see the videcon camera, the 70 mm
        Hasselblad single frame camera, and the 16 mm Milliken
        pin-registry cinecamera.

transmits 30% vertically.   The vertical beam is deflected by a
front surface mirror into a videcon camera (Figs.7,9).   The
various lens systems have been adjusted so that the images on
all cameras are parfocal with that formed on the videcon screen.
Since the videocron screen has no capability of optical accommodation this difficulty has been eliminated.

   With modern fine grain films it is possible to record a
great deal more information on film than can be seen by eye at
that magnification.   It is much more efficient to record photo-
graphically at relatively low magnification and to use further
magnification in viewing the film as long as all of the desired

information is captured on the film, since a doubling of the linear magnification on the film cuts the light intensity by a factor of four.   For survey work, it is theoretically possible to get adequate resolution to identify individual red cells with a magnification on the film of only 20 times, permitting a field of 3 mm diameter to be recorded on the 70 mm Hasselblad format. We have not yet satisfactorily solved the problem of keeping this large an area of living tissue flat enough to remain within the 20 µm depth of field of a lens of N.A. 0.20, which gives a nominal resolution of approximately 1.3 µm.   We have solved the problem of focus, however, by projecting the image on an opaque screen, which is parfocal with the camera, and examining the image with a telescope, shown just to the right of the upper mirror assembly in Fig.4.

The TV camera, shown just above the Hasselblad to the left of Fig.9, and all of the lenses for the film cameras are mounted on the optical platform so that the entire optical train is stably interconnected.   The film cameras are suspended from the uppermost platform, which is independently supported on its own tripod support system (Figs.3 and 4).   Any vibration of the cameras relative to the lens system will be recorded without optical amplification.   Since we normally do all photography with a flash duration of less than 40 microseconds, we have never had blurred images due to relative motion of the optics and the cameras.

In Fig.9 we see, from left to right, the videcon camera; a 70 mm Hasselblad electrically advanced single frame camera; and a Milliken 16 mm pin registry cinecamera capable of frame rates up to 500 frames per second.   To the right of Fig.4 we see a 35 mm pin registry Vaught cinecamera, capable of rates up to 96 frames per second.   The Milliken and Hasselblad are both mounted on quick-change dovetail mounts which permit the substitution of a Bolex 16 mm cinecamera for the Milliken and a 35 mm still camera for the Hasselblad without losing parfocality with the TV system.

SOME APPLICATIONS OF THIS SYSTEM

A few specific applications of our intravital microscope system will be described to show how we have taken advantage of its adaptability and flexibility. We have found that it is most convenient to carry out major surgery and other preparatory manipulations completely independent of the animal table of the microscope. We have mounted flat stainless steel tops on hydraulic lifts salvaged from dental chairs. The animal is prepared on one of these tables, along with the necessary life support and tissue maintenance systems; the table is moved next to the microscope table, its height adjusted for easy transfer of the animal and its auxiliary equipment; and this is moved as a unit to put the preparation in register with the optics of the microscope.

Cat mesenteric preparation

Such a unit has been developed for a cat mesenteric membrane preparation used for several investigations in our laboratory. A detailed description of this preparation will appear in the Proceedings of the Conference on the Microcirculatory Approach to Peripheral Vascular Function, Tucson, Arizona, 1971, which will be published in Microvascular Research.

The primary objective of the preparation is to provide access for *in vivo* microscopic examination of the microcirculation in the mesenteric membrane when the membrane is exposed in a fashion such that no external strain is imposed upon the structure. Essential concomitant conditions for the animal are that the body temperature must be maintained, adequate ventilation assured, and that an index of the operation of the cardiovascular system be provided. Essential additional conditions for the tissue specimen are that the local temperature be stable and the osmotic environment be definable and controllable.

Fig.10 :   Special animal carrier for cat mesentery preparation mounted in position for microscopic observation.

The system is shown in Fig.10 mounted on the animal table of the intravital microscope for operation.   The required surgery on the animal can be performed away from the microscope, with the animal mounted on a subassembly which is later moved onto the animal table.   It carries Teflon pads on its underside to permit easy sliding into position, and registry with the optical system is achieved by fixtures on the animal table which mate with cutouts in the carrier.   The entire small intestine is carried in the semicircular transparent chamber shown in the center of Fig.10.   Pressure gauges, a heat lamp, and supplies of temperature and osmotically controlled bathing fluid are carried on the animal table so that they all move as a unit with the animal.

Utilizing this configuration, the ability to move the entire preparation without imposing additional strain on the tissue permits us to make an "aerial survey" at high resolution, as illustrated in Fig.11 illustrating a basic microcirculatory unit in the cat mesentery. The detail available with our Leitz UMK 50/0.60 objective adapted for water immersion is shown in Fig.12, in which a portion of a rat omentum is pictured.

A photometric double slit velocity method has also been employed to study the erythrocyte velocity in vessels of various sizes in the cat mesentery. The photosensitive elements are phototransistors with a sensitive area 1 mm square. Two such elements are mounted in the screen shown at the left of Fig.5. An enlarged image of the vessel or vessels to be studied is projected onto the screen and as the image of a stream of red cells passes across one of the phototransistors it modulates the light level. The other member of the phototransistor pair, located just downstream from the first, will receive an almost identical signal, but displaced in time by the transit time of the cell pattern over the distance equivalent to that between the leading edges of the phototransistors. From this time and this distance, the velocity of the red cell motion can be calculated. A study of the velocities measured in a cat mesentery is shown in Fig.13.

Basically the same mesenteric preparation has been used for studying the mechanical response of the tissue to stretch. A special clamp and stretching mechanism have been developed by Dr. Billie Mae Chu for these studies which is shown mounted on the microscope table in Fig.14, so that the movement of the clamps relative to the animal is completely independent of the movement of the stage. This permits microscopic examination of parts of the stretched tissue without interfering with the

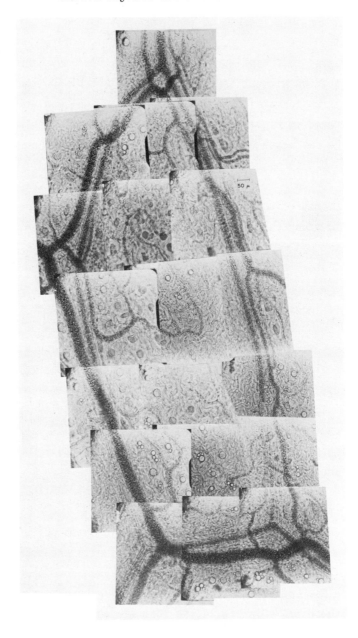

Fig.11 : High resolution "aerial survey" of basic mesenteric
flow unit of cat.   (Reprinted from Microvascular
Research by permission.)

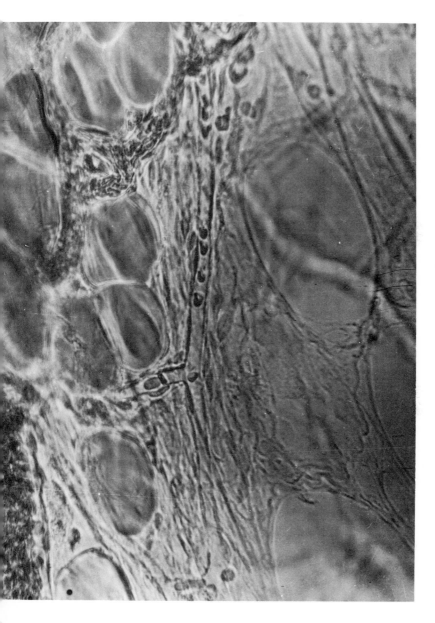

Fig.12 : Deformed red cells in rat omentum, showing detail
available with Leitz UMK 50/0.60 objective modified
for water immersion.

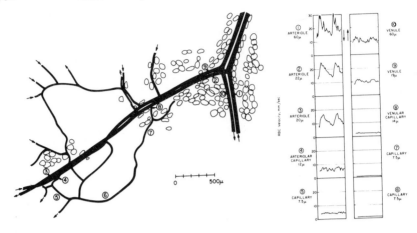

Fig.13 : Corner of mesenteric unit in cat with mid-line red cell velocities in vessels of various sizes.

Fig.14 : Mesenteric stretching unit mounted on intravital microscope table.

stretching process. Fig.15 shows a photomicrograph of cat
mesentery in a relaxed state, in which the collagen fibers are
seen as wavy lines. In Fig.16 the tissue has been stretched
close to the point of irreversible damage, and we see the collagen fibers have been straightened out.

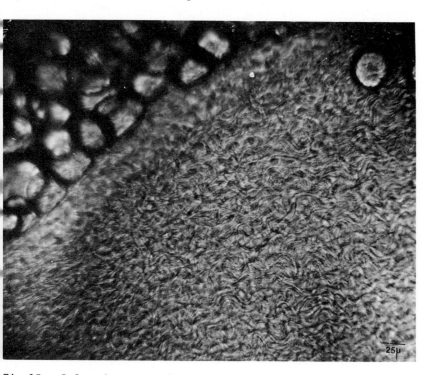

Fig.15 : Relaxed mesenteric membrane. Note wavy pattern of
collagen fibers.

Cat heart studies

Dr. Richard Bing and his team from the Huntington Institute
for Applied Medical Research have been working with our instrument to study the flow of blood in the microcirculation in the
left atrium of the cat's heart.

For this preparation the cat is anesthetized, the trachea
intubated, artificial respiration started, and then the chest is
opened. The cat is laid on its side and an incision is made in

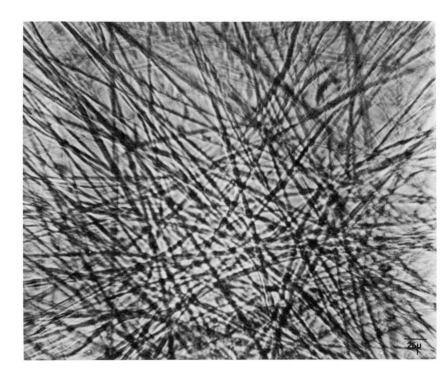

Fig.16 : Stretched mesenteric membrane.   Note that the col-
        lagen fibers are now straight.

the left atrial appendage.   A hollow glass tube, 6 mm in diameter, with the distal end sealed is inserted through the incision to the atrium and held in place with a ligature.   The
top of this tube is flattened for about 1 cm at its distal end,
and this flattened portion is positioned against the part of
the atrium to be studied.   A light pipe can then be inserted
into the tube to transilluminate the desired field.   For a
light source we have used a 100 watt xenon arc with an internal
semi-ellipsoidal reflector.   The arc itself is placed at one
focus, while the second focal point is outside the envelope of
the arc.   This is shown to the left of the animal of the Fig.
17, which shows the overall arrangement for this experimental

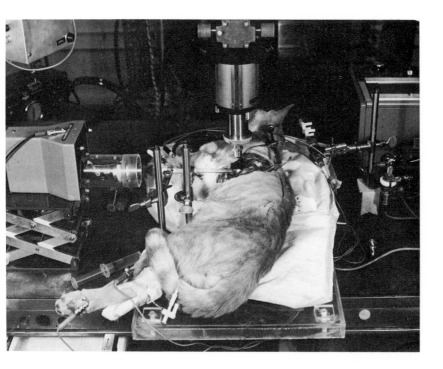

Fig.17 : Cat in position for transillumination of atrium of the
heart by light pipe illumination.

procedure.

For dynamic studies, motion pictures are taken up to 400
frames per second, and flow velocities obtained from frame by
frame analysis.   Care must be taken not to put too much pressure
against the atrial wall, or flow may be impeded.   Consequently
it is impossible to keep the tissue in focus throughout the entire heart cycle.   By focusing at different levels and taking
a burst of pictures at each level it has been possible to obtain
quantitative data at all parts of the heart beat.

THE BLUE-LIGHT LUMINESCENCE

TECHNIQUE AS AN INTEGRATING

VITALMICROSCOPIC METHOD:

CAPACITY AND LIMITATION

G. HAUCK and H. SCHRÖER

Institute of Physiology, University of Würzburg, Germany

Among the vitalmicroscopic methods, luminescence microscopy represents a relatively young technique based on tagging
the plasma with so-called fluorochromes, which produce a secondary luminescence of the plasma when irradiated with light of
the proper wave length. Since, in accordance with the Stokes'
rule, the wave length of the excited luminescence cannot be
longer than the wave length of the excitation light, activation
of the dyes used - such as trypaflavin, thioflavin S, aesculin
or fluorescein - was only possible by application of powerful
ultraviolet light because of the spectral absorption properties
of these fluorochromes and the quality of the available primary
and secondary filters. The alteration of the highly sensitive
small vessel wall due to such induced mobilisation of histamin
substances and by the local heat radiation was an unavoidable
consequence. Therefore, with rapid elimination of these fluorescent dyes, a small molecular size occurs and a bright and
homogeneous luminescence of the extravascular space occurs in a
short time, accentuated by the high visibility of these dyes

even in small quantities.   Under such conditions, a detailed observation of the capillary bed is impossible and therefore the luminescence technique seemed to be unsuitable for vital-microscopic studies.

A change of the status of this method in vitalmicroscopy resulted from the technical progress in three areas:

First, the development of new powerful excitation light sources with a well defined emission spectrum between 280 and 600 nm.   One class is the high pressure mercury lamp, especially the HBO 200 watt lamp.   The main peaks of the non-continuous emission spectrum are localized at 365 and 435 nm in the harmless blue-violet spectral zone (Fig.1).   Further, the

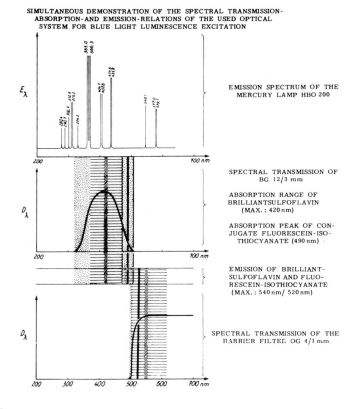

SIMULTANEOUS DEMONSTRATION OF THE SPECTRAL TRANSMISSION-
ABSORPTION-AND EMISSION-RELATIONS OF THE USED OPTICAL
SYSTEM FOR BLUE LIGHT LUMINESCENCE EXCITATION

E_λ

EMISSION SPECTRUM OF THE
MERCURY LAMP HBO 200

D_λ

SPECTRAL TRANSMISSION OF
BG 12/3 mm

ABSORPTION RANGE OF
BRILLIANTSULFOFLAVIN
(MAX.: 420 nm)

ABSORPTION PEAK OF CON-
JUGATE FLUORESCEIN-ISO-
THIOCYANATE (490 nm)

EMISSION OF BRILLIANT-
SULFOFLAVIN AND FLUO-
RESCEIN-ISOTHIOCYANATE
(MAX.: 540 nm/ 520 nm)

D_λ

SPECTRAL TRANSMISSION OF THE
BARRIER FILTER OG 4/1 mm

Fig.1

xenon lamps, characterized by a near-continuous emission spectrum, are especially suitable for excitation of a blue-light fluorescence. And finally the low voltage halogen lamps are worth consideration.

Second, the development of new excitation and barrier filters which are quickly interchangeable and possess excellent selection properties. In particular, the blue glass filter BG 12/1.5 - 3.0 mm (by Schott & Gen.) has its spectral transmission maximum at 400 nm and therefore permits a splendid utilization of the blue-violet radiation of the mercury and xenon lamps for an application to blue-light fluorescence (Fig. 1). A small red light transmission is eliminated by the blue glass filter BG 38. The combination of the primary filters BG 12 and BG 38 with the orange-yellow barrier filter OG 4/1 mm (by Schott & Gen.) marked by a very steep transmission slope at 500 nm (Fig.1) guarantees a sufficient contrast or darkness of the background and an optimal utilization of the fluorescent emission, especially of a fluorochrome new to the vitalmicroscopy - brilliantsulfoflavin - which is subsequently discussed. Another major factor was the development of an interference excitation filter - KP 490 - and a new barrier filter - K 510 - by Leitz and Balzers (3) especially adapted to the absorption and emission properties of the fluorochrome fluorescein-isothiocyanate, which is characterized by the close proximity of the absorption and emission maxima at 480 and 520 nm respectively (Fig.1). Because of the extreme steepness of the transmission slopes between the absorption and emission peaks, an excellent utilization of the emitted luminescence of fluorescein-isothiocyanate results. The heat-absorbing filter KG 2 (by Schott & Gen.) gives sufficient protection against the heat produced.

Third, the introduction into vitalmicroscopy of two fluorochromes well known in bacteriology: brilliantsulfoflavin and fluorescein-isothiocyanate has permitted a re-evaluation of luminescence methods as useful intravital microscopic tools.

Brilliantsulfoflavin, a derivative of naphthalic acid with a molecular weight of 404.3 was introduced into vitalmicroscopy by Witte (7). It possesses favorable absorption and emission properties with respect to the application of blue-light fluorescence. The main absorption peak is localized at 420 nm and the emission maximum at 550 nm (Fig.1). There exists, therefore, an optimal relationship between the blue-violet emission peaks of the mercury lamp HBO 200, the primary filter BG 12, the barrier filter OG 4 and the absorption and emission properties of brilliantsulfoflavin.

The second fluorochrome fluorescein-isothiocyanate was successfully used in vitalmicroscopy (6) and was found to be primarily suitable as an albumin conjugate. For this, the filter combination of BG 12 and OG 4 does not result in an optimal efficiency of the blue-light fluorescence excitation. The use of the new FITC-filter KP 490 and the barrier filter K 510 is preferable, as already mentioned.

Assuming an optimal utilization of the stimulated luminescence, the best vitalmicroscopic results indicate the trans- and epi-illumination in transparent tissues as mesentery, omentum or thin muscles under the conditions of a light path is essentially in a homogeneous or quasi homogeneous medium (Fig.2). This requires the use of apochromatical water immersion objectives and the observation of the living subject in the bath fluid. Further, for transillumination, water immersion of the condenser front lens on the lower side of the observation chamber is desirable. We use a bright- and dark-field fluorescence and phase contrast condenser by Zeiss with a numerical aperture of 1.4. The optimal resolutions are achieved in the range of magnification between 60 and 100 times. In the case of incident light application, the use of the dark field illumination is occasionally preferable, especially for investi-

PRINCIPLE OF THE LIGHT PATH

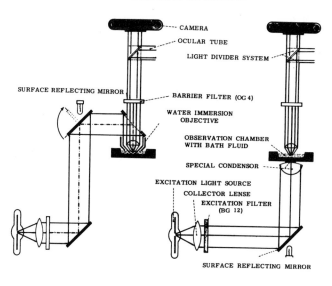

Fig.2 EPI-ILLUMINATION TRANS-ILLUMINATION

gations in which the organ surfaces are highly reflective.
Here we have successfully used the Ultropak system by Leitz.
For an optimal efficiency of excitation, the new fluorescence-
opak system by Leitz, based on the vertical illuminator by
Ploem (5), primarily developed for bacteriological and histological investigations, represents an important development. The
application of quickly interchangeable dichroic reflecting
mirrors with interference properties guarantees utilization of
90% of the stimulating light and an optimal transmission of the
emitted luminescence. Simultaneously the reflecting mirrors
have a prebarrier filter function.

With respect to the use of fluorochromes as indicators even
a very limited movement of the dye across small vessel walls can
be observed. This is due to the high visibility of the fluorochromes even in small quantities (in the extravascular space)
permitting their use when conjugated to proteins or other complexes. Brilliantsulfoflavin shows a quantitative affinity to

the albumins and β-globulins within the physiological range of
the pH-values with a saturation concentration of the plasma at
30 mg% (1,7).   In contrast to fluorescein-isothiocyanate,
brilliantsulfoflavin is water soluble to 0.1% and therefore
directly injectable intravascularly.   A concentration of 10 to
15 mg% of saturation is sufficient for excellent intravascular
luminescence in the plasma for 30 to 45 min (in contrast to the
unphysiologically high plasma concentrations of 150 to 200 mg%
(or more!) required for simple dyes).   In the case of a direct
intravenous injection, the elimination of brilliantsulfoflavin
is characterized in the early stage by dye penetration in the
arterial capillary side because of the presence of unconjugated
free dye molecules.   After 10 to 15 min the maximum perivascular luminescence appears in the venular part of the capillary
bed because of a progressive intravascular complex formation
with the proteins, which can permeate only here.   After injection of a homologue plasma labelled with brilliantsulfoflavin,
however, the dye penetration remains restricted to the venous
side and is completely identical with the elimination pattern
of albumin-conjugated fluorescein-isothiocyanate (2).   This is
a vitalmicroscopic demonstration of the intravascular complex
formation of brilliantsulfoflavin with the plasma proteins and
underlines the suitability of this fluorochrome for vitalmicroscopic studies.   The use of fluorescein-isothiocyanate, however, is dependent on a conjugation procedure and is preferred
for exact quantitative investigations, which are already possible in luminescence vitalmicroscopy.   In our experiments the
albumin conjugation was prepared in accordance with the instructions by Nairn (4) and the albumin-conjugated fraction was separated by use of a sephadex G-25 gel filtration.

In fact, three vitalmicroscopic criteria can be examined
by means of the blue-light luminescence method: the blood flow
pattern in small vessels because of excellent contrast between

the unmarked red cell column and the fluorescent plasma space, the blood distribution pattern within the terminal bed and the dye penetration pattern as evidence of the local vessel wall permeability; furthermore, the pattern of the extravascular dye spreading and drainage will be delineated.

A limitation of the blue-light luminescence method for detailed vitalmicroscopic studies in the capillary bed - not for angiological studies - is given by investigations in organ surfaces using the incident-light principle. This is primarily because the three-dimensional microvasculature shows increasing diminution of the luminescence-free tissue spaces important for a contrasting effect, essential for the luminescence method.

SUMMARY

The blue-light luminescence method permits a successful application of luminescence techniques in the vitalmicroscopy. The efficiency is dependent on the spectral matching primarily of the blue-violet radiation zone of the light source, the transmission properties of the excitation and barrier filters and the absorption and emission range of the fluorochrome employed. An optimal correspondence exists for the system using a high pressure mercury lamp HBO 200, the excitation filter BG 12, the barrier filter OG 4 and the luminescence properties of the fluorochrome brilliantsulfoflavin. Good results are also obtained using the same light source, the FITC-excitation filter KP 490, the barrier filter K 510 and the fluorochrome fluorescein-isothiocyanate. Brilliantsulfoflavin, water soluble and directly injectable intravascularly, forms complexes with the plasma albumins and β-globulins. The dye penetration pattern under these conditions is identical with the pattern of the albumin-conjugated fluorescein-isothiocyanate, which is only applicable in conjugated form. A major improvement in the incident luminescence excitation is obtained with the vertical

G. Hauck and H. Schröer

illuminator by Ploem. Three essential vitalmicroscopic phenomena can be readily observed: the capillary blood flow, the actual blood distribution and the local permeability to the dye penetrating into the extravascular space. A limitation of this method for investigations in organ surfaces is pointed out. The best results are obtained in transparent tissues.

REFERENCES

1. Hauck, G. Zur Frage der Existenz eines "gradient of vascular permeability" an der Endstrombahn. *Arch. Kreisl. Forsch. 59*: 197 (1969).

2. Hauck, G. and Schröer, H. Vitalmikroskopische Untersuchungen zur Lokalisation der Eiweisspermeabilität an der Endstrombahn von Warmblütern. *Pflügers Arch. ges. Physiol. 312*: 32 (1969).

3. Kraft, W. Ein neues FITC-Erregerfilter für die Routinefluoreszent. Leitz-Mitt. *Wiss. u. Techn. 5*: 41 (1970).

4. Nairn, R.C. Fluorescent protein tracing. E. & S. Livingstone Ltd. Edinburgh and London (1964).

5. Ploem, J.S. The use of a vertical illuminator with interchangeable dichroic mirrors for fluorescence microscopy with incident light. *Z. wiss. Mikr. 68*: 129 (1967).

6. Schiller, A.A., Schayer, R.W. and Hess, E.L. Fluoresceinconjugated bovine albumin. Physical and biological properties. *J. Gen. Physiol. 36*: 489 (1952/53).

7. Witte, S. Eine neue Methode zur Untersuchung der Capillarpermeabilität. *Z. ges. exp. Med. 129*: 181 (1957).

ADVANCES IN

SPECTROPHOTOMETRIC METHODS

AS APPLIED TO LIVING TISSUE

D.W. LÜBBERS and R. WODICK

Max-Planck-Institut für Arbeitsphysiologie, Dortmund, Germany.

Spectrophotometric observations have led to the discovery
of interesting and important substances in biology. For
example, the respiratory enzymes, the so-called cytochromes,
were first seen spectroscopically by MacMunn (9) and were later
studied by Keilin (3,4) in the same way.

Although the human eye has a high photosensitivity and the
photographic emulsion is a powerful instrument for long term ex-
position, the progress in the construction of photomultipliers
and in electronics has so much improved the photometers that
today the spectrophotometer is able to measure these substances
in living tissue. Figure 1 shows the extinction spectrum of the
blood and of the heart of the turtle. The upper spectrum with
two peaks is the spectrum of the oxygenated hemoglobin, the
spectrum with one peak corresponds to the de-oxygenated hemoglobin. The points of intersection are called "isosbestic
points". At these points the extinction does not change if the
degree of oxygenation changes. The isosbestic point is suitable for measuring the total hemoglobin concentration in a probe,
the wavelengths of the peak for measuring the degree of oxygen-

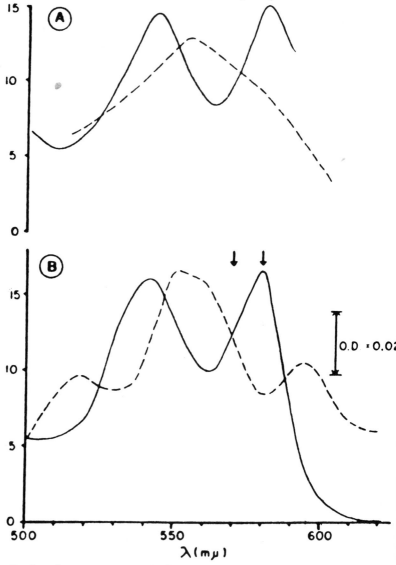

Fig.1 : Extinction spectrum of blood and heart of the turtle.
(A) Extinction spectrum of oxygenated (solid line) and
de-oxygenated (broken line) blood.   The arrows mark
isosbestic points.
(B) Extinction spectrum of an oxygenated (solid line)
and a de-oxygenated (broken line) heart.   In the
oxygenated form the two peaks of oxygenated myoglobin

can be seen; in the de-oxygenated and reduced spectrum
the band at ca. 600 nm corresponds to reduced cytochrome (a_3 + a), the peak at 550 nm to reduced cytochrome c. The shoulder is caused by reduced cytochrome b and de-oxygenated myoglobin.

---

ation. The area between both spectra represents the difference spectrum.

The existence of isosbestic points allows the measurement
of a two-component system by using two suitable wavelengths.
This two-wavelength photometry was developed in the thirties by
Kurt Kramer, Karl Matthes and G.A. Millican (17), and was improved greatly by Britton Chance (1) in the fifties. The two
wavelengths can be obtained by filters or by two monochromators.

The double wavelength method has proved to be very useful,
but if the components of a system have overlapping or similar
spectra, then additional information is necessary. The
measurement has to be carried out in a larger spectral range
(multiwavelength method). Working with living tissue *in situ*
the normal recording spectrophotometers are too slow due to the
movement of the tissue. Therefore, we constructed together
with Niesel and Koehler, a rapid-recording spectrophotometer
(5,10), which is now commercially available as Rapid-Spectro-
meter (Kieler Howaldts-Werke, Kiel). It is a two beam photometer which scans the spectrum 100 times per second. Each
spectrum consists of about 250 measurements and reference points.
This corresponds to a measuring rate of 50,000 points per second,
so that each value is available for about 20 μs only.

The scanning frequency of the spectrum and the splitting
frequency of the double beam are brought about by oscillating
mirrors, which oscillate in their resonance frequencies, one
mirror of 50 by 70 mm at 50 to 60 cycles per second, and the
other one of 1 by 3 mm at 20 to 25 kcycles per second. The
reference beam is used to pick out the different light sens-

D.W. Lübbers and R. Wodick

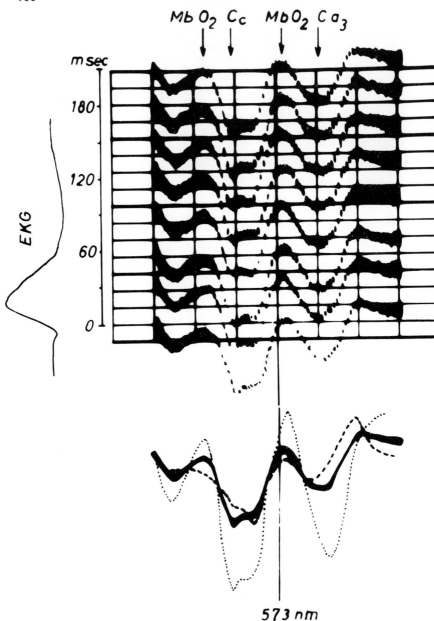

Fig.2 : Difference-reflection spectra of a beating, hemoglobin-
free perfused rabbit heart.
(Abscissa: wavelength; ordinate: extinction; Parameter (ordinate): time in msec.) The difference is

tivities of the photomultiplier and the emission spectrum of the lamp by regulating the diode voltages of the photomultiplier. The signal is logarithmically transformed and then the reference and the measuring pulse are subtracted. The so-formed extinction spectrum is recorded on an oscilloscope screen or transformed to an analog-digital converter in order to evaluate the signals by a digital computer.

The photomultiplier is covered by an integrating sphere with 4 holes. The incident light penetrates the integrating sphere, enters a light pipe and reaches the object. The diffusely reflected light reaches the photomultiplier via the light pipe and the integrating sphere.

Figure 2 shows the difference spectrum of a beating rabbit heart *in situ*, which was measured by means of the rapid spectrometer. The spectra were triggered by the R-peak of the electrocardiogram which is recorded on the left side (scanning time 10 ms frequency about 30 spectra per second (2)).

The quantitative evaluation of the recorded reflection spectra has been studied in cooperation with Schwickardi and Knaust (12). We recorded first the spectrum of the Hb-free perfused cortex of a guinea pig. Then the measured tissue sample was taken out and homogenized. By consecutive dilution of the homogenate the cytochrome concentration of the sample can be measured in transmitted light. Knowing the concentration in the tissue, the results of the measurements using reflected and transmitted light can be compared. Taking the published spectra of the measured concentrations of the cytochrome, we summed

---

measured against reduced heart homogenate. If the heart is reduced and de-oxygenated, then the "zero line" is reached (broken line), the dotted line corresponds to O_2 saturation and total oxydation of the heart. In the lower part of the figure the spectra are drawn one over the other: there is no detectable change of the spectra during heart action.

D.W. Lübbers and R. Wodick

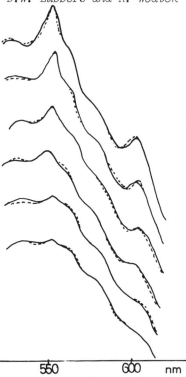

550   600   nm

Fig.3 : Reflection spectra of a hemoglobin-free perfused guinea
pig brain.
(Abscissa: wavelength; ordinate: extinction - the
peak height of cytochrome c corresponds to ca. 0.03
OD). The lowest spectrum corresponds to a normal
oxygen supply, the uppermost spectrum to anoxia. The
solid lines are the measured spectra, the dotted lines
the superposed spectra. The superposition was established by using the published spectra, the real concentration values (measured in homogenates), and a correction for the isosbestic points. Both spectra fit very
well.

up the components to a superposed spectrum. Using the isosbestic points of the reflected spectrum that are situated at other
wavelengths and have a higher extinction than the isosbestic
points of the published spectra, we obtained an excellent fit
between the measured and the superposed spectrum, as shown in
Fig.3. This allows the conclusion that, for the reflected
spectra of the brain, the Bouguer-Lambert-Beer (BLB) Law can be

applied, if the light path is corrected by an experimentally
measured factor and the spectra are corrected for the displacement of the isosbestic points (16).

For the evaluation of the small concentration of the components, we used in this case the changes of the angle of the
peaks; this method has a total error of 10%.

If the spectra of the components are *similar*, then another
evaluation method must be applied.   Knowing the spectra of a
multi-component system, their actual combination in the mixture
can be calculated applying the least squares method of Gauss
(6).   This method looks for the concentration combination for
which the difference between the calculated and the measured
spectrum is a minimum.   The method could be improved by introducing a weight factor (13).   By this weight factor, the extinction values will be weighted corresponding to their information content about the composition of the mixture.   In a
region where all extinctions are similar, the weight is low,
but in regions where the extinction of the component is different, the weight is high.

We tried this method for the analysis of a mixture of
nucleotides.   There is no characteristic peak and there are
regions in which the spectra are very similar.   Wodick was
able to show that there exists an optimal weighting function
for each system.   Figure 4 shows the weighting functions for a
mixture of four nucleotides.   In order to take advantage of the
multi-component analysis, a photometer of high reproducibility
is necessary.   We used for these measurements the DMR 21 of
Zeiss which was provided with an automatic data recording system.
The measuring of a system lasted over 7 minutes.   The wavelength
reproducibility was checked by a holmium oxide filter.   Between
2400 and 3000 Å, 2500 values have been recorded for the wavelength as well as for the extinction.   The wavelength had a
reproducibility of ±0.5 Å and the extinction values changed

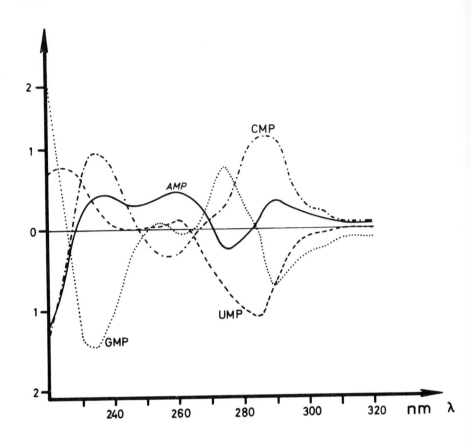

Fig.4 : Weight function for the least-squares analysis of a
        nucleotide system.
        AMP = Adenosine-3-phosphate.
        CMP = Cytidine-3-phosphate.
        GMP = Guanosine-3-phosphate.
        UMP = Uridine-3-phosphate.

only in the range of 10^{-4}.   The extinction, however, had to be
recalibrated.   (We should like to thank the factory for their
help, especially Mr. Franke.)   Measuring the components and the
mixture by the same instrument, the overall error for this multi-
component system was below 1%.   This demonstrates how powerful
this method is.   It has the advantage that the multi-component

analysis can be applied to intact systems without isolating the components.

If we try to measure the oxygenation degree for hemoglobin at the surface of the blood-perfused brain or in the skin, then the application of the previously-mentioned methods gives erroneous results. In order to find the reason for this failure, we have to know the degree of oxygenation in the skin. A complete oxygenation of the hemoglobin in the skin can be brought about by breathing oxygen during skin hyperemia, since then the oxygen tension at the surface of the skin rises to pO_2 values higher than 400 Torr. If we now relate the extinction value of a 100% saturated $Hb-O_2$ solution (abscissa) to the corresponding values of a 100% saturated $Hb-O_2$ in the skin (ordinate), we should find a straight line, if the BLB Law were valid. We actually find a curved line which also shows that for some extinction values of the $Hb-O_2$ spectrum two extinction values exist in the skin spectrum dependent on the wavelength (Fig.5).

How can we explain this complicated curve? On the picture of the blood perfused brain we can easily distinguish two compartments: one filled with blood of red colour and the other consisting of rather white tissue. These two compartments seriously effect the extinction curve obtained from the tissue. This can be simulated by varying the position of the measuring cells. If they are situated one behind the other and one is filled with hemoglobin and the other with solution, each part of the light beam has to pass the same amount of hemoglobin and solution molecules. The light will be weakened in the first cell to $L_1 = L_0 T_1$ and after the second cell to $L = L_1 T_2$. The total weakening amounts to $L = L_0 T_1 T_2$. This is the product of the weakening; such a process therefore is called *multiplicative mixing*. If they are situated side by side, the light passes two different compartments. The weakening for the compartment of the fractional area a filled with hemoglobin is

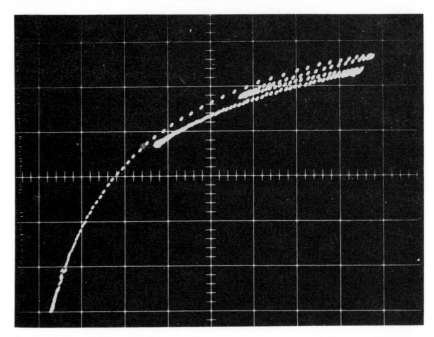

Fig.5 : Transfer function of oxygenated hemoglobin in the skin.
(Abscissa: extinction values of an oxygenated hemoglobin solution in a cell; ordinate: extinction values
of the oxygenated blood in human skin (reflection spectrum)).

$L' = aL_0T_1T_2 = a(L_0T')$ and for the pure solution of the fractional area $(1-a)L'' = (1-a)L_0T_3T_4 = (1-a)L_0T''$.   The total
weakened light $L = L' + L'' = L_0(aT' + (1-a)T'')$ equals the sum
of the weakening which takes place in both compartments.   This
process therefore is called *additive mixing*.

An additive mixture of 40% white and 60% oxygenated hemoglobin considerably disturbs the extinction spectrum, as shown
in Fig.6.   This disturbance can be described as a transformation function of the two compartment systems shown in Fig.7.
If this function for an additive system is known, then the additive mixture can be analyzed.

Reflection in a tissue can be simulated by the additive
mixture in a staircase cell (Fig.8A).   The total amount of light

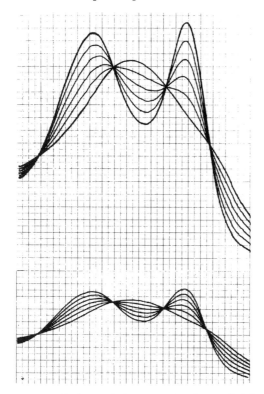

Fig.6 : Effect of additive mixing on the hemoglobin spectrum.
(Abscissa: wavelength; ordinate: extinction; the
two peaks of oxygenated hemoglobin are at λ = 543 nm
and λ = 578 nm.)   Upper part: 100%, 80%, 60%, 20% and
0% oxygenated hemoglobin. Lower part: the same
oxygenation degrees, but with an additive mixture of
40% white (= no absorption) and 60% hemoglobin.

which leaves the cell is given by the sum of the light which
leaves each compartment. The light which is reflected from
the surface corresponds to a cell with no absorption; the light
which is scattered and does not reach the photomultiplier corresponds to a cell compartment which is very thick.   Generally,
reflection spectrophotometry can be understood as a result of
additive and multiplicative mixing.

    In the tissue itself the light path is much more compli-
cated.   If we count how often a certain light path can be found,

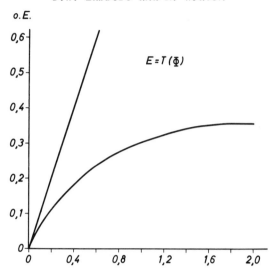

Fig.7 : Transfer function of a two compartment system.
(Abscissa and ordinate: extinction; abscissa: single
compartment system; ordinate: two compartment system.)
If no additive mixing takes place, one obtains the
straight line, with additive mixing the curved one,
similar to the transfer function of the skin (Fig.6).

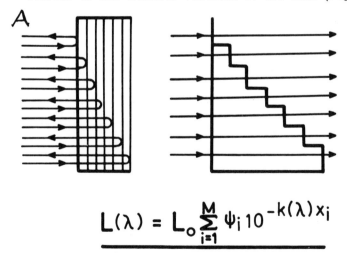

$$L(\lambda) = L_o \sum_{i=1}^{M} \psi_i \, 10^{-k(\lambda)x_i}$$

Fig.8 : Simulation of the reflection spectra by a staircase cell.
(A) Reflection (left side) can be simulated by a cell
with different light paths (staircase, right side). The
transmitted light is the sum over each part of the cell
as the formula shows.

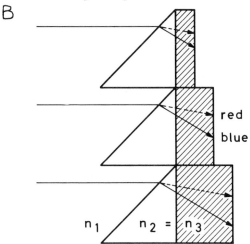

Fig.8 : (B) The influence of different refraction indices can be demonstrated by mounting a prism in front of the cell. Then the lightpath in the cell depends on the wavelength.

then we obtain the *probability density* of the different light-paths in the tissue. Knowing this function, as well as the spectral components of the system, it should be possible to calculate the transformation. As an approximation, the probability density function for brain homogenate has been experimentally determined by adding different amounts of hemoglobin to the homogenate.

After a decrease the measured curve shows a second increase. This second increase is due to multiple reflection. At the moment we cannot completely simulate our skin-hemoglobin system, but we hope we shall be able to in the near future.

If the dependence of the lightpath on the wavelength is small and if the total extinction is small, then a new method can be applied in order to obtain the fractional distribution of two substances, as for example oxygenated and de-oxygenated hemoglobin in additive mixtures. We observed that the distance of the hemoglobin maxima, which change dependent on the degree of oxygenation, is influenced by additive mixing (7). Realizing

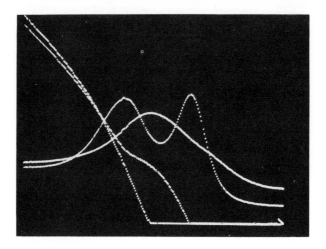

Fig.9 : "Queranalyse", method to analyse the degree of oxygenation.   (Abscissa: wavelength; ordinate: (a) spectra:
extinction; (b) Queranalyse: wavelength distance (in
arbitrary units).)   The wavelength distances between
equal extinction values are measured from shorter to
longer wavelength.   The zero point corresponds to the
maximum of the extinction curve, for de-oxygenated Hb
at 555 nm, for oxygenated Hb at 578 nm.   The curves of
"Queranalyse" intersect: isosbestic point of second
order (independent of the oxygenation degree).

that such invariants exist, we could prove the existence of
other invariants.

Generally, in a spectrum with one maximum or minimum the
distance of two wavelengths having the same absorption is invariant against additive mixing.   The distance can be drawn as a
function of the shorter or longer wavelengths.   In such a way
we obtain the "Queranalyse" (analysis of the wavelength distance
across the spectrum with similar absorptions or extinctions) of
a spectrum (14,15).   Figure 9 shows the "Queranalyse" for oxygenated and de-oxygenated hemoglobin.   The measurements start at
the left and measure for HbO_2 the wavelength distance to the
second peak at 578 nm (ignoring the first one).   As can be seen,
there exist points which are independent of the oxygenation.   We

call these points isosbestic points of the second order.

For analyzing the degree of oxygenation by means of the "Queranalyse" one can use calibration curves or calculate by a computer the best fit for different degrees of oxygenation using the least squares method. There is a pronounced minimum corresponding to the correct oxygenation. If the minimum is not close to zero, this is a sign that an unknown disturbance function is existent and the analysis, therefore, may give wrong results. The two main reasons for such disturbances are:

(1) the wavelength-dependent lightpath. A model for such a lightpath is shown in Fig.8B. By refraction in a medium with higher refraction index the lightpath for red light is shorter than that for blue light. In this case the *quotient* $\Delta E/E$ is an invariant against different lengths of the lightpaths; and

(2) an unknown component.

If a disturbance function exists and cannot be cancelled out by compensation, sometimes another method can be helpful. We remember that in additive mixing the extinctions are in the exponent of a sum. So the calculation of the logarithm does not result in the extinction values. In order to avoid these calculations, we measure the different light intensities between three wavelengths (8). If at the three wavelengths the disturbances are similar, they are cancelled out by subtracting the light intensities $L_1 - L_2 = {}_aL$ and $L_3 - L_2 = {}_bL$ (Fig.10). The degree of oxygenation g can be obtained as

$$g = \frac{a\Delta_b L - b\Delta_a L}{(a-u)\Delta_b L + (v-b)\Delta_b L}$$

If the degree of oxygenation is known, the product of the concentration and the thickness of the layer can be obtained in a similar manner.

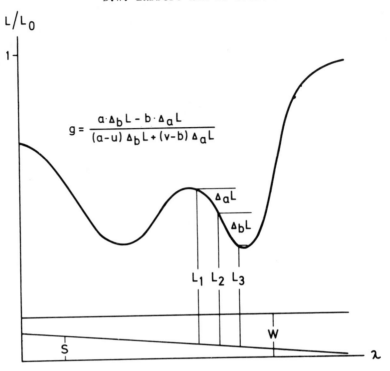

Fig.10 : Measuring the oxygenation degree of hemoglobin by means
of two wavelength differences (3-method).
(Abscissa: wavelength; ordinate: absorption.)   The
spectrum of oxygenated hemoglobin is disturbed by an
additive factor (w) which is slightly wavelength
dependent (s).   If all the absorption values L_1,
L_2, L_3 have the same disturbances, the disturbances
can be cancelled out by forming the differences
L_1 - L_2 and L_3 - L_2.   The formula shows how the degree of oxygenation g can be calculated.

In conclusion we can state, that the analysis of reflection
spectra is complicated by the combination of additive and multiplicative mixing as well as by the wavelength-dependent lightpath.   This has been demonstrated by the transformation of the
cell hemoglobin spectrum into the hemoglobin spectrum of the
skin.   For the multiplicative mixture the multicomponent analysis can be improved considerably by introducing a weight functio
in the least squares method.   The new method of "Queranalyse"

using invariants against additive mixing is a helpful tool for the analysis of reflection spectra. The final goal, to know the transformation that allows us to calculate the concentrations of the components in complicated photometric systems, could be approached by calculation and measuring the probability density function for brain homogenate. Rapid recording spectrophotometers have proved a helpful tool for analysing complicated photometric systems.

REFERENCES

1. Chance, B.   Spectrophotometry of intracellular respiratory pigments, *Science 120*, 767-775 (1954).

2. Fabel, H. and Lübbers, D.W.   Measurements of reflection spectra of the beating rabbit heart *in situ*. *Biochem. Z. 341*, 351-356 (1965).

3. Keilin, D.   On cytochrome, a respiratory pigment, common to animals, yeast and higher plants. *Proc. Roy. Soc. B 98*, 312 (1925).

4. Keilin, D. and Hartree, E.F.   Cytochrome and cytochrome oxydase. *Proc. Roy. Soc. B 127*, 167 (1939).

5. Lübbers, D.W. and Niesel, W.   Der Kurzzeit-Spektral-analysator.   Ein schnellarbeitendes Spektralphotometer zur laufenden Messung von Absorptions- bzw. Extinktions-spektren. *Pflügers Arch. ges. Physiol. 268*, 286 (1959).

6. Lübbers, D.W. and Wodick, R.   The examination of multi-component systems in biological materials by means of a rapid scanning photometer. *Appl. Optics 8*, 1055-1062 (1969).

7. Lübbers, D.W., Piroth, D. and Wodick, R.   Bestimmung der Sauerstoffsättigung des Hämoglobins bei inhomogener Farb-stoffverteilung. *Naturwissenschaften 57*, 42 (1970).

8. Lübbers, D.W. and Wodick, R.   Quantitative Auswertung von Spektren inhomogen verteilten Farbstoffes Beispiel Hämo-globin. *Naturwissenschaften 58*, 321 (1971).

9. MacMunn, C.A.   Researches on myohaematin and histohaematins. *Phil. Trans. Roy. Soc. 177*, 267 (1886).

10. Niesel, W., Lübbers, D.W., Schneewolf, D., Richter, J. and Botticher, W.   Double beam spectrometer with 10-msec recording time. *Rev.-Scientific Instr. 35*, 578-581 (1964).

11. Schneider, St., Lübbers, D.W. and Wodick, R.   (in process of publication).

12. Schwickardi, D. and Knaust, K.   Konzentration und Kinetik der Atmungsfermente am isoliert perfundierten Meerschweinchengehirn *in vivo* und in Hypothermie von 18°C. (no journal cited).

13. Wodick, R.   Die Auswertung von Mehrkomponentensystemen der Kurzzeitspektrophotometrie.   Inaug.—Diss. Marburg (1968).

14. Wodick, R.   Neue Auswertverfahren für Reflexionsspektren und Spektren inhomogener Farbstoffverteilung, dargestellt am Beispiel von Hämoglobinspektren.   Inaug.-Diss. Marburg (1971).

15. Wodick, R. and Lübbers, D.W.   Reflexionsphotometrische Analyse von inhomogen verteilten Farbstoffgemischen an lebenden Geweben mit Hilfe von Lichtweginvarianten. *Pflügers Arch.* *319*, R 60 (1970).

16. Wodick, R., Schwickardi, D. and Lübbers, D.W.   Konzentration und Kinetik der Atmungsfermente im Meerschweinchengehirn *in vivo*.   *Pflügers Arch. ges. Physiol.* *291*, 25 (1966).

17. Zijlstra, W.G.   Fundamentals and Applications of Clinical Oximetry.   Verlag van Gorcum - Assen, Netherlands (1953).

L A S E R S

I N   B I O M E D I C A L   R E S E A R C H

LEON GOLDMAN

Laser Laboratory, Medical Center of the University of
Cincinnati, U.S.A.*

For all disciplines of biology, the laser has much to offer
now and for the future. There is much more to this new technology than just a study of the reactions of so-called non-
ionizing irradiation on living tissue. For the biomedical applications of the laser (1,2,3) in the field of biology there
have been great advances, much more perhaps than in the field of
the medical applications with the exception of laser ophthalmology. As new systems and new techniques develop, the range
of applications will extend. Some obvious chief deterrents to
progress in biomedical research and development are inadequate
funds (as contrasted with industry and the military), the continuing need for expensive, sophisticated instrumentation, multi-
disciplined cooperation, the necessity of adequately planned
safety programs and the frequent lack of communication between
the various divisions in biology and medicine (4,5).

   It must be remembered that there are many different types
of lasers with frequencies varying from ultraviolet to the far
infrared, with pulse duration varying in the pulse laser systems
from milliseconds to nanoseconds and even picoseconds. There

---

* Laboratory established by the John A. Hartford Foundation.

Leon Goldman

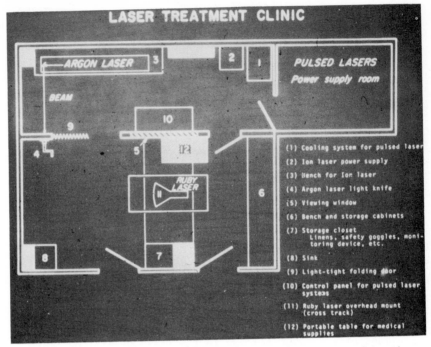

Fig.1 : Laser treatment Out-Patient Clinic established by the
John A. Hartford Foundation at the Medical Center of
the University of Cincinnati, U.S.A.

are also continuous wave laser systems (CW).   All the previous
laser systems, the ruby, 694.3 nanometers, the neodymium,
106.0, the neodymium YAG, 106.0, ultraviolet from harmonics,
the helium neon, 6324.8, and the carbon dioxide, 1060.0, have
all been used in biology.   Ultraviolet lasers are now being
used for research in biology.   Higher outputs of ultraviolet
lasers are usually obtained with an ultraviolet doubler, as
from a neodymium YAG laser system.   Now the fascinating new so-
called tunable dye lasers with the ability to portray almost the
entire spectrum offer a tremendous opportunity for investigations in biology.   These make for differences in biological
reactions not only because of their different frequencies but
also because of their different energy and power characteristics

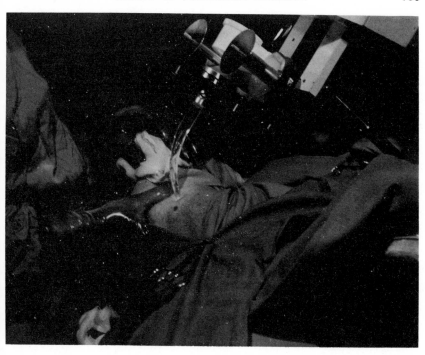

Fig.2 : Ruby laser treatment of cancer of the skin with beam
transmitted through curved tapered quartz rod.

Miniature semi-conductor laser systems will be able to be used
deep in tissue and in body cavities.   Research still continues
for the development of an X-ray laser and this will not be too
far away.

Energy and power densities of lasers may vary.   Transmission techniques, quartz rods, mirrors, prisms, fiber optics,
etc. also affect output.   New fiber optics systems will be used
for transmission of laser beams into cavities, blood vessels,
heart and deep into the lungs.   Finally, the optical characteristics of the living tissue target area itself affects the reaction in terms of reflectance, absorption and transmission.

SAFETY PROGRAM

It must be remembered that most fields of laser research
in biology come under the category of high risk exposure to
lasers. This is so in any type of R & D work (Research and
Development) with lasers. Therefore, the safety program must
be planned carefully in advance. One must refer constantly to
current information about laser safety. In the United States,
ANSI (American National Standards Institute) is setting up
these basic requirements on laser exposure through its Z-136
Committee (6). It is hoped that there will be international
agreement of the standards of maximum permissible exposure
levels and not the confusion of safety standards as holds now
for microwaves. University laboratories are no hallowed
places with special privileges, as regards safety, for students,
fellows, or staff who may be working in them. The same rules
and restrictions for laser safety apply to them as to anyone in
industry, biomedical work, etc. If the laser area is extensive, one individual designated as the Laser Safety Officer
should have responsibility. Again, if laser systems are to be
used with different systems in an area, then prior construction
of the laboratory area control must be so constructed for safety.
If it is possible for laser systems to be enclosed, such as in
animal exposure boxes, computer technology, etc., then there is
very little concern about hazards. If these laser systems must
be open such as in micro-irradiation and for some features of
micro-emission spectroscopy and Raman spectroscopy, human and
animal laser surgery, then one must be concerned about the
hazards of the reflected beams. Hazards to the eyes are most
important, not only for acute burns, but for chronic exposure
and special protective glasses for each respective system should
be used. Warning signs in the area should indicate that the
laser is being employed. An international laser radiation sign

is now offered.   Chronic exposure of the skin should be avoided.
Air pollution should be considered depending upon the impact on
the target area of such viable organisms, cancer tissue fragments,
toxic chemicals, metals, ionized air, etc.   The safety programs
should be reviewed and revised from time to time.

MICRO-IRRADIATION

In the micro-irradiation procedures, biologists are familiar with ultraviolet systems.   Now with a coherent form of
irradiation, micro-emission can be much more precise and the
beam width can be less than one micron.   This introduces the
possibility of micro-surgery not only on cell tissue, but also
on organelles.   Refinement of this technique has been recently
done (7,8,9) and has produced micro-irradiation of the large
chromosome of the salamander.   Some recent studies have related also to DNA and histone.   As Scheidt and Traut (10) of
the University of Münster have done with X-rays on the chromosomes of the broad bean, *Via faba*, scanning electron microscopy
(SEM) will aid considerably in the detailed study of these
"gaps" in chromosomes impacted by lasers.   The newly available
ultraviolet lasers can be used for precise studies of the repair
of chromatid breaks.   Attempts to have the laser identify differences in chromosomes have not been successful, unlike quinacrine fluorescence microscopy.

There has also been extensive interest in the application
of laser impacts to the development of laser micro-emission
spectroscopy.   Some difficulties were experienced initially
with quantitation.   Recent developments (11,12) have shown that
for micro-sampling analysis, excellent quantitation can be obtained with immediate readout with computer attachment.   This
enables studies to be done of cations of the nucleus, cytoplasm,
selective studies on mitochondria stains for increased absorption of laser, trace metals analyses, sequential analysis in

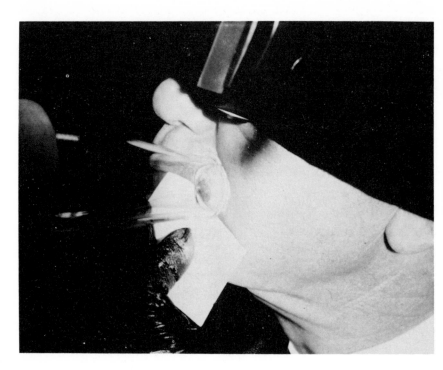

Fig.3 : Laser safety arrangement for treatment of cancer of the
face; eye protection with black cloth and protective
glasses; laser head with plume trap; white cardboard
to protect lips; gloved hand of operator to protect
against chronic exposure to lasers.

forensic medicine, etc.   Recently, we have done studies of
periodic micro-emission spectroscopy of hairs in children to
evaluate exposure to such environmental pollutants as lead, mercury, cadmium.   With Brech (12) and Allemand (11), we have done
experiments on calcium, gold in the skin, nails, and trace metal
in bone marrow smears of a patient who developed severe anemia
after gold.   In the bone marrow study, crater sizes of the impact varies from 10 to 50 microns.   Control studies of the
carriers were made.   Ca, C, Mg, Fe and S were detected but no
gold.   After the spectroscopy, the bone marrow smears were
stained and the laser impact areas found.   Micro-emission spec-

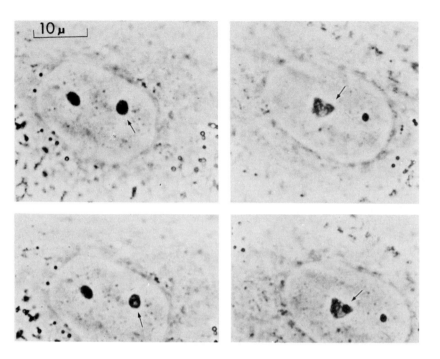

Fig.4 : Argon laser micro-irradiation of human skin fibroblasts
with 0.005% quinacrine; production of nucleolar lesions
(Donald Rounds - Pasadena Tumor Institute).

troscopy, then, can be done with proper controls, on smears,
tissues, etc.   Then these are fixed and stained and the structures impacted can be identified on the fixed and stained tissue.   In other words, there is no complete destruction of the
specimens for analysis.

Desmond Smith (13) and his associates at Heriot-Watt University, Edinburgh and Bell Telephone Laboratories have been
working on tunable infrared lasers for infrared spectroscopy for
rapid chemical analyses for the spectroscopy of excited chemical
states.   Robinson, Christian, Guillardo and Katayama (14) have
used infrared laser systems for continuous monitoring of the
atmosphere for environmental pollutants.   Goodman, Seliger and

Minkowski (15) have considered the use of Q-switched laser for selectively stimulating chemical reactions by absorption of infrared photons by "non-thermal" enhancement. As an experimental model, they have selected the hydrolysis of adenosine triphosphate (ATP) in H_2O and D_2O.

In addition to Raman spectroscopy by Q-switched neodymium laser for studying air pollution, tunable semi-conductor diode lasers have been used for remote or long range sensing and also for point sampling of molecular pollutant gases.

LASER PHYSIOLOGICAL CHEMISTRY

Laser technology is now developing in the field of chemistry and a special division of laser chemistry has been developed in connection with the Laser Laboratory at the University of Cincinnati under the directorship of R. Marshall Wilson (16). This is for important studies in Raman spectroscopy, the detection of potential phototoxic and photosensitive agents, singlet oxygen, formation of dimers, photolysis, etc. Raman spectra can be demonstrated only by a few milligrams of powdered materials. Crystals are to be used in the future. With lasers, it is possible to measure even low Raman shifts. As indicated, Raman spectroscopy can be combined with infrared spectroscopy. Recently, Brueck and Mooradian (17) have reported the "efficient operation of a CW spin-flip Raman laser in Insb". The CO_2 laser was used as a pump. This can be used in high resolution infrared spectroscopy. A study of interest to physiologists is the use by Eisenthal, Chuang, Rehm, Smith and Drexhage of IBM of neodymium glass laser (106.0 nm) in picosecond pulses to study orientational relaxation of Rhodamine 6G dye molecules. In the chemistry laboratory, glassware may produce hazardous reflectant laser beams when open systems are used.

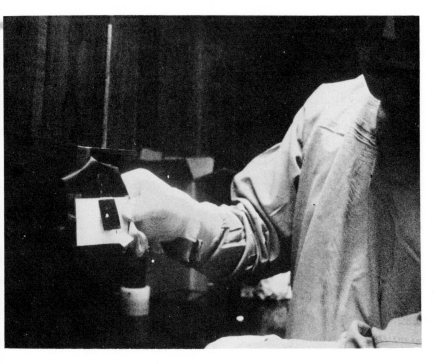

Fig.5 : Exposure of Mycobacterium kansasii to ruby laser impacts
to study effect on chromogenic mycobacteria.

EXPERIMENTAL EMBRYOLOGY

Collimation possible by the laser beam has made for selective localization of irradiation not only in chromosomes, as
mentioned previously, but also on the chick embryo, ova and
testis.   Not only is there selective and precise micro-surgery
for cells and tissues but here also a technique for comparing
non-ionizing to ionizing irradiation.   Will this precise technique be of any value in the future to modify the embryo with
genetic defects before replacement and implantation (18)?

Leon Goldman

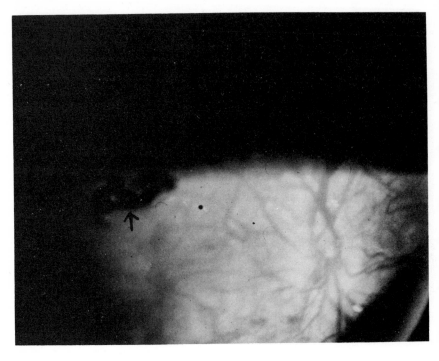

Fig.6 : Removal of hemorrhage in eye with argon eye laser.

THROMBOSIS AND THROMBOLYSIS

With the absorption of the laser in red cells and impacts on blood vessel walls, thrombosis has been studied with scanning electron microscopy by Naprstek (19) and Vahl (20) in our laboratory.   Striking pictures of clumped red cells, without platelets have been seen.   Vahl (21) has done detailed studies with scanning electron microscopy.   Thrombolysis has been attempted by canulization of thrombi and vaporization of thrombi by laser impacts.

HAIR BULB EXPERIMENTS

The hair bulb is a sensitive test model for studying the effects of irradiation.   With instrumentation developed in connection with the American Optical Company, Charles Koester (22), it has been possible to deliver selective irradiation to the hair bulb. Our initial goals were attempts at epilation and to produce localized changes in the individual bulb, similar to the more diffuse changes observed in irradiation of animal hair. We studied also the regrowth of hair after laser treatments of tattoo marks.

EXPERIMENTAL MICRO- AND MACRO-LASER SURGERY OF BIOLOGICAL INTEREST

In addition to extensive clinical studies by our Laboratory on laser surgery of the eye, skin, maxillofacial area and abdominal viscera (29), described later, we have done experiments on laser impacts of mouse brain as an experimental model for studying cerebral edema and on bone impacts of the rabbit as a test model for the development of fat embolization. Hogberg, Renius, Stahle, Vogel and Wallen, at the University of Uppsala (23,24) have produced, by the laser, minute localized lesions of the inner ear of great significance for physiological investigations and for possible therapeutic measures.

HOLOGRAPHY

Holography by the laser has also great potential for research in physiology. Stress patterns can be visualized as our laboratory has shown with bone. Raoul Van Ligten (25) has shown the possibilities of microholography in biology. This technique would provide excellent third dimensional patterns for study of the growth of a single living cell. Bell Telephone Laboratories (26) have developed a system of peripheral holography so that the whole 360° aspect of the object is covered. For those interested in the physiology of respiration, TRW Instruments have proposed a reference grid for the hologram. This can be used to study breathing excursions of the chest by providing an illuminated

**3 Laser Beams
Generated Here**

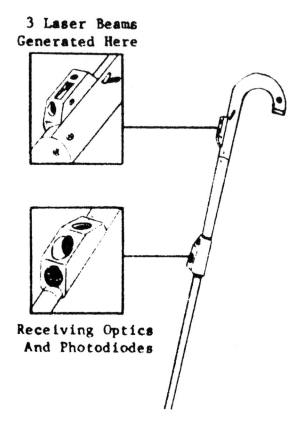

**Receiving Optics
And Photodiodes**

Fig.7 : Laser cane for the blind;   laser radar by junction diode
lasers.

reticule over the reconstructed holographic image.   In this
manner, the detailed surface interferogram of respiration may
be visualized.   Double exposure holography may be of value in
the study of heart action.

With laser techniques, George Stroke (27), one of the great
pioneers in laser holography, has increased the resolution of
Crewe's electron microscope (28) by the technique of "holographic
image deblurring".   This has made possible the actual visuali-
zation of the doubled helical structure of a virus.

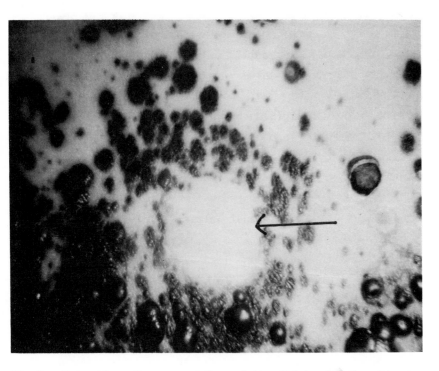

Fig.8 : Metastic melanoma nodules of the thigh with localized
        clear area; free of melanoma one year after high out-
        put ruby laser impact.

Some additional fields of interest are concerned with ad-
juvant effects of ultrasonics and laser in the fascinating new
field of acoustical holography pictures resulting from the combined effects of lasers and microwaves.   This is acoustic
imaging by holography.   Acoustical holography can be used to
detect tumors in soft tissues.   It may be possible to use
holography also in transillumination.   We have been using
laser transillumination to study masses in soft tissues.

MISCELLANEOUS

We have used laser surgery with high output CO_2 lasers on
such highly vascularized organs as the liver and lungs. As in-

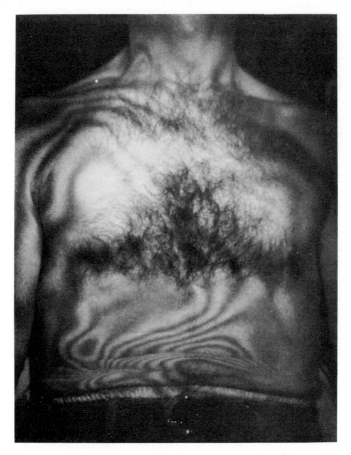

Fig.9 : Laser holography; interferograms made with Ruby Laser
Holocamera of TRW Systems; patient had left lung removed; inhalation with about 150 microseconds between
exposures (S.M. Zivi-TRW Instruments).

dicated, there are fundamental problems of tissue cutting, the
extent of areas induced coagulation necrosis, blood loss, liver
function studies and healing. The biological effects of trans-
versely excited atmospheric pressure (TEA) CO_2 lasers are now
under study. Laser surgery is important now in ophthalmology
and will be important soon in other fields of surgery. Over
the past ten years with various high output laser systems, we
have treated hundreds of patients with accessible malignancies,

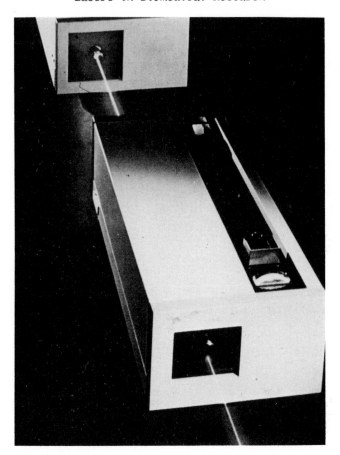

Fig.10 : Tunable lasers of great promise for research in biology, an example is Nd YAG, pumping fixed frequency UV doubler; UV beam made visible by fluorescent system (Chromatix).

birthmarks, tattoos and warty growths. Planned programs of laser safety and the development of flexible and reliable instrumentation in a comprehensive laser medical installation have all made this possible.

Some of our current research programs include studies of the effect of laser irradiation on seed and plant growth (30,31), pattern and pigment changes in bacteria and fungi (32). Studies have been done in absorption and heating effects of lasers on

ferrofluids introduced in tissues.   Air pollution from materials working in plastics, metals, have also been studied.
The effect of the thermal reaction in tissue from laser impacts
as related to tumor immuno-biology is also under investigation.
The impact of the laser on the fields of communication in
science as regards laser computer technology for laser printers,
video cassettes, establishment of centers for immediate dissemi-
nation of scientific information, etc. have been under study by
the Laser Laboratory.   We are looking to the future for the
good of man.

CONCLUSIONS

These are but brief and scarcely scanned introductions of
only some of the applications of modern laser technology in the
field of physiology.   The challenge is there to you.   The instrumentation is still not a routine product of the assembly
line save for certain the more commonly used laser systems.   The
laser systems are still sophisticated and expensive.   With the
increase in basic knowledge of the effects of lasers on living
tissue, there will be many more developments in the fields of
biology and medicine.   The imagination, interest, goals and
support for the investigator will determine the range of applications.

REFERENCES

1.   Goldman, Leon.   Biomedical Aspects of the Laser - The
     Introduction of Laser Applications into Biology and Medi-
     cine.   Springer-Verlag, New York, 1967.

2.   Goldman, Leon and Rockwell, Jr. R.J.   Lasers in Medicine.
     Gordon & Breach Science Publishers, Inc. (in press).

3.   Lasers - Second Conference on the Laser (Leon Goldman,
     Chairman).   *Annals of The New York Academy of Sciences
     168*: February 10, 1970.

4.   Goldman, Leon.   Biomedical Laser Technology: A Challenge
     to the Engineer.   *Biomedical Engineering 6*: 22, January,
     1971.

5. Goldman, Leon.   Snags in Laser Biomedicine.   *Biomedical Engineering 6*: 9, January, 1971.

6. First International Conference on Non-Ionizing Radiation, Cincinnati, March, 1971.

7. Berns, Michael W. and Rounds, Donald E.   Laser Microbeam Studies on Tissue Culture Cells.   Second Conference on the Laser.   *Annals New York Academy of Sciences 168*:   February 10, 1970, p.550.

8. Berns, Michael W.   Personal communication.

9. Rounds, Donald E.   Personal communication.

10. Scheidt, W. and Traut, H.   *Mutation Research 11*:   p.253. Abst. *New Scientist and Science Journal 29*:   April, 1971, p.250.

11. Allemand, Charly.   Advanced Laser Microprobe.   Personal communication.

12. Brech, Frederick.   Advanced Laser Microprobe.   Personal communication.

13. Smith, Desmond.   *New Scientist and Science Journal*: April, 1971, p.6.

14. Robinson, Jas. W., Christian, Charles, Guillardo, John and Katayama, Nibuki.   Infrared Laser Systems.   *C & EN*: May 24, 1971, p.11.

15. Goodman, M.F., Seliger, H.H. and Minkowski, J.M.   On the Possibility of Infrared Laser-Induced Chemistry.   *Photochemistry and Photobiology 12*:   355, 1970.

16. Wilson, R. Marshall.   Personal communication.

17. Brueck, S.R.J. and Mooradian, A.   Efficient Single Mode, CW, Tunable Spin-Flip Raman Laser.   Lincoln Laboratory, MIT, 1971, Sect.1, p.12.

18. Edwards, Robt. G. and Sharpe, David J.   Social Values and Research in Human Embryology.   *Nature 231*:   87, May 14, 1971.

19. Naprstek, Zdenek.   Personal communication.

20. Vahl, Joh.   Personal communication.

21. Vahl, Joh.   Analyse von laser-bestrahlten Zahnschmelz und Zahnfullungsmaterial.   *Laser und angewandete Strahlen technik 1*, 1971.

22. Koester, Charles.   Personal communication.

23. Hogberg, Lars, Renius, Staffan, Stahle, Jan, Vogel, Klaus and Wallen, Gunnar.   Laser Microsurgery upon Inner Ear and

Myelinated Nerves, Vestibular Function on Earth and in Space. Wenner Gren Symposium. No.15. Pergamon Press, Oxford and New York, 1970.

24. Hogberg, Lars and Stahle, Jan. The Laser as a Microsurgical Tool. March, 1971, Forsoarets Forskningsanstalt, Avdelning 2, Stockholm.

25. Van Ligten, Raoul. Personal communication.

26. Peripheral Holography. *Camera*, January, 1971. Abst. Photographic Applications in Science, Technology and Medicine.

27. Stroke, George. *New Scientist and Science Journal*, May 6, 1971, p.304.

28. Crewe, Albert V. A High Resolution Scanning Electron Microscope. *Scientific American*: 1971, p.26.

29. Goldman, Leon *et al*. Some Parameters of High Output CO_2 Laser Experimental Surgery. *Nature 228*: No.5278, pp.1344-1345, December 26, 1970.

30. Paleg, L.G. and Aspenill, D. Field Control of Plant Growth and Development through Laser Activation of Phytochrome.

31. Parr, Wordie. Personal communication.

32. Schwarz, Jan. Personal communication.

QUASI-ELASTIC PHOTON

SCATTERING IN BIOPHYSICS

F.D. CARLSON

Department of Biophysics, Johns Hopkins University, Baltimore, U.S.A.

The use of light scattering techniques to determine the size and shape of particles in solution is well known in physics, chemistry and biophysics. A measurement of the concentration and angular dependence of the absolute intensity of light scattered from a solution of macromolecules permits one to determine: molecular weight, the virial coefficient and the radius of gyration of the scattering particle. In such classical light scattering studies the incident light used is normally quasi-monochromatic but it is not coherent. If, however, one uses highly coherent monochromatic laser radiation as the incident beam it is possible to obtain some additional information about the scattering particles, namely information about their translational and rotational dynamics in solution. The scattering of laser light from fluids and solutions is referred to as quasi-elastic photon scattering in order to differentiate it from classical light scattering of incoherent light. We shall develop here: the physics of quasi-elastic photon scattering by particles in solution, the techniques for detecting the spectral changes in coherent light following scattering and applications of quasi-elastic photon scattering to the study of some problems of biophysical and biological interest.

THEORY

Consider a solution containing N identical scatterers in a volume V which is illuminated by a plane parallel beam of light coming from a stabilized continuous wave laser. Such a beam will be plane polarized and highly monochromatic. For example, light from a typical gas or ion laser with a frequency of approximately 3×10^{15} Hz will show a variation of less than 100 Hz. The scattering geometry for such a situation is shown in Fig.1. The electric vector, $\bar{E}$, of the incident coherent radiation can be represented by:

$$\bar{E} = \bar{E}_0 \, e^{i\omega_0 t}$$

We associate a momentum vector, $\bar{K}_0$, with the incident radiation, its direction is that of the radiation and its amplitude is $2\pi n_0/\lambda$ where n_0 is the refractive index of the medium and λ the wavelength in vacuum. The light scattered by a single particle at an angle, ψ, to the incident beam has a momentum $\bar{K}_s$ where $|\bar{K}_s| = 2\pi n_0/\lambda$ also.

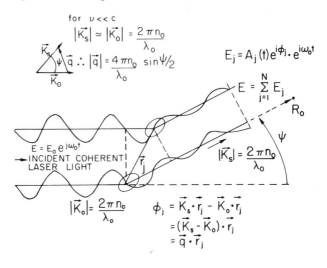

Fig.1

The distance from the origin of the system to the jth scattering particle is designated by the position vector, $\bar{r}_j$. The amplitude of the scattered light at R_0 will depend on the scattering characteristics of the particle and on its orientation in space. Generally as the particle rotates and changes its orientation in space the amplitude of its scattered field will change in time. This time dependent amplitude factor is designated by $A(t)$. The phase of the radiation scattered by the jth particle will be shifted relative to the phase of the radiation scattered from the same particle in the same orientation located at the origin of the system. The phase shift, designated by ϕ_j, depends on the position of the scatterer relative to the origin and on the scattering angle, ψ, and it is due to the difference in path length travelled by light scattered from a particle located at the origin and light scattered from a particle located at a distance $F_j \bar{r}_j$ from the origin. This difference is given by the difference between the projections of the position vector of the particle on the momentum vectors of the scattered radiation and the incident radiation (see Fig.1). That is,

$$\phi_j = \bar{K}_\delta \cdot \bar{r}_j - \bar{K}_0 \cdot F_j$$

$$= (\bar{K}_\delta - \bar{K}_0)\, \bar{r}_j$$

$$= \bar{q} \cdot \bar{r}_j$$

where: $|\bar{q}| = (4\pi n_0/\lambda)\sin \psi/2.$

Finally, the frequency of the scattered optical field is the same as that of the incident optical field. Giving for the field at a distant point R_0 due to scattering by a single particle:

$$E_j = A_j(t) \cdot e^{i\phi_j} \cdot e^{-i\omega_0 t} .$$

The total field at the distant point R_0 due to light scattered from all N particles in the volume is:

$$E_s = \sum_{j=1}^{N} E_j = \sum_{j=1}^{N} A_j(t) \cdot e^{i\overline{q}\cdot\overline{r}_j} \cdot e^{-i\omega_0 t} .$$

We turn next to calculating the power spectral density (power spectrum of the optical field from the expression for the total field). To do this, we make use of the Wiener Khintchine theorem which states:

$$I(\omega) = \frac{1}{2\pi} \int_{-\infty}^{\infty} C(\tau) \cdot e^{i\omega\tau} \, d\tau$$

where: $\quad C(\tau) = \langle E_s^*(t) \cdot E(t + \tau) \rangle$

and: $\quad C(-\tau) = C^*(\tau).$

According to this theorem the distribution of power in the frequency domain and the autocorrelation of the time dependent optical field are related as a Fourier transform pair. The power spectrum of the scattered field can be calculated with the aid of this theorem as follows.

$C(\tau)$ for E_s is:

$$C(\tau) = \sum_{j=1}^{N} A_j^*(t) \, e^{-i\overline{q}\cdot\overline{r}_j(t)} \, e^{i\omega_0 t} \qquad \text{multiplied by}$$

$$\sum_{j=1}^{N} A_j(t+\tau) \, e^{i\overline{q}\cdot\overline{r}_j(t+\tau)} \, e^{-i\omega_0(t+\tau)}.$$

For identical, statistically independent scatterers, cross terms $(j \neq 1)$ vanish. Further we assume no correlation between orientation and position; then:

$$C(\tau) = N \, e^{-i\omega_0 t} \langle A^*(t)\, A(t+\tau)\rangle \langle e^{-i\overline{q}\cdot\overline{r}(t)} \, e^{i\overline{q}\cdot\overline{r}(t+\tau)} \rangle \, d\tau$$

and

$$I(\omega) = N \frac{1}{2\pi} \int_{-\infty}^{\infty} e^{i(\omega-\omega_0)\tau} \left[C_A(\tau) \right] \left[C_\phi(\tau) \right] d\tau \quad.$$

where:

$$\left[C_A(\tau) \right] = \langle A^*(t)\, A(t+\tau) \rangle$$

$$\left[C_\phi(\tau) \right] = \langle e^{-i\overline{q}\cdot\overline{r}(t)} \, e^{i\overline{q}\cdot\overline{r}(t+\tau)} \rangle$$

where angular brackets designate the time average.

The autocorrelation function has two components, one indicated by $\left[C_A(\tau) \right]$, arises from the time correlation of the orientation of the particle and the other $\left[C_\phi(\tau) \right]$ arises from the time correlation of the spatial position of the particle.

It is instructive to consider the form of the autocorrelation function given by various types of particle dynamics.
For particles whose dimensions are much less than those of the
wavelength of the incident light, Rayleigh scatterers, the intensity of the scattered field is spherically symmetric and
there is no fluctuation in intensity with changes in orientation
of the scatterer. For Rayleigh scatterers, therefore, the
orientation part of the autocorrelation function is a constant
equal to 1.

Table I summarizes the autocorrelation functions and the
power spectra for three kinds of particle dynamics: stationary
particles with random positions in space, particles all moving
with the same constant velocity, v, and particles freely dif-
fusing in solution. For the case of the stationary scatterers
the power spectrum is the same as the spectrum of the incident
radiation. That is, there is simply a single sharp line. For
the case of the scatterers moving with constant velocity, the
power spectrum again shows just a single sharp line but its frequency is shifted by an amount equal to $\bar{q} \cdot \bar{v}$. This corresponds
to the well known Doppler shift. For particles executing
simple Brownian diffusive motion, the situation is more complicated. The spectrum is centered on the frequency of the incident radiation but we no longer have a sharp line - the line has
been broadened and the shape of the broadened line is dependent
upon both the translational diffusion constant of the scattering
particles and the scattering vector, $\bar{q}$. This function which
describes the shape of the spectrum of the scattered radiation
is called a Lorentzian function. The width of the Lorentzian
at half its maximum value is given by:

$$\Delta W_{1/2} = D_t \, |q|^2$$

TABLE I

SUMMARY OF POWER SPECTRA FOR VARIOUS PARTICLE DYNAMICS

Spherical Scatterers

$$\left[C_A(\tau)\right] = |A|^2 = \frac{4\pi I_0 V^2 n_0 (n - n_0)^2}{R_0^2 \lambda_{vac}^4}$$

$$I(\omega) = N |A|^2 \frac{1}{2\pi} \int_{-\infty}^{\infty} e^{i(\omega-\omega_0)\tau} \left[C_\phi(\tau)\right] d\tau$$

1. Fixed at random positions:

$$\left[C_\phi(\tau)\right] = 1 \quad ; \quad I(\omega) = N|A|^2 \delta(\omega - \omega_0)$$

2. Constant velocity $\bar{v}$:

$$\left[C_\phi(\tau)\right] = e^{i\bar{q}\cdot\bar{v}\cdot\tau} \quad ; \quad I(\omega) = N|A|^2 \delta(\omega - \omega_0 + \bar{q}\cdot\bar{v})$$

3. Translation diffusion:

$$\left[C_\phi(\tau)\right] = e^{-D_T q^2 \tau}$$

$$I(\omega) = N|A|^2 \frac{D_r q^2/\pi}{(\omega - \omega_0)^2 + (D_T q^2)^2}$$

$$\Delta\omega_{1/2} = D_T q^2$$

The case of diffusive motion is of considerable interest
for it corresponds to the situation it obtains when we have
macromolecules self-diffusing in solution at equilibrium. From
a determination of the spectral line width of coherent light
scattered from such a solution one can evaluate the diffusion
constant of the macromolecule directly.

The case of constant velocity scatterers considered above
is a trivial one and not of particular interest biologically.
However, it can be readily generalized to cover the situation
where the scatterers have different velocities, that is, they
are travelling with different speeds and in different directions
such as one might observe in a test tube containing motile bacteria.   If the particles, even though they do not all have the
same velocity, hold essentially a constant velocity during a
period of observation, then as Nossal (6) has shown the spectral
half width depends directly on $\bar{q}$.

For the case of diffusive motion, however, the half width
depends on $|\bar{q}|^2$.   This means that since $\bar{q}$ contains the
$\sin \psi/2$ Doppler scatterers will give spectra whose widths vary
linearly with the $\sin \psi/2$ while Brownian scatterers give spectra whose half widths vary as the $\sin^2 \psi/2$.   Scatterers which
are stationary in space but rotating give spectra which show no
dependence on ψ.

LIGHT BEATING SPECTROSCOPY

In the preceding section we showed that light scattered
from a solution of moving particles has its spectral characteristics broadened.   The spectral broadening due to scattering
from particles of biological interest ranges from 10 Hz to 10^6
Hz.   The light produced by a helium-neon laser has a frequency
of- 3×10^{15} Hz.   The spectral changes of interest produce
changes of only 1 part in 10^{14} to 1 part in 10^9.   Such
small spectral changes require an instrumental resolution beyond

the capability of the best optical spectroscopic methods. There
is, however, another technique which is ideally suited for
measuring such small changes in spectral characteristics. The
technique is called "optical beating" and it operates on the
same principle that allows one to detect modulated radio waves
with a non-linear detector. The detector used in "optical
beating" is the photomultiplier, a device which produces a current that is proportional to the intensity, hence the square of
the incident optical electric field. Since the photomultiplier
current is proportional to the square of the optical field, it
will contain components whose frequencies are equal to the sum
and difference frequencies of the components present in the
optical field. As we have noted, the line widths due to
scattering from moving scatterers are of the order 10 to 10^6
Hz. The beating of the various components within a line will
produce spectra centered at 0 Hz in the audio frequency range
and having widths ranging from 20 to 2×10^6 Hz. The power
spectrum of a photomultiplier current will then have the same
functional form as the power spectrum of the spectrally broadened
line and it will be centered at 0 Hz. One need only measure
the audio frequency power spectrum of the photomultiplier current
in order to determine the shape of the spectral line of the optical field.

Another way of describing the same phenomenon is in terms
of the intensity of the optical field. The fluctuations in the
intensity due to the dynamic characteristics of the scatterers
in the solution produce fluctuations in the photomultiplier current. If the scattered light incident upon the photomultiplier
is not purely monochromatic there will be rapid fluctuations in
the intensity in the optical field, and hence in the photomultiplier current. However, as the field becomes more nearly purely
monochromatic, the intensity fluctuations become less and less
rapid until a purely monochromatic beam is obtained and there are

no intensity fluctuations at all.   At this point only the shot
noise fluctuations remain in the photomultiplier current.

When the various frequency components of the optical field
beat with one another to produce a fluctuating photomultiplier
current the process is referred to as homodyne spectra. Heterodyne spectra can be obtained by providing a local oscillator in
the optical field which has the same frequency as the incident
laser radiation but has an intensity much stronger than that of
the scattered radiation.   The technique produces spectra which
have the identical functional form as the spectrum of the optical field.   Unlike homodyne spectra the frequency dependence
for heterodyne spectra is not multiplied by a factor of 2.
Heterodyne technique while technically difficult to use sometimes has the advantage that it greatly increases the sensitivity
of the spectrometer and also improves the signal-to-noise ratio.

It can be shown that the photomultiplier current power
spectrum that arises from the quasi-elastic photon scattering of
Rayleigh scatterers executing diffusive Brownian motion is given
by the following:

$$P_i(\omega) = \frac{e \langle i \rangle}{\pi} + \langle i \rangle^2 \, \delta(\omega) + 2 \langle i \rangle^2 \, \frac{(2\Gamma_0/\pi)}{\omega^2 + (2\Gamma_0)^2}$$

and the corresponding power spectral density for the case of
heterodyne detection is

$$P_i(\omega) = \frac{e \langle i_{L0} \rangle}{\pi} + \langle i_{L0} \rangle^2 \, \delta(\omega) + 4 i_{L0} \, \frac{(\Gamma_0/\pi)}{\omega^2 + \Gamma_0^2}$$

The various terms in these expressions correspond to the following contributions to the total spectrum observed.   The first
term in each expression, $e \langle i \rangle / \pi$, is the direct current component of the spectrum and normally this is not recorded because

one uses a spectrum analyzer sensitive only to the alternating current components in the photomultiplier current.   The second term, $\langle i \rangle^2 \delta(\omega)$,   corresponds to the shot noise which is generated by the photomultiplier detection process itself, and the spectrum of this process is always present in the photomultiplier current.   The third term,

$$2 \langle i \rangle \quad \frac{(2\Gamma_0/\pi)}{\omega^2 + (2\Gamma_0)^2} \quad ,$$

corresponds to the spectrum of the scatterers and shows the typical Lorentzian dependence on frequency.   It is possible to calculate the photomultiplier current spectrum that would be obtained from scatterers executing other kinds of motion, but we shall not do so here.   It is, however, worthwhile to point out that for the case of more than one kind of dissimilar scattering particle (say two particles of different sizes) the homodyne power spectrum of the photomultiplier current becomes complicated because there is not only the contribution due to the self beating of the components of each particle's spectrum but the components of the spectrum of one particle also beat with those of the dissimilar particle to give a "mixed" component.   Consequently the photomultiplier current contains not just two Lorentzian components, one associated with each of the particles, but there is also a mixed component which is associated with a mixture of the spectra of the two particles.   This is not the case for heterodyne detection of a two particle system where only two Lorentzian components are found in the photomultiplier current.

EXPERIMENTAL TECHNIQUES

Figure 2 depicts schematically a spectrometer that can be used for the study of quasi-elastic photon scattering for macromolecules or from particles of biological interest.   The incident laser radiation is typically provided by a helium-neon con-

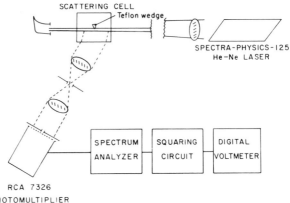

Fig.2

tinuous wave laser; for example, Spectra Physics Model 125 is
quite suitable. The highly collimated laser radiation is
focused onto the scattering cell by means of a simple lens. The
scattering cell is equipped with a light stop to reduce unwanted
scattered radiation and there is a collecting system consisting
of a collimator mounted on a goniometer in such a way as to enable one to collect scattered light at any desired angle. The
collected scattered light is imaged on the photocathode of the
photomultiplier. The photomultiplier current is amplified and
then squared and then analyzed by a conventional audio frequency
spectrum analyzer to obtain the audio frequency power spectrum
of the current.

   Spectrometers of this general type can be made to study
spectra with half widths ranging from 20 Hz to 10^6 Hz. With
reasonable care such spectrometers are capable of measuring diffusion coefficients of typical biological macromolecules to a
precision of 1% or so. Heterodyne detection can be achieved
by introducing into the scattering cell a small Teflon wedge
which interrupts the laser beam and scatters it strongly in all
directions. The Teflon wedge behaves as a stationary diffuse
scatterer and produces no spectral shift but merely provides a
very strong component in the scattered light which has the same

frequency as the incident light. It behaves then as a local oscillator that beats with the light scattered by the moving scatterers in solution that are within a coherence length of the Teflon wedge.

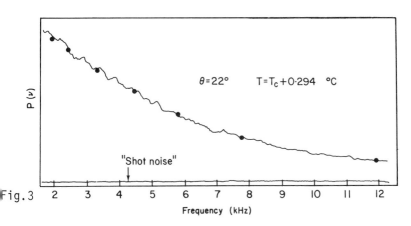

Fig.3

Typical recordings of homodyne spectra are shown in Fig.3. Figure 4 is a plot of the half width of the power spectrum at half maximum as a function of $\sin^2 \psi/2$ obtained and a solution of polystyrene latex spheres. As predicted for Brownian scattering the spectra of the half widths vary linearly with $\sin^2 \psi/2$ and the heterodyne spectrum is half as wide as the homodyne spectrum.

Recently another technique for measuring spectral line widths of scattered light has been developed. The technique known as digital autocorrelation spectroscopy actually allows the direct real time measurement of the autocorrelation function which as we have already seen contains all of the information that the power spectrum contains.

This technique makes use of the fact that the electron avalanche which occurs when a photon hits the photocathode follows a sharp pulse. These pulses can be made into discrete pulses of uniform amplitude and duration much like the pulses shown in Fig.1. In this one discrete series of pulses can generate cor-

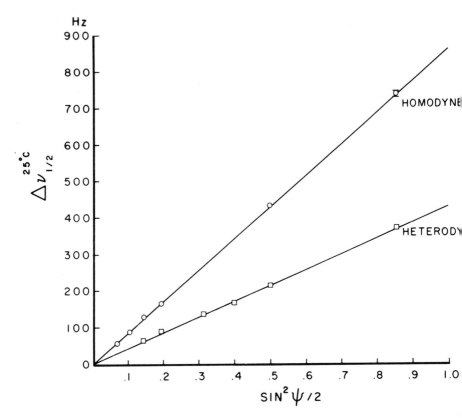

Fig.4

responding to the arrival times of scattered photons at the
photocathode surface.   It is then possible, with the aid of a
digital autocorrelator, to autocorrelate the discrete time
series continuously as the experiment proceeds.   The main advantages of digital autocorrelation are that it is fast and
extremely accurate.

Finally, for those cases which produce extreme broadening
of the spectral line, that is 10^8 Hz   or greater, one can re-
sort to conventional optical spectroscopic devices for resolving
them.   A high quality Fabray Perot Etalon will resolve line
widths of 10^8 cycles or greater.

BIOLOGICAL APPLICATIONS OF QUASI-ELASTIC PHOTON SCATTERING

So far major applications of light scattering spectroscopy to biological problems has been in the field of biophysical chemistry. Several laboratories have reported measurements of diffusion constants of various viruses. Diffusion constants of large viruses are so small that they are extremely difficult to measure accurately by conventional techniques. Already the diffusion constants of several coliphage viruses and tobacco mosaic virus have been measured by light scattering with a precision of 1%. The diffusion constant data together with a determination of the sedimentation coefficient has allowed a more accurate estimate of the molecular weight for these viruses than has hitherto been possible.

Several proteins have also had their diffusion coefficients determined by the technique of quasi-elastic photon scattering. Among those reported in the literature are bovine serum albumen, lysozyme, squid haemocyanin and myosin.

One of the most interesting possibilities which the technique has to offer the biophysical chemist is as a probe for the study of protein interactions with other molecules and with themselves in solution. The sub-unit structure of proteins and the alosteric changes which occur upon protein interaction with other molecules are two of the most active fields of modern molecular biology. Since size and shape changes reflect themselves in changes in the diffusion constants, photon scattering offers the possibility of studying association reactions and alosteric changes of proteins.

In our own laboratory we have made a careful study of the self association of the myosin molecule to form a dimer. The suggestions that such an association reaction might occur with myosin first came to us from an experiment in which we measured the diffusion constant as a function of protein concentration.

A very large decrease in the diffusion constant with increasing concentration of myosin suggested that the myosin molecule was associating into some higher polymer. From an analysis of the absolute intensity and spectral distribution of the light scattered from solutions of myosin we were able to demonstrate that a monomer-dimer exists in solution. It was also possible to measure the equilibrium constant of the reaction, the radius of gyration of the monomer, the diffusion constant of the monomer and dimer, the molecular weight of the monomer, and hence that of the dimer, the virial coefficient of the monomer and the coefficient of concentration dependence of the frictional factor for the monomer. These results are summarized in Table II. Recently we have completed a more thorough analysis of our earlier data and have been able to conclude that the myosin dimer is of the parallel form, that is the head tail directions of both of the sub-units of the dimer are the same.

Quite recently there have been some preliminary reports that lysozyme also shows a dependence of the diffusion constant on concentration which strongly suggests an association reaction

Although as yet unexplored but possibly by far one of the most potentially interesting applications of the technique is the study of the motility of micro-organisms and sperm. Berge and Volochine (1) reported the results of an experiment in which they measured the spectrum of light scattered from living and dead rabbit spermatozoa. Their spectra clearly showed a broadening which is almost certainly due to the motility of the living sperm. Recently Nossal (6) has developed the theory of spectral broadening from light scattered from moving motile organisms. His results provide a theoretical basis for velocity distribution of motile organisms from their scattered light spectrum. Also, Nossal and Chen (7) have reported measurements of bacterial velocity distributions. They confirmed the linear sin $\psi/2$ dependence of spectra obtained from motile bacteria and have thus

TABLE II

PHYSICAL CONSTANTS OF MYOSIN FROM LIGHT SCATTERING MEASUREMENTS

(in 0.5 M KCl, 0.2 M PO_4, 0.001 M EDTA, pH = 1.3)

Mol. weight of monomer	466.000 ± 14.000
Radius of gyration of monomer	468 ± 32 Å
Virial coefficient Bm = Bd	3.21 dl/g
Monomer-dimer Equil. constant	10.6 ± 1.3 dl/g
Diffusion constant monomer dimer	1.24 × 10^{-7} cm^2/sec 0.84 × 10^{-7} cm^2/sec
Frictional factor virial	3.27 ± 0.40 dl/g
Lengths: monomer dimer	1481 ± 130 Å 2121 ± 180 Å

shown that the bacteria behave as Doppler scatterers.

Although I know of no attempts to measure protoplasmic streaming or blood by the method of quasi-elastic photon scattering there is no reason why these phenomena cannot be studied in this way. In the field of fluid dynamics the technique goes under the name of laser Doppler shift velocimetry. For the last 6 or 7 years fluid dynamicists have used it to measure local velocities in laminar and turbulent fluids.

Finally there are other biological phenomena which could in principle at least be studied fruitfully by this technique. The movement of organelles both inside and outside cells would produce a spectral broadening of light scattered from them. The beating movement of flagella and cilia, for example, could well be studied by the technique. One kind of intracellular movement currently being studied in our laboratory is the movement of the myosin cross bridges that is thought to occur during muscular contraction.

In this presentation I have focused upon one aspect of quasi-elastic photon scattering, namely spectral broadening associated with the scattering of coherent light from particles moving in solution. There are processes which occur in solids, within liquid crystals, and within membranes that also give rise to spectral broadening on light scattering. Physicists and chemists interested in these processes have just begun to use light scattering techniques and there is a rapidly developing technology and theory in this field. It would be surprising indeed if these techniques were not applied to the study of biological phenomena which occur in condensed phases similar to the solid, liquid crystalline and membrane state.

REFERENCES

1. Berge, P., Volochine, B., Billard, R. and Hamelin, A. *Compt. Rend. 265:* 889 (1967).

2. Cummins, H.Z., Carlson, F.D., Herbert, T. and Wood, G. *Biophys. Journ. 9:* 518 (1969).

3. Cummins, H.Z. and Swinney, H. Light Beating Spectroscopy. *In:* Progress in Optics, Vol. VIII, Ed. Wolf, E. (Amsterdam, Netherlands: North Holland Publ. Co.) (1970).

4. Dubin, S., Lunacek and Benedek, G.A. *Proc. Nat. Acad. Sci., USA, 57:* 1164 (1967).

5. Herbert, T.J. and Carlson, F.D. Biopolymers. (In press).

6. Nossal, R. *Biophys. Journ. 11:* 341 (1971).

7. Nossal, R. and Chen, S.H. CNRS Colloquium on Light Scattering in Fluids. Paris 17-19 July 1971.

THE ANALYTICAL

LASER MICROPROBE*

DAVID GLICK

*Histochemistry Division, Pathology Department,
Stanford University Medical School, Stanford, California, U.S.A.*

Use of the laser microprobe for analysis of elements in
microscopic samples of biological materials, such as selected
structures in a frozen-dried microtome section of tissue or
chosen cells in a fresh air-dried smear or tissue imprint, for
quantitative cytochemical studies has been the basis of our
interest in developing this analytical approach. (Aside from
biological uses, the technique can also be applied very broadly
to non-biological materials as well.)   In addition to the direct
analysis of elemental components, possibilities exist of extending the application to organic constituents by means of chemically specific metal or metal-containing stains for these substances.

The principle employed is placement of the dry sample in the
optical axis on the stage of a microscope, vaporization at a very
high temperature with a Q-switched ruby laser flash sent through
the microscope, collection and transmission of the light of the
incandescent vapor to a spectrograph and finally photographic or
photomultiplier-tube recording of the spectral line intensities

* Supported by research grants GM 16181, HE 06716 and 5K6AM18,513
  from the National Institutes of Health, United States Public
  Health Service.

of selected elements in the sample.

This short presentation will be confined to an outline of the developments in our laboratory and not of the subject as a whole.   The latter would require a much more extensive treatment than is practical for the present purpose.   The new instrumental design and subsequent improvement, studies to establish optical and electronic parameters for optimal analytical use, investigation of the influence of the organic matter in the sample on the spectral emission, determination of effects of the atmosphere in which the sample is placed and of other conditions that affect analytical capability, all have been the result of our group effort involving a number of investigators. The individuals concerned and their particular contributions are indicated by the names and titles in the publication list appended.

Our first reports (1,2,3) pointed out the state of development at the time and analytical potentialities of the use of the laser microprobe, problems to be considered in this use, and some early thoughts on sample preparation.   Subsequently, more general reviews of initial histochemical application were also given (4,5,6), and though these were essentially repetitive they were presented to widely different audiences.   It should be stated that in one of these reviews (6, Tables 1 and 2) estimation of analytical detection limits of elements gave values that have since been shown to be too low.   Recently, direct experimental tests showed that, defined as the amount of element giving a signal twice the standard deviation of that of the background, the detection limit in the presence of an organic (gelatin) matrix is typically of the order of 10^{-14} g (7).   Actual values for a number of elements will be reported.

Our first apparatus was an early model Jarrell-Ash instrument employing a Wadsworth 1.5 m, f/22 spectrograph that was soon replaced by a more sensitive Czerny-Turner 0.75 m, f/6.3 spectrograph. It had become apparent from the start that an essential need was instrumental improvement, particularly to increase the control and reproducibility of the laser energy at every firing. This proved to be an extensive task but was successfully accomplished by effective team work with a group at the Stanford Research Institute. The design and performance of the improved instrument was described in a joint publication (8).

There still remained however the need for further development of certain instrumental components. One of these was the electric spark attachment of Brech and Cross, supplied with the original Jarrell-Ash system, that was used to intensify spectral emission of the incandescent vapor by having the latter set off a spark discharge when the vapor shot up between the tips of a pair of carbon electrodes positioned just above the sample. A voltage of 1-2 kV was used across the electrodes and the spark that was set off not only intensified the spectral emission (especially important for photographic recording which is less sensitive than photomultiplier-tube recording of the spectral lines), but it also burned a zone around the chosen sample in the focus of the microscope. This burning added unwanted vapor to that of the sample and in addition to the contamination it reduced the sampling resolution. Improvement of this resolution to 10-25 μ from 50-250 μ was accomplished by further tapering of the electrode tips, shortening the gap between them and reducing the energy of discharge. Obviously, it would be preferable to eliminate the spark excitation entirely to avoid any effect on the sampling, and this has been the aim of later work in which the sensitivity necessary for adequate measurement at high resolution was increased electronically and, to some extent, optically.

Another attachment was designed to permit automatic correction for the immediately adjacent background of the phototube measurement of intensity of a spectral line. The mechanical

construction and circuitry of this polychromator was described
in detail (10).

Finally, another laser microprobe system was constructed
based on experience with the previous one (11). The chief improvements made were: (a) increase in energy stability and
homogeneity of the laser beam, (b) further control of beam diameter, (c) addition of binocular and phase optics and a camera to
the microscope, with a safety device for protection of viewer
and microscope condenser from the laser beam, (d) increase in
stability of the laser beam monitor with addition of an indicator to show the number of pulses delivered, (e) reduction of
spherical and chromatic aberration and increase in efficiency of
the lens system collecting flight from the plume, and (f) addition of an attachment for selection of the portion of the plume
to be used for the spectral analysis.

The well known suppression of spectral emission of metallic
elements by organic material in the sample obviously required
investigation for the quantitative use of the technique.
Accordingly, a study of this matrix effect was carried out in
model systems and it was observed that a threshold concentration
of matrix existed, above which the emission decreased.   The magnitude of the threshold varied with the nature of the emitting
element and of the matrix itself (12).   In spite of the suppression of the emission, which is appreciable in biological samples,
the useful signal obtained could still be exploited for analytical purposes.

From earlier work, particularly by Vallee and co-workers at
the Biophysics Research Laboratory and Department of Biological
Chemistry, Harvard Medical School, Boston, the influence on the
intensity of spectral emission of the atmosphere in which the
incandescent plasma is generated by D.C. arc has been described.
Since properties of plasmas formed by electric arcs and laser
beams have some distinct differences, it was necessary that at-

mosphere effects on the laser plume be established to determine what influence on spectral analysis was involved. An investigation was carried out of signal-to-background ratios of iron and magnesium lines in metallic and biological samples in air, argon, helium, nitrogen, oxygen and vacuum employing laser energy levels of 1.2 - 8.0 mJ (13). Although both the nature of the atmosphere and the laser energy used had similar effects on the signal intensities from metallic or biological samples, and although it was observed that some advantage could follow use of certain atmospheres in certain analyses, the general conclusion drawn was that the possible gain was not sufficient to merit the complication of a change from the use of air at normal pressures.

An important advance for the analytical effectiveness of the laser microprobe was time differentiation of the signal derived from the light emitting plasma (14). The duration of the Q-switched laser flash is usually less than 100 nsec and emission from the vapor usually persists for several μsec. To separate useful emission from the non-specific background, photo-electric measurement of the emission at particular times proved effective. Oscillograph tracings of the photomultiplier tube output from spectral lines revealed an initial large contribution to the emission from background which fell off steeply with time while the useful signal decreased more slowly. Thus signal-to-background ratio increased with time to a maximum and then decreased. The time taken to reach the maximum varies with laser energy, the emitting element and the nature of the sample. The experiments performed showed, in general, that the electronic time differentiation, while reducing the useful signal to about 40%, could increase the signal-to-background ratio up to 2000%. A statistical study of relationships between spectral line and background intensities and laser energy was also carried out (15).

It is apparent that very considerable effort has gone into design, construction, testing and optimization of the apparatus and the study of parameters significant for analytical performance. With this extensive background our group has begun work on biomedical applications, e.g. iron in single red blood cells is being measured, and analysis of magnesium and calcium in fine droplets (12 nl) of blood serum is being carried out (16). With collaboration of Drs Mary M. Herman and Klaus G. Bensch, uptake of gold by individual cultured fibroblasts treated with gold thioglucose is being determined (17).

Of course applications of the analytical laser microprobe in biology and medicine can be extensive, and it is our hope that the beginning we have made, though modest in spite of the time, effort and expense involved, will provide a basis for future development.

REFERENCES

1. Rosan, R.C., Glick, D. and Brech, F.   Progress in laser microprobe emission spectroscopy. *Federation Proc. 24*: No.2, Pt.I, 542 (1965).

2. Rosan, R.C., Brech, F. and Glick, D.   Current problems in laser microprobe analysis. *Federation Proc.* Suppl. 14, *24*: No.1, Pt.III, S.126-S.128 (1965).

3. Rosan, R.C.   On the preparation of samples for laser microprobe analysis. *Appl. Spectr. 19*: 97-98 (1965).

4. Glick, D. and Rosan, R.C.   Laser microprobe for elemental microanalysis, application in histochemistry. *Microchem. J. 10*: 393-401 (1966).

5. Glick, D.   The laser microprobe, its use for elemental analysis in histochemistry.   *J. Histochem. Cytochem. 14*: 862-868 (1966).

6. Glick, D.   Cytochemical analysis by laser microprobe-emission spectroscopy. *Annals New York Acad. Sci. 157*: 265-274 (1969).

7. Treytl, W.J., Marich, K.W., Saffir, A.J., Orenberg, J.B. and Glick, D.   Laser microprobe optical emission detection limits for certain elements in an organic matrix.   In preparation.

8. Peppers, N.A., Scribner, E.J., Alterton, L.E., Honey, R.C., Beatrice, E.S., Harding-Barlow, I., Rosan, R.C. and Glick, D.   Q-switched ruby laser for emission microspectroscopic elemental analysis. *Anal. Chem. 40:* 1178-1182 (1968).

9. Beatrice, E.S., Harding-Barlow, I. and Glick, D.   Electric spark cross excitation in laser microprobe-emission spectroscopy for samples of 10-25 μ diameter. *Appl. Spectr. 23:* 257-259 (1969).

10. Beatrice, E.S. and Glick, D.   A direct-reading polychromator for emission spectroscopy. *Appl. Spectr. 23:* 260-263 (1969).

11. Marich, K.W., Carr, P.W., Treytl, W.J., Beatrice, E.S. and Glick, D.   Improved Q-switched ruby laser microprobe for emission spectroscopic elemental analysis.   In preparation.

12. Marich, K.W., Carr, P.W., Treytl, W.J. and Glick, D. Effect of matrix material on laser-inauced elemental spectral emission. *Anal. Chem. 42:* 1775-1779 (1970).

13. Treytl, W.J., Marich, K.W., Orenberg, J.B., Carr, P.W., Miller, D.C. and Glick, D.   Effect of atmosphere on spectral emission from plasmas generated by the laser microprobe. *Anal. Chem.*   In press.

14. Treytl, W.J., Orenberg, J.B., Marich, W.J. and Glick, D. Photoelectric time differentiation in laser microprobe optical emission spectroscopy. *Appl. Spectr. 25:* 376-378 (1971).

15. Saffir, A.J., Treytl, W.J., Marich, K.W., Orenberg, J.B. and Glick, D.   Statistical relationships between spectral line and background intensities and laser energy in performance of a Q-switched ruby laser microprobe.   In preparation.

16. Marich, K.W., Orenberg, J.B., Treytl, W.J., Saffir, A.J. and Glick, D.   Biomedical application of the Q-switched ruby laser microprobe.   In preparation.

17. Herman, M.M., Bensch, K.G., Marich, K.W. and Glick, D.   The effect of gold thioglucose on mouse L-strain fibroblasts, morphological and laser microprobe studies.   In preparation.

INVESTIGATIONS

TOWARD A LASER-MICROPROBE*

F. HILLENKAMP, R. KAUFMANN and E. REMY

*Gesellschaft für Strahlen- und Umweltforschung mbH. München,
Germany;
Physiologisches Institut der Universität Freiburg/Breisgau,
Germany;
Infratest-Industria, München, Germany.*

In the biological and medical sciences there is a growing
interest in the investigation of subcellular structures and
their functional interactions in the natural integrated living
material.   The methods so far applied, e.g. autoradiography,
staining techniques etc., have turned out to be of rather
limited value particularly in the determination of water-soluble
compounds.

There is therefore a demand for new methods and several new
technologies, most of them originally developed for materials
testing, such as the ion-microprobe and the electron-microprobe,
have therefore been tried in biological research and will be discussed.

We have looked into the possibility of using a laser as the
probing instrument and a mass-spectrometer for the analysis of
the probed material.   The principal concept of this method is
to irradiate a small area, preferably of diameter of 1 μ or
less, with high enough an intensity to get the material of this

* This project is sponsored by the Stiftung Volkswagenwerk.

area ejected from the sample. The at least partially ionized
material will be ejected directly into a mass-spectrometer
which analyses its constituents according to their atomic or
molecular mass. The laser is well suited for such purposes
because of its exceptional photometric brightness, i.e. the
amount of energy a radiating area emits at a given wavelength
into the unit solid angle within unit time and which may exceed
the brightness of the sun by several orders of magnitude. The
size of the irradiated spot is - at its lower limit - in principle limited by the effect of diffraction to a diameter 0.5 -
1 wavelength of the light used, i.e. in our case to 0.5 - 1 μ.
With the lasers now available evaporation and even ionization
of the irradiated volume of material can be achieved within a
time as short as 10^{-11} sec.

We have chosen a ruby laser for our experiments because it
is a very well developed sytem and such a laser was immediately
available to us, though almost every pulsed laser system could
in principle be used. Blood cells smeared out on cover slides
and fixed in methylene were taken as test objects for most of the
experiments. The actual investigations with the complete instruments will be done on deep frozen, non-dried sections of
biological specimen.

Figure 1 shows a schematic diagram. On the right is the
microscope with the illumination system consisting of the light
source, L_1, M_1 and the condenser C, the objective O and
the binocular eyepiece BE. Through the partially transparent
mirror PM_2 photographs could be taken with the camera L_2-F.
Condenser and objective allowed the observation in bright field
and phase contrast. The microscope used was an Orthoplan of
the Ernst Leitz KG.

The partially transparent mirror PM_1 was added to deflect
the laser beam into the microscope. The ruby laser on the left
consists of a 3" ruby crystal pumped by a linear flash lamp, a

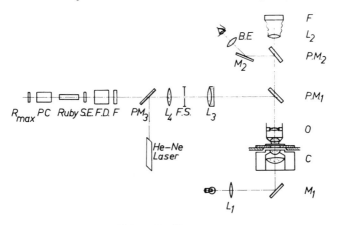

Schematic Diagram

Fig.1

Pockels cell PC, a fully reflecting dielectric mirror R_{max} and a sapphire etalon SE as the coupling mirror. Through a rotation of the ruby through 90° about its axis the laser could be changed from the normal mode of operation, emitting pulses of about 1 msec in duration, to the Q-switched mode of operation with pulses of about 20 nsec duration. This way light of λ = 694 nm wavelength, which is in the red part of the visible spectrum was available. For some of the experiments a nonlinear crystal was put into the beam which acted as frequency doubler FD, thus making light of wavelength λ = 347 nm in the near ultraviolet available. The filter F then eliminates the remaining red light.

The laser beam is focused by the lens L_4 onto a small field stop FS and is then focused into the object plane by the achromatic lens L_3 and the objective O. A small CW Helium-Neon laser was added for the purpose of aligning the whole system and aiming the ruby laser pulses onto the required area.

Figure 2 is a drawing of the specimen support. A microscope slide S with a center hole of about 1 cm in diameter is put on the microscope table MT. A cover slide CS is put

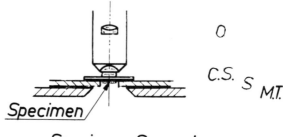

Specimen Support

Fig.2

over the hole with the cells on the lower surface facing the condenser. This arrangement allowed a free evaporation of the material into the space underneath the cover slide and still used the oil immersion objective in an optical arrangement of optimal correction.

Figure 3 shows a sample of red blood cells which have been irradiated with ruby laser pulses of about 1 msec duration. Going down from the top the applied energy has been increased in

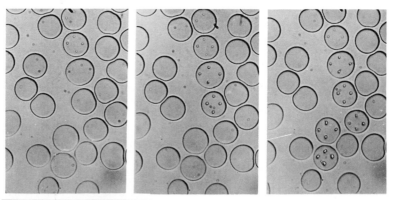

Red Blood Cells
λ=694nm; T=1msec.

Fig.3

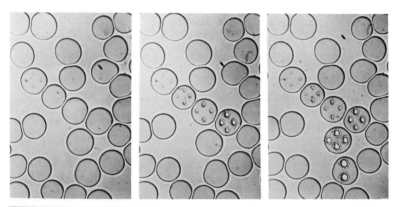

Red Blood Cells
λ=694nm; T=20nsec.

Fig.4

a 1 - 2 - 5 - 10 sequence.   The energy span between the uppermost cell and the lowest one in the right picture is then   100.
It is obvious that near the threshold the affected area is
indeed limited by diffraction.   These holes are about 0.5 μ
in diameter as can be easily deduced from the 6 - 7 μ's of the
cell diameter.   It is remarkable that even at 100 times the
threshold energy, the hole diameter does not increase beyond
about 1 μ.

Figure 4 again shows red blood cells, but irradiated with
Q-switched laser pulses of about 20 nsec duration.   At
threshold the diameter of the holes is again about 0.5 μ but
it appears to increase faster with rising energy as compared to

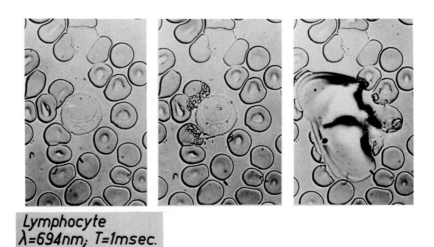

Lymphocyte
λ=694nm; T=1msec.

Fig.5

the non Q-switched case.   In addition the cell surface seems to
be more distorted at energies above 10-times threshold.   The
damage threshold for Q-switched pulses is lower than that of normal pulses by at least a factor of 100.   Similar observations
have been made by other investigators for damage thresholds of
the retina.   The very high intensity of the Q-switched pulses is
certainly an important reason for this, but the complete mechanism of interaction is not yet fully understood.

Attempts to perforate leukocytes in the same manner with
normal or Q-switched pulses rendered to our surprise only little
success.   Figure 5 shows a lymphocyte unirradiated on the left,
with a few holes in the nucleus in the middle and after a cata-

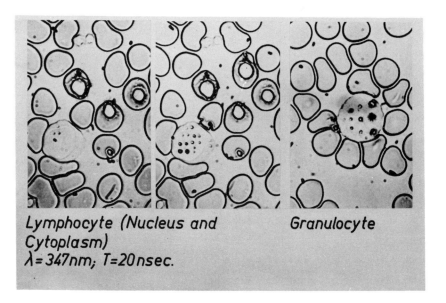

Lymphocyte (Nucleus and   Granulocyte
Cytoplasm)
λ=347nm; T=20nsec.

Fig.6

strophic explosion on the right. This was however a lucky
case, because in most cases the white blood cells show either no
effect at all or explode if the energy reaches a certain
threshold which is at least three orders of magnitude higher
than the value for perforating red blood cells. This very sharp
threshold makes red light unsuitable for our purposes.

When working with ultraviolet light, white blood cells could
as readily be punched as the red ones. Figure 6 shows a lymphocyte with just a few holes in the left and a large number of
holes in the middle. No difference in reaction between the
nucleus and the cytoplasm could be detected. The threshold of
the leukocytes is somewhat higher than that of the red blood

cells - the large holes in the red cells on the upper right were
done with the same energy as was used for the small holes in the
lymphocyte.   On the right is a granulocyte with three columns
of holes of three different energies.   We cannot give a sound
explanation for these observations as yet.   At the ruby laser
wavelength in the red both types of cells show over 99% transmission and no difference in residual absorption could be detected with the microspectrophotometer available to us.   The
existence of such differences which would have to be in the
range of much less than 1% total absorption can not however be
excluded.   The different structure of the cells may be at least
partially responsible for the observed effect.

The small holes shown are definitely not fully resolved by
the light microscope and even the appearance of the large holes
is not really conclusive as to whether the cells have been fully
perforated or the proteins just got coagulated creating a change
in refractive index and/or absorption.   To resolve this question
pictures were taken of irradiated cells first in the light microscope with a magnification of 10^3 and then with a scanning
electron microscope with a magnification of 10^4.   Figure 7
shows four pairs of pictures, each pair showing a cell as it
appears in visible light (on the left) and as traced out by the
secondary electrons (on the right).   The pictures with visible
light were taken from the side where the laser beam entered the
cell; the electron microscope pictures were taken from the side
where the laser beam left the cell.   Three observations are
worth noticing:

1.   Every time the irradiated area appears as a dark dot only
     in the light microscope, the cells turn out to be not
     fully penetrated (left column); every time small dark
     circles appear in the light microscope, the cells are fully
     punched (right column).

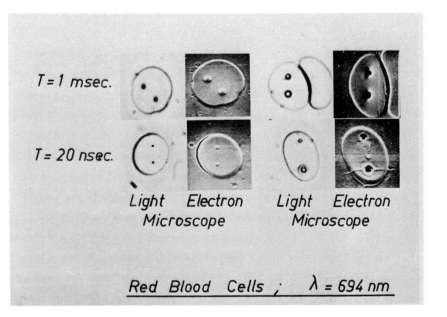

Fig.7

2. True holes of diameters as small as 0.5 μ can reproducibly be achieved.

3. The rim of the holes punched with normal mode lasers appears smooth, whereas in the case of Q-switched pulses material (most probably of the cell membrane) appears to be vigorously thrown out of the cell and scattered over the surrounding area. Further experiments are necessary for a complete understanding and explanation of this effect.

All three observations have been made consistently with a large number of cells.

As a demonstration of the fact that the method is not restricted to blood cells we have, more or less at random, chosen two other objects. Figure 8 shows a fixed and air dried Mouse-L cell in phase contrast with three holes done with ultraviolet light. The holes are somewhat elliptical because no field stop was used in this case. Figure 9 shows a bull spermatozoon with

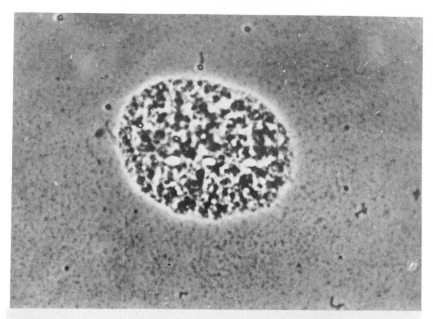

Mouse L-Cells (Phase-Contrast)
λ=347nm; T=20nsec.

Fig.8

a hole in the head and one in the neck.

We would like to thank the Stiftung Volkswagenwerk for sponsoring these investigations and the Ernst Leitz KG which very generously supplied us with the microscope and the necessary accessories. We are grateful to the Gesellschaft für Strahlen- und Umweltforschung which allowed us to conduct the experiments in their laboratories.

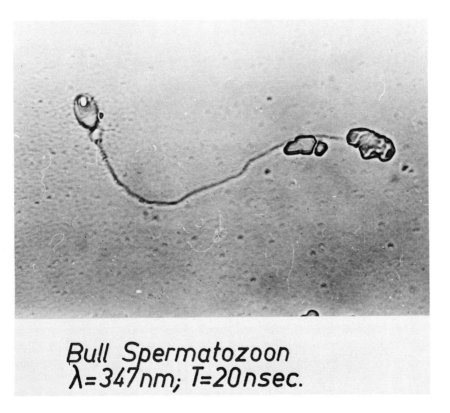

Bull Spermatozoon
λ=347nm; T=20nsec.

Fig.9

ACCESS TO METABOLIC PROCESSES

IN LIVING MATTER

MADE POSSIBLE BY THE LASER

NILS KAISER

Max-Planck-Institut für Plasmaphysik, München, Germany

Most of the methods now being used to investigate metabolism
are based on chemical analysis.   For this purpose the test
specimen has first to be removed from its physiological environment and prepared.   For many metabolic investigations, however,
it would be better to record these processes under physiological
conditions, that is, conduct experiments *in vivo.*
Chemically speaking, biological metabolism constitutes
changes in molecular structure.   All molecules have characteristic resonance absorption in the frequency band.   The resonance
frequencies of the molecules of metabolic interest are primarily
in the near and far infrared.   Changes in molecular structure
are accompanied by changes in the resonance frequencies of the
molecules.   This suggests the use of infrared spectroscopes for
these experiments.   The devices available hitherto are unsuitable, however, since the generators used with them do not have
sufficient transmitting power to measure aqueous solutions, water
having pronounced intrinsic absorption in the infrared and being
an unavoidable component of experiments *in vivo* (1,2,3,4).
A survey of the absorption values of pure water is presented
in the first figure;   these are compiled from data given in the

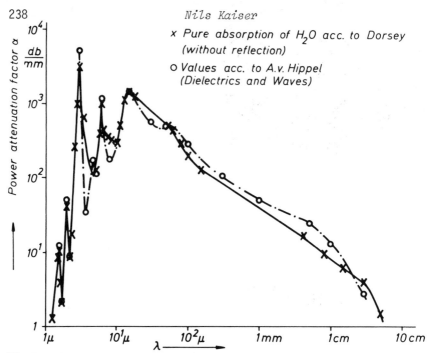

Fig.1 : Absorption of water in the range 1 μ to 10 cm.
Compiled from values given in the literature.
10000 Å = 1 μ = 10⁻⁴ cm.

literature (Fig.1).

As can be seen from Fig.1, the maximum attenuation of water at a wavelength of about 5 μ is about 7 . 10^3 dB/mm. However, the attenuation maximum above 10 μ is only slightly less, and our measurements were made at this wavelength.

With the rapid advance of laser technology in recent years, generators with sufficient power are now available in almost the entire near and far infrared regions. Before the proposed experiments could be prepared, the following questions first had to be clarified:

(1) What layer thickness of water, aqueous solution, or blood can be irradiated and still readily be measured?

(2) Is it possible at this layer thickness to maintain circu- lation of the test specimen through the cell by pumping, for

example, a bacteria culture or tissue-suspension and to
operate as well an extracorporal circulation system for experiments *in vivo*?

(3) What generator power is required for such measurements?

(4) To what extent is the test specimen heated as a result?
The answer to the last question is of major importance biologically owing to the risk of protein denaturation. The problem
was first tackled in a simple experiment consisting of a
generator-specimen-detector array set up by K. Plank. A CO_2
laser (wavelength 10.6 μ) was chosen as generator since it is
known from the literature that pure water has a pronounced absorption band at this wavelength. Measurements at this wavelength would therefore show to what extent investigations are
possible in the entire infrared region despite the high attenuation of water.

First pure water was investigated, the results being as
follows:

(1) With a layer thickness of 0.1 mm the attenuation was
    about 25 dB.

(2) At this layer thickness a flow rate of 30 cc/min could
    readily be maintained. This is roughly equivalent to that
    of an extracorporal system in an animal experiment.

(3) The CO_2 laser used yielded an output power of 2W in CW
    operation.

(4) No critical heating of the test specimen was observed.

As the energy balance shows, with no heating expected: the
attenuation of 25 dB due to the test specimen practically corresponds to total absorption; an input intensity of 2W is
equivalent in 1 min to 120 watt seconds = 29 cal; at a flow
rate of 30 cc/min this corresponds to a temperature rise in
the test specimen of about $1^{\circ}C$.

As such a temperature increase is not critical biologically,
animal experiments were subsequently conducted in collaboration

with K. Messmer of the Institute of Experimental Surgery at the
University of Munich.   An arterio-venous bypass was inserted to
divert the circulation system of a heparinized dog through a
flow cell with adjustable layer thickness and with Irtran as
window material.   This extracorporal system functioned perfectly
for several hours at a layer thickness of 0.1 mm.   An attenuation of about 25 dB was again measured at this layer thickness.

From these results and the data in the literature it can be
assumed that laser experiments can be made in aqueous solutions
on living biological specimens, and extracorporal systems in the
entire infrared region (5,6).

In investigations of metabolic processes it is largely the
absolute values of or variations in very small concentrations
that have to be recorded.   Instruments for such purposes must
therefore have a high measuring sensitivity.   A suitable interferometer arrangement developed in collaboration with K. Plank
is shown in Fig.2.

The coherent wave train coming from the laser is split into
two equal components by the first beam splitter.   One beam
passes via a mirror to the test vessel, irradiates the test speci
men, and joins the second beam, which has passed through the
reference cell.   The two beams then form an interference pattern
This interfering wave train impinges on a screen provided with a
slit behind which the detector is placed.   The interferometer
can be balanced by the germanium mirror mounted on a piezoelectri
crystal at the top left.   For this purpose a voltage is applied
to the crystal, and this shifts the mirror to an angle of 45° to
the beam axis. After the identical test and reference cells have
been filled, e.g. pure water in the reference cell and aqueous
solution in the test cell, an interference maximum is set on the
bolometer by means of the adjustable mirror.   If a reaction then
takes place in the test cell, and the refractive index for the

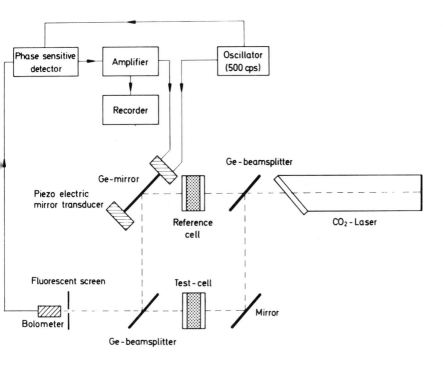

Fig.2 : Basic circuit of the laser interferometer for single-
frequency and swept-frequency operation with simultaneous recording of attenuation and phase for both
modes.

test wavelength changes as a result, the bridge becomes unbal-
anced and the bolometer current changes. The phase-sensitive
detector tries via the amplifier to follow the crystal mounted
mirror, so that a maximum is again set on the bolometer. If
the bolometer current and control voltage for the mirror are
now recorded together on a two-channel recorder, the variations
in attenuation and phase of the test specimen during a reaction
are determined as functions of time.

As previous microwave measurements have shown, either the
attenuation or the phase may yield more significant values if
the refractive index of the test specimen varies (7,8,9). It is

therefore advantageous to measure these two values simultaneously

Besides being suitable for the single-frequency measurements described above, the measuring setup can also be used for measurements with swept frequency. All that is needed for the purpose is to make the optical paths between the first and second beam splitters equal in length. A tunable laser has then to be used as generator; such lasers have been available for some time. In this case the attenuation and phase can be determined as functions of the wavelength. The simultaneous recording of these two values affords improved detection of resonances particularly in multi-component systems or biological specimens (10). The reference cell is incorporated to make the optical path lengths of the two beams equal and extract disturbing substances, e.g. water. The use of such identical test cells in the reference beam of dual-beam spectroscopes is familiar from infrared technique. It is intended first to use the tunable laser to detect on the wavelength-band the resonances of interest for certain metabolic processes and then to follow the development of these absorption bands by means of a very sensitive single-frequency instrument. The theoretical sensitivity of the single-frequency apparatus is very high. At present it depends primarily on the sensitivity of the detectors.

For information purposes a few measurements were made by G. Venus on mixtures of water and ethanol with a simple irradiation device. Here the transmitted test signal was merely modulated and then amplified. This simple setup allowed changes in concentration of as little as ±0.5% to be determined with good reproducibility. Further improvement of this display sensitivity by about a power of ten is possible without major technical outlay. Achieving this order of magnitude would be most interesting for biological applications.

In conclusion, it can be stated that the use of lasers as light sources has made it possible to conduct spectroscopic in-

vestigations on aqueous solutions and living biological specimens without having to extract samples. The investigations
apparently do not destroy biological specimens. The apparatus
can theoretically be elaborated to provide a very high degree of
accuracy. It is expected that this method will provide more
knowledge of intermolecular interactions in aqueous solutions
and of the reaction kinetics of metabolic processes.

REFERENCES

1. Kaiser, N. Lab. Report Max-Planck-Institut für Physik
   und Astrophysik, 8046 Garching-München, Germany, 1.7.
   (1962).

2. Kaiser, N. Proc. "16th Colloquium of the Biological
   Fluids" Vo.16, 81-86, Bruges, Belgium (1969).

3. Kaiser, N. Third International Biophysics Congress,
   August-September 1969, Cambridge, Massachusetts.

4. Kaiser, N. Zeitschrift *"Laser"* Nr.4, 56-57 (1969).

5. Kaiser, N., Casimir, W.v., Gehre, O., Plank, K. and
   Schneider, W. Max-Planck-Institut für Plasmaphysik, 8046
   Garching-München, Germany: IEEE Transactions (in process of
   publication).

6. Kaiser, N. Symposium Israel Laser Group, Tel Aviv,
   1. July 1970.

7. O'Brien, B.B. IEEE Transactions on Instrumentation and
   Measurement Vol. IM-16, Nr.2, 124 (1967).

8. Casimir, W.v. Disseration, T.H. München, January 1968.

9. Schneider, W. Disseration, Universität München, January
   1969.

10. Gehre, O. IEEE Transaction on Instrumentation and
    Measurement Vol. IM-19, Nr.1, S. 14-17, February 1970.

BIOELECTRODES

GEORGE EISENMAN

Department of Physiology, UCLA,
Los Angeles, California, U.S.A.

I INTRODUCTION

The subject of "Bioelectrodes" encompasses such a wide range of phenomena that it would be difficult to do justice to it even in a lengthy book, much less within the space of a single chapter. However, the *sine qua non* of a useful bioelectrode is its ability to respond selectively to a particular ion species in the presence of other species which are often very similar to it in the usual chemical sense; and it is to this narrower subject of the molecular basis of electrode selectivity that this chapter is chiefly directed.

Certain fundamental processes are common to all electrodes; and our theoretical understanding of the origin of electrode potentials has progressed to the point that it is now possible to assess the common features, as well as crucial differences, among the various electrode types. In this way, one is able to reduce the problem of understanding the selective discrimination of apparently diverse kinds of electrodes to that of understanding the selectivity of a few elementary processes common to all electrodes. These processes involve the relative differences in the affinities of ions for the various molecular binding structures in the membrane and their mobilities (whether bare, hydrated, solvated or complexed) within the membrane.

Electrodes having useful selectivity to particular cations
and anions may be conveniently classified into four general
types according to their mechanism of operation.

Class A consists of membranes made from solid ion exchangers, of which H^+-selective (11), cation-selective glass
electrodes (34) and anion-selective collodion membranes (63)
are typical examples. In glass the membrane is relatively permeable to monovalent cations as a result of a combination of
their high ion-exchange affinities for fixed membrane sites and
favorable mobilities within the membrane. Such electrodes are
useful for all monovalent cations and are at present the electrodes of choice for H^+, Na^+, Li^+, Ag^+, and even K^+ and
NH_4^+ in certain applications (17,18,19).

Class B comprises liquid ion exchanger membranes (see (92)
for review of early literature and (23), and (84) for discussion
of mechanism) which yield useful electrodes for divalent cations
such as Ca^{++} and Mg^{++} (82), monovalent cations such as K^+
(107) and acetylcholine$^+$ (1), and anions such as Cl^-, NO_3-
and ClO_4^- (85).

Class C encompasses the recently developed electrodes (100,
59,61,106,27,38) based on lipophilic neutral ion binding molecules such as the antibiotics, valinomycin and nonactin, and
certain synthetic cyclic polyethers. These molecules can form
coordination complexes with monovalent cations, replacing the
water of hydration by ligand oxygens of the complex (73). Since
the exterior of the complex is hydrophobic, the charged complex
is unusually soluble in media of low dielectric constant and
such molecules act as carriers of cations across the solvent of
the membrane (75,29,30,101). This class of electrodes offers
the greatest selectivity presently available for ions such as
K^+ and NH_4^+. Although the precise mechanism by which these
antibiotics produce their effects on thick commercial electrodes
is just beginning to be understood (61), it is already known

TABLE I

COMPARISON OF POTENTIAL SELECTIVITY CONSTANTS FOR THICK ELECTRODES, THIN BILAYER MEMBRANES, AND EQUILIBRIUM 2-PHASE SALT EXTRACTION

NONACTIN

	Bilayer membranes[1]	Thick electrodes[2]	Equil. constants for 2-phase salt extraction[3]
Li	<.001	.00056	.00026
Na	.0071	.0067	.017
K	1.0	1.0	1.0
Rb	.58	.42	.47
Cs	.033	.031	.061
NH$_4$	8.0	2.5	47.
H	–	.018	–

VALINOMYCIN

	Bilayer membranes[4]		Thick electrodes[5]		Equil. constants for 2-phase salt extraction[6]
Li	<.012[a]	.0025[b]	<.00021[a]	.0021[b]	.000007
Na	<.014[a]	.0035[b]	<.00026[a]	.00065[b]	.000017
K	1.0[a]	1.0[b]	1.0[a]	1.0[b]	1.0
Rb	1.9[a]	2.3[b]	1.9[a]	2.0[b]	1.95
Cs	.44[a]	.53[b]	.48[a]	.32[b]	.62
NH$_4$	–		.012[a]	–	.19
H	10.8		.000056[a]	–	–

Data sources:

[1] Szabo *et al.* (101), Table 5; and Eisenman *et al.* (31), Table 3
[2] Pioda and Simon (74); Simon (95), in Najol-2 octanol
[3] Eisenman *et al.* (30), Table 15 for CH$_2$Cl$_2$
[4] a. Lev and Buzhinsky (59), b. Mueller and Rudin (68)
[5] a. Pioda and Simon, in Diphenyl ether; b. Lev *et al.* (60,61) in Heptane
[6] Eisenman *et al.* (31), Table 1 for CH$_2$Cl$_2$

(see Table I) that the selectivities in bulk electrodes are the
same as those in the well-understood bilayer membranes for which
comprehensive theoretical treatments have been developed (29,4,55
and verified (101,31,98).

Class D encompasses the so called "solid-state" crystalline
electrodes, useful for ions such as F^- and $S^=$ (35,84). The
crystal-impregnated rubber electrodes (77) and classical solubility-product electrodes (e.g. Ag; AgCl) may conveniently be re-
garded as members of this class also.

A universal feature underlying the ion selectivity in all
the above electrodes will be seen below to be the competition
between the *equilibrium* free energies of hydration (for measurements in aqueous media) vs. the equilibrium interactions with the
particular ligand groups in the membrane. *Non-equilibrium*
(diffusional) processes will also be shown to contribute to electrical properties, with the particular properties of each class
reflecting the relative roles of equilibrium interactions and
ionic mobilities. To anticipate the reasons the selectivity
differs from class to class, it is worth pointing out here that
one of the principal restrictions on the selectivity of solid
ion exchangers results from the opposing effects of mobilities
and affinities (e.g., the poor selectivity for divalent cations
is a consequence of their low mobilities). Part of this limitation can be overcome by making the ion exchanger a liquid and
thus allowing the ion exchange sites to move; but an ever
greater freedom is attained in the third class of electrodes,
where the mobile entity bears the electrical charge of the com-
plexed-ion, which is merely solubilized in the membrane by the
neutral carrier molecule. Class C is unique in that strong
binding affinities can be utilized without concomitant restrictions on the mobility of the charged species. This is because
this class merely loads and unloads its ions at the interfaces
whereas the solid electrodes, and to some extent the liquid ion

exchangers, must do this throughout the membrane.

Before proceeding with the examination of the individual classes of electrodes, it should be noted that their range of selectivity can be greatly elaborated if they are used as secondary sensors for the product of a specific (e.g. enzyme) reaction. Thus, Updike and Hicks and Guilbault and his colleagues (40,41,42) have pioneered the use of specific ion sensors of the above types in conjunction with appropriate immobilized enzyme systems using, for example, a cation-responsive glass electrode to measure 1-amino acids or urea by sensing the ammonium ion produced by 1-amino acid oxidase or urease, respectively.

It should also be noted that certain physical features of liquid-phase electrodes lend themselves to easier fabrication of microelectrodes than the more conventional glass membrane microelectrodes (3,46,58) since merely filling a pipette with the liquid sensor suffices to make such a microelectrode (70,105).

The following section contains a characterization of the electrode behavior in ionic mixtures for these four classes of electrodes. Since the mathematical description of the electrode potential in ionic mixtures is a prerequisite for the definition of selectivity, this is considered in some detail except for class D, whose selectivity can generally be ascertained from elementary solubility-product considerations. Moreover, since an understanding of the origin of the electrode potential is needed for the comparison of selectivity among the various classes of electrodes, this subject is also developed. Section III compares the factors determining the selectivity of each of these electrode classes and shows why each class has certain advantages, as well as limitations, compared to the others. Section IV summarizes, through a series of empirically-deduced (but theoretically-based) selectivity patterns, the principal selectivity data for those classes of electrodes (chiefly A and C) where this is presently feasible. These patterns contain

all the information needed to decide on the electrode type having
the optimal selectivity for the ions of interest and the appli-
cations-oriented reader will find this section, together with
Section III, to contain the chief practical results.   The fifth,
and concluding section analyzes the molecular basis of ionic
selectivity emphasizing the common features underlying the equilibrium component of selectivity for all classes of electrodes.

II   GENERAL PROPERTIES OF THE FOUR CLASSES OF ELECTRODES

A. *Solid ion exchangers (e.g. the glass electrodes)*
(1)   <u>History</u>.   Cremer (9) first observed that the electrode
potential of a glass membrane differed in solutions having differing H^+ contents.   Three years later, Haber and Klemensiewicz
(35) showed that this response was in fact the Nernstian response
expected for an ideal hydrogen electrode

$$V = const. + \frac{RT}{F} \, ln \, C_{H+} \tag{1}$$

where C_{H+} is the H^+ concentration or, more precisely, the H^+
activity.

It was soon noted that in alkaline solutions a deviation from
this equation, or "alkaline error", was present (48) which was
more prominent in the presence of small amounts of Al_2O_3 or B_2O_3
(47,90).   These observations were extended by Lengyel and Blum
(57), who demonstrated that the introduction of Al_2O_3 or B_2O_3
(or both) into the glass caused its potential to become strongly
dependent upon Na^+, as well as H^+ and, indeed, established
that one could thereby obtain a Nernstian response to Na^+ as
well as to H^+.   Unfortunately, they concluded, incorrectly as
it turned out later, that further variation of glass composition
beyond introducing 10% Al_2O_3 or B_2O_3 had no important effects
so that it was not for several decades that it was established by
Eisenman, Rudin and Casby (26) that not only was a useful respons

to Na^+ developed by the addition of Al_2O_3, but that reproducible responses to the other alkali metal cations could also
be developed by systematic variations of the composition of such
glasses. In particular, these authors established, through a
systematic study of three-component glasses containing only
Na_2O, Al_2O_3, and SiO_2, that the selectivity among the different group Ia cations was a systematic and reproducible function
of the Na_2O, Al_2O_3 ratio in these glasses and discovered the
existence of particular glass compositions, which had sufficiently
high selectivities for K^+, as well as Na^+, to become the
first selective K^+ electrodes and Na^+ electrodes. Subsequent studies by Eisenman (17,18) laid down the detailed rules
concerning the relationship between glass composition and the
selectivity for the ions H, Li, Na, K, Rb, Cs, Ag, Tl,
HN_4, Mg, Ca, Sr, Ba, and certain substituted ammonium ions;
and presently manufactured ion-selective glass electrodes are
made from these compositions, although this is obscured by the
fact that most manufacturers do not disclose the composition of
the glass in their electrodes. A description of the responses
of certain commercial electrodes, as well as an extensive
characterization of the electrochemical properties and selectivities to other cations than those of group Ia (e.g. Ag^+,
Tl^+, NH_4^+ and various substituted ammonium ions will be found
in the monograph, "The Electrochemistry of Cation-Sensitive
Glass Electrodes" (18), as well as the recent volume on "Glass
Electrodes for H^+ and Other Cations" (19,20,21).

The above inorganic glasses are not the only examples of
useful electrodes based on solid ion exchangers. Thus, dense
organic oxygen polymers (e.g. collodion membranes (97)), particularly when densely compacted by drying (64,17-Fig.8A,63), can
give very selective electrodes for such cations as Cs^+ and Rb^+
(as can be seen from the data labelled "collodion" in Fig.1) and
such anions as SCN^-, NO_3^-, and I^- (63-Fig.10-6).

(2) Equations describing electrode potential and definition of "Selectivity". To examine the properties of these electrodes further, it is first necessary to have a quantitative description of the electrode potential and a definition of selectivity. The first equation for the glass electrode which was capable of describing not only the H^+ response but also the Na^+ "error" of glass electrodes was due to Nicolskii (69), whose equation was of the form:

$$V = const. + \frac{RT}{F} \ln (a_{H+} + K^{pot} a_{Na+}) \qquad (2)$$

where K^{pot} is a "potential selectivity constant", to be discussed more extensively below. (The constant voltage sources due to internal filling solutions of the electrode, internal references, etc. are all lumped in the term labelled *"const."*.)

Eisenman *et al.* (26) confirmed the essential correctness of this equation, but found that in order to describe a wider range of data, it had to be elaborated to the form:

$$V = const. + \frac{nRT}{F} \ln \left[a_i^{1/n} + (K^{pot}_{ij} a_j)^{1/n} \right] \qquad (3)$$

containing an additional parameter, n (which is a constant specific for the non-ideal behavior of a given glass for a given pair of cations). Eisenman *et al.* established that Eq. (3) not only described the potential in H^+-cation mixtures, but that it described the potential for mixtures of any pair of alkali metal cations at constant pH. For such pairs as Na^+ and K^+, n was usually 1, indicating an ideal behavior, whose basis has been considered elsewhere (17, pp. 281-282). In the case of such ideal behavior Eq. (3) reduces to the simpler form

$$V = const. + \frac{RT}{F} \ln (a_i + K^{pot}_{ij} a_j) \qquad (4)$$

The "selectivity" of glass to various cations is defined by K_{ij}^{pot} of Eqs (3) and (4), since K_{ij}^{pot} is a weighting factor characteristic of the selectivity of a given membrane composition between a given pair of ions. Thus, a value of $K_{ij}^{pot} = 10$, means that ion J^+ is ten times as effective as ion I^+ in determining the electrode potential.

3) Theory of the origin of the glass electrode potential and the contributions to electrode selectivity of both equilibrium ion exchange selectivity and non-equilibrium mobility.

The classical controversy (11) as to whether the origin of glass electrode potential was due to diffusion or to phase boundary processes was finally resolved by the development of a theory of the glass electrode potential by Karreman and Eisenman (51) and its extension by Conti and Eisenman (6,7). These authors suggested (and it was later verified experimentally (23,13,49,37, 38,39)) that the glass electrode was simply a perfect cation-exchanger membrane whose electrode potential was expected to represent a straightforward sum of contributions from both diffusional and phase boundary processes. This theory indicated that the potential selectivity constant of Eqs (3) and (4) is defined by:

$$K_{ij}^{pot} = \frac{u_j^*}{u_i^*} K_{ij} \tag{5}$$

and depends upon the relative mobilities of the ion species within the glass, u_j^*/u_i^*, as well as the conventional ion exchange equilibrium constant K_{ij} of

$$J^+(\text{aqueous}) + I^{+*}(\text{membrane}) \xrightleftharpoons{K_{ij}} J^{+*}(\text{membrane}) + I^+(\text{aqueous}) \tag{6}$$

where asterisks have been used to denote quantities characteristic of the membrane phase. (This convention will be adhered to henceforth; and quantities pertaining to the aqueous phase

will either be unmarked or designated by (') and ('') when
it is necessary to distinguish between the solution on two sides
of the membrane.)

The usefulness of the theoretical relationship, Eq. (5), in
dicating the roles played by mobility as well as ion binding in
controlling the electrode selectivity will be made apparent in
Section III, where it will be shown, by comparison with the corresponding equations for liquid ion exchangers and neutral carriers, how it is that the mobility restrictions for solid ion
exchange membranes are circumvented in liquid systems.

Note that reactions (6) can be decomposed into two (hypothetical) sub-reactions

$$J^+ \text{ (aqueous)} \xrightleftharpoons{K_j} J^{+*} \text{ (membrane)} \tag{7a}$$

$$I^{+*} \text{ (membrane)} \xrightleftharpoons{1/K_i} I^+ \text{ (aqueous)} \tag{7b}$$

for which

$$K_{ij} = \frac{K_j}{K_i} \tag{8}$$

Definition (8) will be useful when comparing the selectivity of
solid vs. liquid membranes in Table II.

With this definition, Eq. (5) can be written in an alternative form

$$K_{ij}^{pot} = \frac{u_j^*}{u_i^*} \frac{K_j}{K_i} \tag{9}$$

B. *Liquid ion exchangers*

Membranes made from a water-immiscible solvent interposed
between two aqueous solutions were among the first electrode
systems whose properties were characterized (12,71,92). When

the solvent of such an electrode is "doped" so as to contain an
appreciable concentration of a suitable organophilic ion (e.g.
a fatty acid or an aliphatic amine), it becomes a liquid ion
exchanger (50,81,14,2,83,84,22). Such an exchanger has been
postulated to differ from a solid ion exchanger only insofar as
its ion exchange sites are free to move within the membrane
phase (8,87,88). The properties of such liquid ion exchangers
have been discussed by Eisenman (22), Ross (84) and by Sandblom
and Orme (89) in a recent definitive review.

(1) Case of complete dissociation (e.g. solvents of high dielectric constant. In the limiting, but probably rarely encountered, case of a solvent having sufficiently high dielectric
constant for there to be negligible association between the site
and the counterion, Conti and Eisenman (8) have shown that the
membrane potential is given (for ideal behavior, where n=1) by
an equation of identical form to Eq. (3), namely:

$$V = const. + \frac{RT}{F} \, ln \left(a_i + \left[\frac{u_j^* k_j}{u_i^* k_i} \right] a_j \right) \qquad (10)$$

The bracketed set of parameters define the electrode selectivity
but their physical meaning is different from the case of the
solid ion exchanger discussed in the previous section. In particular, for the liquid ion exchanger, u_i^* is the mobility of
the dissociated KI^+ species in the solvent of the membrane
and k_i is its partition coefficient between water and the solvent, corresponding to the reaction

$$I^+ \, (aqueous) \xrightarrow{\;\; k_i \;\;} I^{+*} \, (membrane) \, . \qquad (11)$$

Thus, for such a liquid ion exchanger the selectivity between
species I^+ and J^+ is determined by the partition coefficients

and mobilities of the dissociated (*sic*) ions instead of the affinities of these ions for the fixed sites and their mobilities
of jumping from site to site. The selectivity in this limiting
case for a liquid exchanger is thus solely characteristic of ion-
solvent interactions, and no specific interactions with the exchanger molecule are expected to manifest themselves. This is
not surprising, since the dissociation between the liquid ion
exchanger molecules and their counterions is complete in this
limiting case

(2) Case of significant association between counterion and
site. More realistically for the low dielectric constant solvents usually used for liquid ion exchange membranes, some association is expected to occur within the membranes between the
counterion species I^{+^*} and the site species S^{-^*} according to
reaction:

$$I^{+^*} + S^{-^*} \underset{\longleftarrow}{\overset{K^*_{is}}{\rightleftharpoons}} IS^* , \qquad (12)$$

which results in the formation of the neutral ion pairs, IS^*,
which are freely mobile within the membrane. For such a system,
Sandblom, Eisenman and Walker (87,88) have shown that the electrode potential is given by the somewhat more complex expression

$$V = \frac{RT}{F}(1-\tau)\ln \frac{a_i + \left[\dfrac{(u^*_j+u^*_S)k_j}{(u^*_i+u^*_S)k_j}\right]a_j}{a''_i + \left[\dfrac{(u^*_j+u^*_S)k_j}{(u^*_i+u^*_S)k_j}\right]a''_j} + \tau \ln \frac{a'_i + \left[\dfrac{u^*_{js}}{u^*_{is}}K_{ij}\right]a'_j}{a''_i + \left[\dfrac{u^*_{js}}{u^*_{is}}K_{ij}\right]a''_j} \qquad (13)$$

(here written explicitly in terms of the activities of the ions
on both sides, of the membrane, as symbolized by the superscripts

(') and (")). The selectivity is characterised by the bracketed sets of parameters, which will be discussed shortly.

Eq. (13) indicates that the electrode potential of a liquid ion exchanger membrane generally is expected to consist of two logarithmic terms (each term, however, being of the same simple form of the preceding electrode equation). The relative contributions to the potential of these terms is governed by the parameter:

$$\tau = \frac{u_s^* (u_{js}^* K_{js}^* - u_{is}^* K_{is}^*)}{(u_i^* + u_s^*) u_{js}^* K_{js}^* - (u_j^* + u_s^*) u_{is}^* K_{is}^*} \quad ; \quad 0 \leqslant \tau \leqslant 1. \tag{14}$$

Inspection of the bracketed selectivity constant for the first term of Eq. (13)

$$\left[\frac{(u_j^* + u_s^*) \, k_j}{(u_i^* + u_s^*) \, k_i} \right]$$

indicates that it depends on the product of the partition coefficient ratio of the dissociated ion species (k_j/k_i) and the ratio of the summed mobilities of the dissociated ions and the dissociated sites $[(u_j^* + u_s^*)/(u_i^* + u_s^*)]$. The former quantity (k_j/k_i) is independent of the chemical nature of the ion exchange sites and depends solely on the solvent; while the latter depends on the ion exchange site only through its contribution to the summed mobilities of the dissociated sites and counterions. Thus the selectivity of the first term of Eq. (13) depends chiefly on the properties of the solvent and only secondarily on the chemical nature of the liquid exchanger.

On the other hand, the selectivity in the bracketed parameter of second term of Eq. (13)

$$\left[\frac{u^*_{js}}{u^*_{is}} K_{ij}\right]$$

is determined by the product of the mobility ratio of the associated ion pairs (u^*_{js}/u^*_{is}) and the ion exchange equilibrium constant (K_{ij}) of the reaction:

$$J^+(\text{aqueous}) + IX^*(\text{membrane}) \underset{}{\overset{K_{ij}}{\rightleftharpoons}} JX^*(\text{membrane}) + I^+(\text{aqueous}) \qquad (15)$$

K_{ij}, of course depends not only on the properties of the solvent but also importantly on those of the organophilic ion exchanger, as will be considered further in Section III.

The value of τ, which lies between 0 and 1, depending on the properties of the solvent and of the ion exchanger, determines the relative importance of the two logarithmic terms of Eq. (13). Fortunately, for certain systems of practical interest, Eq. (13) is largely determined by either the first or the last term alone. For example, if the association constant for the I^+ species is much larger than that for J^+ (and also if the mobilities of all neutral pairs are approximately equal so that $u^*_{is} \simeq u^*_{js}$), Eq. (14), reduces to:

$$\tau = \frac{u^*_s}{u^*_j + u^*_s}$$

from which τ is identifiable as the transference number of the dissociated site species relative to the most dissociated counterion species, and $(1-\tau)$ is seen to be simply the transference number of the most dissociated counterion species. This means that the second term will determine the potential predominantly whenever the dissociated site species is more mobile than the most dissociated counterion.

C. Neutral lipophilic cation-binding molecules which act as carriers of ions

(1) <u>Description.</u> Following the discovery that certain neutral macrocyclic lipophilic cation-binding molecules, such as the antibiotics valinomycin and monactin (68,59) and the synthetic cyclic polyethers (29), made extremely selective electrodes for ions such as K^+ and NH_4^+ in ultra-thin (ca. 50 Å thick) phospholipid bilayer membranes, two lines of work on these molecules have developed, to some extent independently. In one line, the mechanisms of action on thin (bilayer) membranes have been carefully characterized (101,98) and detailed theories developed (4,55,5) in terms of the underlying chemical equilibria (30) as well as certain important rate processes (98). In the other line, a number of practical electrodes have been devised utilizing thick electrode phases and electrode designs similar to those used with liquid ion exchangers. The first of such antibiotic-based thick electrode systems was described by Stefanac and Simon (99) even before the effects of the antibiotics on thin membranes had been observed; and although the initial electrodes had markedly sub-Nernstian responses, modification of solvents and improved electrode design soon verified the potential utility of such electrodes (60,74, 91). Indeed, thick electrodes of this type have now been developed to the point of practical usefulness (35,38,33), and even Ba^{2+} specific electrodes have recently been described by Levins (62).

These macrocyclic ion binding molecules are devoid of charged groups but contain an arrangement of oxygens energetically suitable (through ion-dipole interaction) to replace the hydration shell around cations. Such lipid-soluble molecules are thus able to solubilize cations in organic solvents, forming mobile charged complexes with the cations therein, and in this way providing a mechanism for cation permeation across the

membrane.   An electrode based on such molecules therefore resembles a liquid ion exchanger in its fluid nature but differs in
that the complexed cation species (IS^+) bears the charge of
the sequestered cation instead of being electrically neutral.
The chemistry and structure of these molecules has been studied
extensively by increasingly sophisticated techniques (73,52,36,
72,76,45,15) and the underlying chemistry is well characterized
both at equilibrium (30,31) and even as to certain aspects of
their kinetics (15).

(2)   Theory of potential.   Although the theory of thick membranes is still in a rudimentary state (but see (61)), the
theory of thin systems has been completely developed and verified.   However, it has been recognized (31) that the electrode
selectivity of bilayers and thick membranes are identical for
the two important molecules, valinomycin and monactin.   The
close correspondence between the selectivity conferred by these
molecules on bilayers and on bulk electrodes is illustrated in
Table I.   This correspondence is easily understandable in terms
of the theoretical conclusion, to be discussed below, that the
selectivity seen in bilayers reflects essentially the simple
equilibrium affinities of these molecules for the ions.   It is
not difficult to conceive of a variety of ways in which this
would also be true for thick membranes as well, but I will res-
trict theoretical considerations here to the electrode potential
of thin bilayer membranes since these considerations must apply
at least formally to the thick membranes as well.

The electrode potential of bilayer membranes is expected
theoretically (30,4), and has been shown experimentally (101) to
be described by an equation of the same form as Eq. (4) or Eq.
(10), namely

$$E = \frac{RT}{F} \, ln \; \frac{a'_i + \left[\dfrac{u^*_{js} \, K_j}{u^*_{is} \, K_i}\right] a'_j}{a''_i + \left[\dfrac{u^*_{js} \, K_j}{u^*_{is} \, K_i}\right] a''_j} \tag{16}$$

where K_j/K_i is the ratio of the equilibrium constants of the salt extraction reactions

$$I^+ + X^- + S^* \overset{K_i}{\rightleftharpoons} IS^{+^*} + X^{-^*} \tag{17a}$$

$$J^+ + X^- + S^* \overset{K_j}{\rightleftharpoons} JS^{+^*} + X^{-^*} \;, \tag{17b}$$

which measure the extent to which the neutral molecules, S, can extract the salt, I^+X^-, from water into an organic solvent phase (30). The potential selectivity constant in this case is:

$$\left[\frac{u^*_{js} \, K_j}{u^*_{is} \, K_i}\right]$$

and it is worth anticipating an important result of Section III here. Namely, that there are a number of observed instances in which the size and charge of the complexed-cation is virtually independent of the particular cation sequestered so that the ratio $(u^*_{js}/u^*_{is}) = 1$. In these cases we have the very simple and important result that the electrode selectivity depends solely on the ratio of equilibrium constants, K_j/K_i. This suggests that one can hope to make electrodes as selective as one can tailor the binding of a molecule. This should be true as long as the binding is not so strong that, for example, the rate of dissociation of the complex becomes rate-determining.

D. "Solid-state" crystalline electrodes

As discussed by Ross (84), certain crystalline materials (e.g. LaF_3, Ag_2S) are known which exhibit ionic conductivity at room temperature and have sufficiently low solubility that a thin section of crystal can be used as an electrode membrane. A dispersion of such crystals in an insulating matrix can also form the electrode membrane (77). Such membranes can be highly selective electrodes regardless of whether they operate through the lattice-defect mechanism proposed by Frant and Ross (34) or simply as classical solubility-product electrodes of the second kind (e.g. like the Ag; AgCl electrode). The "site" in the lattice for a particular ion is specifically suited with respect to size, shape and charge distribution to favor the normal lattice species with extremely high selectivity. For example, even when mixed lattices are possible (as between AgCl and Ag_2S or between CaS and Ag_2S (86)) the energy differences between the different lattices are so large in the usual electrode sense (being at least tens of kilocalories per mole) that interferences due to foreign ions are negligible and lack of selectivity usually results from seconday chemical reactions (i.e. dissolution and precipitation) at the crystal surface. These reactions, since they involve more complex rearrangements than the simple exchange of counterions characteristic of the preceding three classes of electrodes, are sufficiently dissimilar to make it rather artificial to try to compare this class of electrode precisely with the other three. I will therefore confine this section to noting that solubility-product equilibria for the various possible surface lattices set important restrictions on the selectivity of such electrodes and that such equilibria share a certain formal similarity to the equilibria of the previous three classes of electrode mechanisms. In particular, the selectivity for a solid-state electrode depends on a favorable lattice energy for the preferred ion. This resembles the favorable ion-site interaction energies for ion-exchangers, as will be seen in Section V (cf. discussion of Eq. (23)). The selectivity must also depend on mobility, at least to the extent that an energetically less favored ion, which could exist as an isomorphorous lattice replacement, might conceivably dominate the electrode potential if its mobility were sufficiently larger than that of the energetically favored ion. This is the counterpart of the mobility contributions to the selectivity of a solid ion exchanger such as glass. Indeed, it is probably not too gross an oversimplification to assert that the chief difference between a "solid-state" crystal electrode and a solid ion exchanger is that the detected species is usually needed for insolubility of the lattice in a "solid-state" electrode whereas it is an exchangeable component in the ion-exchanger. Thus the physical existence of the crystal in the "solid-state" electrode depends upon the presence of the detected species (e.g. the F^- ion of the insoluble LaF_3 lattice) through solubility considerations; while the insolubility of the glass lattice is essentially independent of the counterion species, since the oxygen atoms are held in a 3-dimensional framework by the cations of higher valence (e.g. Si^{4+}, Al^{3+}). This distinction is not absolute; indeed, certain deviations of glass electrodes at high pH, most marked in the case of Cs^+ and Rb^+ ions (17), probably reflect a cation-dependent attack of OH^- on the silicate lattice. Lastly, the high selectivity characteristic of the detailed geometry of the site in a particular crystal lattice resembles the precise coordination and size requirements within the sequestration cavity of a neutral carrier molecule like valinomycin.

III COMPARISON OF THE FACTORS EXPECTED THEORETICALLY TO
DETERMINE THE SELECTIVITY FOR EACH OF THESE TYPES
OF MEMBRANES

The factors determining the relative ion selectivities for
the various classes of electrodes can be compared most easily
with the aid of Table II, which summarizes the parameters involved in the electrode selectivity of solid ion exchangers,
dissociated liquid ion exchangers, associated liquid ion exchangers and neutral carriers, respectively. (Unfortunately,
the theory for solid-state crystalline electrodes has not
developed to the point where they can meaningfully be included
in this table.) From the column labelled "Exact Representation" it should be clear that both mobility and equilibrium
factors determine the electrode potential selectivity for each
of these mechanisms. However, from the column labelled
"Approximate Representation" it can be seen that mobility terms
contribute only slightly to the selectivity of certain liquid
ion exchange membranes, and not at all to neutral carrier membranes. The physical reasons for this will become more clear
on examining each of these mechanisms in greater detail below,
while the elementary atomic basis for the free energies underlying the equilibrium ion selectivity factors of these electrodes will be examined in Section V.

A. Solid ion exchangers

Scrutiny of Eq. (5) indicates that the relative effects of
species I^+ vs. J^+ on the electrode potential of a given membrane depend on the relative mobilities of these species in the
membrane (u_j^*/u_i^*) as well as on their ion exchange equilibrium
constant (K_{ij}).   It has been observed (19,23), and is not unexpected, that the more strongly an ion is preferred by such an
ion exchanger, the more poorly does it move within the membrane

TABLE II

THE CONTRIBUTION TO ELECTRODE SELECTIVITY OF MOBILITIES (u^*), EQUILIBRIUM CONSTANTS (K), AND PARTITION COEFFICIENTS (k)

	Exact representation	Approximate representation	Note that:
SOLID ION EXCHANGERS	$\dfrac{K_j u_j^*}{K_i u_k^*}$	$\dfrac{K_j u_j^*}{K_i u_i^*}$	high affinity usually opposed by low mobility
LIQUID ION EXCHANGERS Dissociated	$\dfrac{k_j u_j^*}{k_i u_i^*}$	$\dfrac{k_j u_j^*}{k_i u_i^*}$	u^* and k depend only on solvent and are independent of exchanger species
Associated Poorly mobile sites	$\dfrac{k_j(u_j^*+u_s^*)}{k_i(u_i^*+u_s^*)}$	$\dfrac{k_j u_j^*}{k_i u_i^*}$	selectivity depends on mobility of ions and their partition coefficients
Highly mobile sites	$\dfrac{K_j u_{js}^*}{K_i u_{is}^*}$	$\dfrac{K_j}{K_i}$	selectivity depends on mobility of neutral complexes and the equilibrium selectivity constants
NEUTRAL CARRIERS	$\dfrac{K_j u_{js}^*}{K_i u_{is}^*}$	$\dfrac{K_j}{K_i}$	the mobility is independent of the species of complexed cation so that $u_{js+}^* = u_{is+}^*$

since the more difficult it is for it to jump from site to site.
For this reason, there are opposing effects between mobilities
and affinities of ions in solid ion exchangers, which set a
serious restriction on the selectivity of such systems.   The
relatively low selectivities of glass electrodes for divalent
cations appears to be a consequence of poor mobility of these
species (19, p.168; 102).

Part of this limitation is overcome when one allows the
ion exchange sites to move by making the ion exchanger a liquid
in which the (neutral) complex between the counterion and the
mobile-site can move; but an even greater improvement occurs
for the neutral carrier mechanism, in which the complex between
the counterion and carrier molecule bears the charge of the ion.

B. *Liquid on exchangers*

(1) Dissociated.   Examination of the row labelled "dissociated"
in Table II indicates that for a highly dissociated liquid ion
exchanger, the selectivity among various ions should be completely independent of the chemical properties of the lipophilic
ion exchanger molecule other than the sign of its charge. Thus,
a solvent such as nitrobenzene can be caused to become a cation-
selective electrode by adding to it a suitable lipophilic anion
such as oleic acid or, alternatively, can be made to be anion
selective by adding a suitable lipophilic cation such as dode-
cylamine (22).   Such a membrane can exhibit high ionic selectivities (cf. Fig.7 of (22)) presumably due to ion-solvent vs.
ion-water interactions; and, as expected from Eq. (10), this
selectivity is indeed independent of the lipophilic anion, at
least to the extent that it is the same for aliphatic carboxylic,
sulfonic, or phosphoric acids.

(2) Associated.   For the more commonly-encountered type of
liquid ion exchanger where the dielectric constant of the solvent
is too low for association between counterion and site to be

neglected, the selectivity is expected to be a function of both the solvent and the exchanger as has been illustrated in Table II for two special cases of associated exchangers, one with poorly mobile sites, the other with highly mobile sites.

In the first subcase (labelled "poorly mobile sites" in Table II) for sites which are poorly mobile compared to the counterions, τ becomes zero in Eq. (11). Since the mobility of the site species can be neglected in comparison to that of the counterions in this case $(u_{s-}^* << u_i^*, u_{j+}^*)$, the selectivity becomes approximately the same as that for the previously considered completely dissociated case. Again both u_j^*/u_i^* and k_j/k_i are independent of the chemical nature of the lipophilic exchanger since they reflect solely the interactions with the solvent of the dissociated counterions.

In the second subcase (labelled "highly mobile sites" in Table II) where the sites are highly mobile compared to the counterions $(\tau=1)$, the selectivity of the second term of Eq. (11) is expected to dominate. Since the mobilities of the associated ion pairs might be expected to be nearly independent of the counterion species,[†] it is not too farfetched to assume $(u_{js}^*/u_{is}^* \simeq 1)$ and to represent the approximate selectivity as due to ion-exchange affinities alone, as has been done in the table.

C. *Neutral carriers*

The last row in Table II summarizes the selectivity expectations for the neutral carrier mechanism. Experimental evidence has been obtained for the four macrotetralide actins (101) and for valinomycin (31) that the mobilities of the complexes

---

† This assumption should be excellent for a molecule like Nigericin, which can sequester cations in the center of a spherical lipophilic complex. It will be less good for an exchanger which screens the cation less effectively.

are essentially the same for all cations.   In this case the
mobility ratio (u_j^*/u_i^*) is unity and the selectivity is
governed solely by equilibrium parameters, as indicated in the
Approximate Representation of Table II, where only the ratio
K_j/K_i appears.

It is worth pointing out that, to the extent that the complexes have the same size and shape for all cations, the ratio
K_j/K_i can be measured directly and conveniently by two-phase
salt extraction equilibria (30).   The thermodynamic basis of
this procedure has been discussed in considerable detail elsewhere (31).

IV   DESCRIPTION OF THE QUANTITATIVE PATTERNS OF SELECTIVITY
CHARACTERISTIC OF EACH OF THE THREE CLASSES
OF ELECTRODES

It has been found possible and useful to summarize the selectivity observed among ions for the various types of electrode
systems in terms of quantitative patterns of selectivity of the
type first found to exist for glass electrodes (17) and later
shown to be applicable even to data for cell membranes (18,10)
and certain antibiotics (31).   Such patterns are expected
theoretically whenever one deals with an ion detecting system in
which a single parameter, such as the electrostatic field
strength, dipole moment, or coordination number of ligand groups
is varied in a continuous and systematic manner.   Such systematic patterns are also expected if other molecular parameters
underlying the energies of equilibrium interaction (such as size
and rigidity of sequestration cavity, or the particular spatial
array of ligands) are varied.

A *priori*, we should expect to observe a different quantitative selectivity pattern for each electrode class, and within
each class for every type of molecular detector.   Nevertheless,
we will see below and in Section V that certain aspects of selec-

tivity, such as the observed sequences of selectivity among
cations share certain features in common for all these systems.
Thus, the selectivity of all electrodes observed to date falls
on one of the follow-sequences (or the indicated minor variants
of these);

I	Cs > Rb > K > Na > Li	
II	Rb > Cs > K > Na > Li,	or IIa Cs > K > Rb > Na > Li
III	Rb > K > Cs > Na > Li,	or IIIa K > Cs > Rb > Na > Li
IV	K > Rb > Cs > Na > Li	
V	K > Rb > Na > Cs > Li	
VI	K > Na > Rb > Cs > Li	
VII	Na > K > Rb > Cs > Li,	or VIIa K > Na > Rb > Li > Cs
VIII	Na > K > Rb > Li > Cs	
IX	Na > K > Li > Rb > Cs	
X	Na > Li > K > Rb > Cs	
XI	Li > Na > K > Rb > Cs	

These sequences were initially deduced theoretically to apply to
monopolar ion exchanger sites and found experimentally to account
for the observed selectivities of glass electrode and ion exchangers (16,17,79). The theoretical considerations were later
extended to dipolar sites (25) in order to be applicable to neutral carriers (and, incidentally, the solvent phases of liquid
exchangers), for which this pattern was also found to prevail
over a rather wide variety of geometries.

Indeed, the class of systems in which such patterns have
been experimentally observed now includes such typical neutral
carrier molecules as the macrotetralide actins, where the variable has been the degree of methylation of the molecule (30),
valinomycin, and certain cyclic polyethers. The existence of
this common set of selectivity sequences is understandable as a
consequence of the general features underlying the equilibrium

selectivity common to all these electrodes: namely, the competition between ion-water vs. ion-site interactions, whose energetic dependence on cation size will be shown in Section V to be
sufficiently similar even between monopolar and dipolar sites
for this result to be expected.

Where they have been deduced, such patterns provide not
only a convenient way in which to represent the quantitative
aspects of selectivity for a given electrode system but also, as
I will try to show here, a useful way of comparing selectivities
from one electrode system to another. Thus, despite the qualitative similarities in sequence of selectivity observed,
important quantitative differences exist which, for example,
allow a much higher K^+ over Na^+ selectivity to be observed
with neutral carriers than glass electrodes. Interestingly,
the selectivities of glass electrodes and even collodion, will
be shown to be describable by essentially a single pattern,
which sets serious restrictions on the kinds of selectivities
that can be obtained with such electrodes. By contrast, a
quantitatively different pattern holds for the macrotetralide
actin series of molecules, and a still different one for valinomycin.

A. *Solid ion exchangers (e.g. glass electrodes)*
To define what we mean by an empirical selectivity pattern,
let us begin with the selectivity among certain monovalent
cations characteristic of glass electrodes. Figure 1 (which
reproduces Fig.15 of (18)) demonstrates the experimental evidenc
for the existence of a single quantitative selectivity pattern
of the cations Li^+, Na^+, K^+, Rb^+, Cs^+, and H^+ for alumin
silicate glass electrodes. The ordinate plots the experimental
observed membrane potential difference (in mV) for the various
cations relative to K^+ at pH 7 in 0.1 normal solutions.
(These values correspond to 58 mV times the common logarithm o

the electrode selectivity; a potential difference of 58 mV implies a selectivity of 10, a potential difference of 116 mV implies a selectivity of 100, etc..)

The effects of the individual cations relative to K^+ (the horizontal line on all subfigures) are indicated on each of the subfigures. Thus, the uppermost subfigure gives the selectivity of Li^+ relative to K^+; the next gives Na^+ relative to K^+; and so forth. All data points below the K^+ line represent electrodes for which the electrode is more selective to the indicated cation than to K^+ (and *vice versa*). Data points at the same left to right location along the abscissa of all subfigures represent observations for a given electrode composition, whose left to right position has been fixed empirically by its observed Na^+ to K^+ selectivity, except at the extreme right where the Cs^+ to K^+ selectivity has been used.[†] Once this left to right position is fixed for a given composition, the observed selectivity for each of the other cations is located on the ordinate of each subfigure as an experimentally determined data point. Data for certain non-Na_2O-Al_2O_3-SiO_2 glasses are also presented in Figs 2 and 3.

Examining the pattern which emerges from these data points it becomes apparent that the selectivity for each of the alkali cations can be reasonably well described by a characteristic curve or selectivity isotherm, drawn on the figure as a visual average to the experimental points. (In examining Fig.1, the reader should note that the scatter in the data does not represent experimental error in measuring the individual selectivity values; for the experimental error due to this is no more than

---

† It goes beyond the scope of the present section to discuss the fact that the Na^+ isotherm is not simply a straight line but has a maximum at the right. In this case the data have had to be scaled by the extrapolated Cs^+ isotherm. This is discussed elsewhere (17).

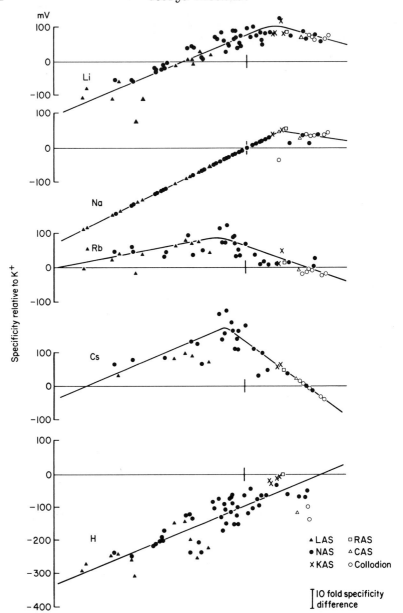

Fig.1 : Selectivities of aluminosilicate glass electrodes (after (18), Fig.15). Described in the text. The constituents making up the various electrodes are indicated by the symbols at the lower right-hand corner of the figure.

A ten-fold selectivity difference relative to K^+ corresponds to 58 mV on the ordinate. The solid curves trace the selectivities characteristic of sodium aluminosilicate glasses after Eisenman (17), Fig.8A. Note that the selectivities for the non-Na_2O-containing glasses, as well as collodion, fall within the range of observations for the Na_2-O-containing glasses; so that their selectivity can also be characterized by the isotherms characteristic of sodium aluminosilicate glasses.

twice the diameter of the plotted points. Rather, the scatter represents the effects of pooling data from widely differing chemical compositions.) It was in this manner that the first set of isotherms was constructed for aluminosilicate glass electrodes (17).

The finding of a single pattern, in which essentially the same isotherms held for electrodes of widely differing chemical composition (e.g. the Na_2O of the glass is replaced by Li_2O, K_2O, Rb_2O, or Cs_2O, as indicated by the designations "LAS", "NAS", "KAS", "RAS" and "CAS"), was surprising and a bit disappointing since these composition changes were made in the hope of producing radically different selectivity differences among the cations, since such changes in composition were expected to change not only the particular site field strength (due to electronegativity and size differences among these cations and consequent changes in lattice structure and screening) but also because they were expected to alter the degree of composition and hydration of the lattice and the inter-site spacing, and therefore the mobility of any given ion within the glass, as well as the magnitude of selectivity for a given site field strength.[†] Indeed, these replacements did lead to electrodes

---

[†] The effects of hydration are discussed elsewhere on both equilibrium selectivity ((17), cf. p.300) and on mobility of ions ((19), pp.170-171). The reader should be reminded that increasing hydration should decrease selectivity magnitude with-

George Eisenman

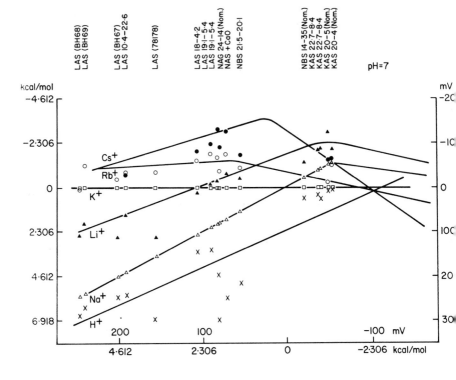

Fig.2 : Selectivities of non-sodium aluminosilicate glasses
(after (17), Fig.8). This figure is plotted in the
same manner as Fig.1; and the curves superpose the
"isotherms" for sodium aluminosilicate glasses drawn on
Fig.1 for comparison with the data points for non-
sodium aluminosilicate glasses. Each vertical row of
data points corresponds to selectivity data for the designated glass composition ("LAS" stands for $Li_2O-Al_2O_3-$
SiO_2 glasses; "KAS" for $K_2O-Al_2O_3-SiO_2$ glasses; "NAG"
for $Na_2O-Al_2O_3-Ge_2O$ glasses; "NBS" for $Na_2O-B_2O_3-SiO_2$
glasses; "NAS + CaO" for $Na_2O-Al_2O_3-SiO_2$ glasses.
Otherwise the nomenclature follows that of Eisenman.

out altering the sequence of selectivity and should also increase ionic mobility drastically, and differently, for different ions.

(17). Notice that the data for the glasses designated "BH68, etc." correspond to those for the EIL series of sodium electrodes; whereas the data designated "78178" are for the Beckman sodium electrode.

of greatly differing selectivities for a given composition (17, 18) but unfortunately (from the point of view of finding new electrodes) did not radically change the selectivity pattern. They merely shift the left to right position in the figure. The isotherms of Fig.1 even held for air-dried collodion membranes ((17), Fig.8A), as indicated by the data points labelled "collodion" in Figs 1 and 2.

Although there is a certain amount of scatter (due, no doubt, to some small influence of the composition-dependent factors discussed above, e.g. degree of hydration), the most striking observation of Fig.2 is that these data are surprisingly well described by the same selectivity isotherms as for the sodium aluminosilicate glasses.

In order to produce an even more drastic change in glass composition and thus, hopefully, to alter the selectivity pattern more radically in the glass electrode system, the aluminum in these glasses was replaced by boron, gallium, indium and antimony, as symbolized by B, G, I, and Sb, respectively in Fig.3 (Eisenman and Ross, unpublished data). The magnitude of the selectivity of a given glass is again ordered from left to right by its Na to K selectivity ratio, and the lines in this figure are the isotherms from Fig.1, but here superposed as in Fig.2 to make easier comparison of selectivity among the cations. Although one might have expected the magnitudes of the selectivity patterns of these glasses to be grossly different from those for the aluminosilicates, in fact one only begins to see significantly different selectivities in the boron and antimony containing glasses (the somewhat compressed selectivities of these glasses perhaps indicating a higher hydration state), and the

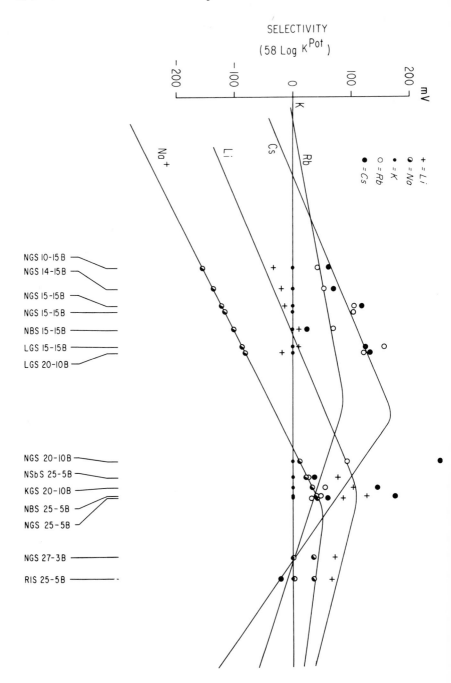

Fig.3 : Selectivity of non-aluminosilicate glasses. The aluminum in these glasses has been replaced by other triply charged ions: Boron (B), Gallium (G), Indium (I), and Antimony (Sb); and the selectivity data (from unpublished observations of Eisenman and Ross) corresponding to each of these compositions has been plotted by the observed data points. For comparison the curves trace the selectivity isotherms characteristic of aluminosilicate glasses from Fig.1. In all cases the glass composition is given as the batch composition, which should correspond rather closely to the final composition. Notice that the most serious deviations from the aluminosilicate pattern occur with the borosilicate glasses; whereas the best correspondence is seen for the gallium and indium glasses.

data points for the gallium and indium glasses correspond surprisingly closely to the average selectivity isotherms for the aluminosilicates.

From this it appears that the selectivity pattern established for aluminosilicate glasses pretty well defines the selectivity of all presently known (or likely to be makeable) silicate glasses. Since it has previously been established that this pattern also holds when Silicon is replaced by Germanium (17), and we have already noted that it holds for carboxylate anions in the nitrocellulose lattice of dried collodion, it can be concluded tentatively that this selectivity pattern is likely to hold for all condensed oxyanion systems.[†] Thus a single selectivity pattern is apparently characteristic of the molecular organization of all condensed oxyanion lattices of the glass type. This implies both a useful simplification and a

---

[†] Incidentally, the selectivity characteristics for certain commercially available Na+ selective electrodes are illustrated in Fig.2 by the data points corresponding to the E.I.L. BH67, BH68 and BH69 and the Beckman 78178 electrode. These are lithium aluminosilicate glasses labelled, "LAS (BH67), etc." and "LAS (78178)" on this figure.

serious restriction on the selectivities that can be obtained
in such glassy systems, for which we now seem to have attained
all the selectivity which is possible for the various ions of
interest.   It therefore appears that if one wishes to trans-
cend this pattern radically within a solid ion exchanger electrode, it will be necessary to change the nearest neighbor to
the cation from oxygen, for example, to sulfur.   By contrast,
the pattern is easily transcended by changing the type of electrode to a neutral carrier and varying the detailed structure
of the molecular detector, as will become clear for the selectivities of valinomycin and the macrotetralide actin to be discussed below.   One can also alter the pattern by varying the
solvent and site of a liquid ion exchanger, as will also be
noted for nitrobenzene shortly, or by finding an appropriate
crystal lattice which is also insoluble.

   For a single pattern to apply so generally to condensed
glassy systems is not expected *a priori*, and its finding deserves brief discussion here.   We have noted that electrode
selectivity depends on such parameters as mobility and ion-
exchange selectivity, each of which might be expected to vary
systematically as a function of composition for a given chemical system, but might be expected to differ for significantly
differing systems in the way that the precise pattern for valinomycin will be shown to differ from that for the Nonactin series
of antibiotics.   While the general selectivity features of
diverse systems might be expected to resemble each other qualitatively (e.g. in the sequences of selectivity) from general
theoretical considerations (17), as is discussed in Section V,
there is no *a priori* reason why the magnitudes of selectivity
for a given sequence should be the same from one system to
another.   Indeed, it has been shown elsewhere that differences
in hydration state are expected to alter radically the magnitude
of selectivity expected for a given type of site ((17), pp.294,

00). That a single pattern is indeed observed suggests in
erms of a theory of selectivity developed elsewhere (16,17)
hat all condensed lattices based on oxygen atoms share a com-
ion density of packing and hydration state for a given field
trength of the sites and that the principal effect of varying
he lattice-forming and modifying ions is to set the field
trength of the ion exchange sites of the glass. The field
trength of these sites, in turn, determines the hydration
tate of the glass and those properties dependent on this state
e.g. (i) ionic mobility and (ii) magnitude of selectivity for
given sequence). The field strength thus appears to be the
rimary independent selectivity variable. With hindsight, this
an be understood since all that a cation (or a water molecule)
ees in a glass is the external field of the oxygens to which
hey can approach closely as modified by the lattice forming the
toms (e.g. Al^{3+}, Ga^{3+}) which are once removed and whose effects
re only felt as modifications of the effective field strength
f the oxygens   Indeed, this situation has a precedent in the
imple rules governing the pK of oxyanions pointed out by
ossiokoff and Harker (53) and Ricci (80).

Thus, the large changes in electrostatic field strength of
he aluminosilicate sites in the glass produced by replacing one
lkali oxide by another, or by replacing aluminum by other triply
harged cations, are accompanied by comparable hydration states
f the glass matrix, so that all glass compositions having sites
f the same "field strength" must also have comparable degrees
f hydration, regardless of the differing chemical compositions
roducing the same field strength in each case (e.g. the selec-
ivity of the Na_2O-containing glass, NAS 10-10 $((Na_2O)_{10}$
$Al_2O_3)_{10}$ $(SiO_2)_{80})$ would correspond to that of the Li_2O-contain-
ng glass, LAS_{25-40} $((Li_2O)_{25}$ $(Al_2O_3)_{40}$ $(SiO_2)_{65}))$ which re-
uires 2.5 times as much Li_2O for the same effective site
field strength" (17,18). This somewhat surprising conclusion

is reached because the degree of hydration is not only expected
to affect the magnitude of ion exchange selectivities ((17),
p.300) but also because the mobility ratios of ions in glass
are also quite sensitive to the degree of hydration ((19), Ch.5)
Therefore if a given site field strength were not also accompanied by a uniquely defined hydration state, a different quantitative selectivity pattern would be expected for Li_2O-based
as opposed to Na_2O-based glasses, and so forth.

The entire glassy system then is one example of a system
in which the spatial arrangements of the nearest neighbors (i.e.
oxygen atoms and water molecules) to the species being detected
are similar, regardless of detailed composition differences.
A comparable situation will again be encountered within a homolo
gous series of cyclic antibiotics (e.g. the macrotetralide
actins) having a given array of ligand oxygens; but it will be
seen that a different pattern will result from a relatively
small change in arrangement of the coordination cage (e.g. in
valinomycin). It is this ability to tailor the relevant detai
of ligand orientation so as to alter ionic selectivity which
makes this latter class of electrodes so potentially useful.

Comparable selectivity patterns exist for Ag^+, Tl^+ and
NH_4^+, as well as for divalent cations. The isotherms for the
important monovalent cations Ag^+, Tl^+ and NH_4^+ are presented i
Fig.4, while those for the alkaline earth cations and certain
amines and amino acids will be found elsewhere ((18), Figs 37-3

Considering together the selectivity patterns of Figs 1 an
4, it can be concluded that the cations H^+, Ag^+, Na^+ and Li
are the most favorable for being selectively measured by glass
electrodes. However, reasonably useful selectivities also
exist for K^+, Rb^+, Cs^+ and NH_4^+. At present, glass electrodes are the electrodes of choice for H^+, Na^+, Ag^+, Li^+
and sometimes K^+ and NH_4^+, for which their specificity is
sufficiently high and their excellent stability characteristics

are desirable.   In addition, since these electrodes function well in organic solvents and are unaffected by organic cations they can be used in non-aqueous media, as well as in the presence of lipid-soluble or surface-active molecules.

It appears not to have been generally appreciated that one can use glass electrodes for any given cation, provided only that the competing cations are present at sufficiently low levels to be unimportant.   What these levels are for a given electrode composition is ascertainable from Fig.1.   The properties of commercially available electrodes can also be easily indicated through Fig.2, and I have labelled certain of these on this figure for the reader's convenience.   Note that the Corning, Na^+ selective electrode No.476710 is an NAS_{11-18} sodium aluminosilicate glass having the composition recommended by Eisenman *et al.* (26), while the Beckman (78178) and E.I.L. Na^+ selective electrodes are lithium aluminosilicates also described by Eisenman *et al.* (26) and labelled "LAS (78178)", and "LAS (BH67, BH68, BH69)" on Fig.2.   The K^+ selective glass electrodes of Corning (476132) and of Beckman (78137 type) are both NAS_{27-4} sodium aluminosilicates as described by Eisenman *et al.* (26).

B. *Liquid ion exchangers*

The most systematic compilation for this system has been given by Sandblom and Orme (89) and by Ross (84); and, as indicated in Eq. (13), such a large variety of parameters can contribute to the selectivity of liquid ion exchangers that it is not surprising that the selectivity of such electrodes cannot be systematized as easily as the above solid ion exchangers.   Therefore, no attempt at deducing general selectivity patterns will be made here.   However, it is worth noting that the basic competition between hydration energies and the energies of interaction with the solvent contributes importantly to the selectivity of such electrodes.   Thus, the ratio of partition coefficient k_j/k_i

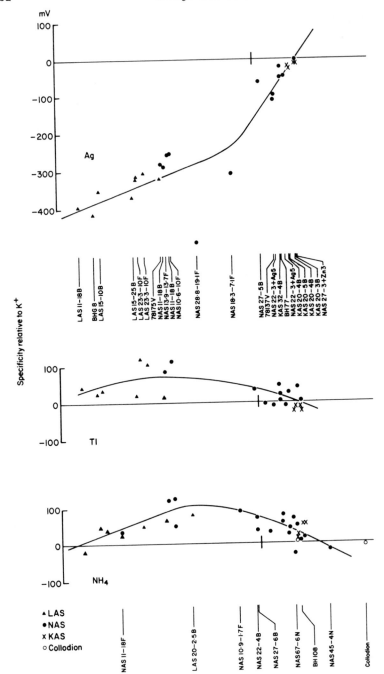

Fig.4 : Selectivity patterns for Ag^+, Tl^+ and NH_4^+ (after (18), Figs 26, 30 and 34). The scales and procedure of ranking are identical to those of Fig.1; and the various glass compositions are indicated explicitly along the abscissa. All selectivities are referred to K^+ (the horizontal line) as in the preceding figures.

for the dissociated ions in a given membrane solvent, which contributes to the selectivities of liquid exchangers (recall Table II), reflects this competition through the relationship, deducible from Eq. (11):

$$RT \, ln \, (k_j/k_i) = (\bar{F}_j^{hyd} - \bar{F}_i^{hyd}) - (\bar{F}_j^{solv} - \bar{F}_i^{solv}) \qquad (18)$$

This illustrates that this selectivity term depends on the differences from ion to ion in their free energies of solvation $(\bar{F}_j^{solv} - \bar{F}_i^{solv})$ vs. their free energies of hydration $(\bar{F}_j^{hyd} - \bar{F}_i^{hyd})$. Recognizing this, the lack of selectivity observed among monovalent cations in liquid ion exchangers where the solvent is an aliphatic alcohol ((22), Fig.5), can be readily understood as expected from the comparable values from cation to cation of the differences of free energies of alcoholation vs. hydration (recall the similarities between the HOH and ROH ligands). By contrast, the high selectivities observed between monovalent cations in nitrobenzene ((22), Fig.7) can be understood as due to the expected differences between solvation energies from cation to cation in this solvent (which is quite dissimilar as a ligand from H_2O) and the hydration energies from cation to cation. The selectivity pattern characteristics of this solvent is compared on Fig.6 with the isotherms for glass electrodes, and the data for a cyclic polyether having the same sequence (I) of ion selectivity. The factors determining the selectivity of the ion-site interactions resemble those for ion-site interactions in solid ion exchangers as well as for neutral carriers.

C. *Neutral carrier molecules*

(1) The macrotetralide actins. A characteristic pattern of selectivity exists for series of the macrotetralide actins which depends in a systematic manner on the increasing methylation of these molecules from non-, to mon-, to din- to trinactin (30,31, 101). This pattern is shown in Fig.5 where the data points have been ranked (using the K^+/Cs^+ selectivity as a common isotherm) as was done for glass electrodes previously; and the selectivities for these molecules can be compared easily with those of glass electrodes traced as fine lines from Fig.1. Notice that the data points for the macrotetralide actins are substantially different in their magnitude from those of glass electrodes. For example their K^+ selectivity relative to Na^+ and Li^+ is much greater and they have a very high NH_4^+ selectivity for reasons which will be discussed later.

Also notice that the selectivity between cations is a function of methylation, the most highly methylated molecule (trinactin) being the least discriminating of this series between K^+ and the smaller cations, Na^+ and Li^+, as would be expected from Section V for its expected higher dipole moment (due to the electron donating effect of the methyl groups). It should be possible to alter radically the selectivity characteristics of these molecules by substituting more powerful electron withdrawing (e.g. Cl) or additional donating groups.

At present the macrotetralide actins are the most selective detectors known for NH_4^+ and also provide excellent K^+ selectivity. Practical electrodes based on these selectivities have indeed been described by Simon (95) and by Garfinkel (38).

(2) Valinomycin. The selectivity pattern characteristic of valinomycin-based electrodes is illustrated in Fig.6. As in Fig.5, these data points can be compared directly with the selectivity isotherms characteristic of glass, traced as fine lines, as well as with the selectivity for the macrotetralides in the

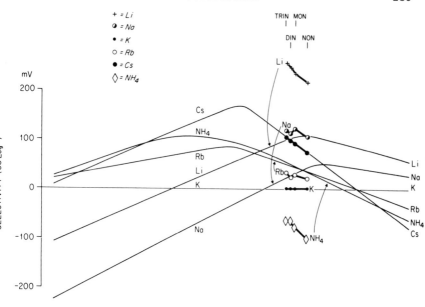

Fig.5 : Comparison of the glass electrode selectivity pattern
with that of the macrotetralide actin antibiotics. The
fine lines are the glass isotherms of Fig.1 (and also
include the ammonium isotherm from Fig.4). The data
points represent the experimentally observed selectivities for the macrotetralide actins: nonactin,
monactin, dinactin and trinaction, each of which contains one more methyl group than the preceding number.
The selectivities are ranked according to the K/Cs selectivity, a procedure necessitated for the macrotetralide actins since their K^+/Na^+ selectivity far exceeds
anything observable in glass. The selectivity data
from salt extraction measurements (25) have been used
rather than those from bilayer membranes since the former have been determined with greater accuracy. The
fine arrows indicate the extent to which the selectivity
characteristic of the macrotetralide actins differs from
that characteristic for glass electrodes.

preceding figure. The selectivity pattern for valinomycin is
different in its details both from that of the macrotetralides
(as can be seen by comparison with Fig.5) and from that of glass
(as can be seen by comparison with the fine lines). This no
doubt reflects the difference in orientation and spacing of the

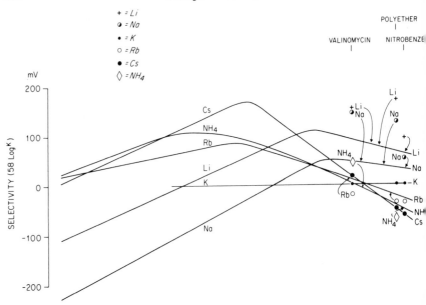

Fig.6 : Comparison of the glass electrode selectivity pattern
with that of: valinomycin, the cyclic polyether
(XXXII), and nitrobenzene.   Three sets of data points
are presented, summarizing from left to right the selectivity pattern characteristic of valinomycin, the
cyclic polyether, and the solvent nitrobenzene, respectively.   All data have been ranked by the K^+/Cs^+ selectivities as in Fig.5.   Differences in patterns are
revealed by deviations of the data for the other cations
as compared to the glass isotherms.   For valinomycin
the pattern is seen to be very different from that of
glass, as indicated by the fine lines connecting the
data points for valinomycin with the isotherms for glass
For the cyclic polyether a similar discrepancy is also
seen, although here the differences are predominantly in
magnitude rather than sequence.   For nitrobenzene the
selectivity deviations are the least and appear to be
chiefly differences in the magnitude of selectivity of
the Li and Na relative to K.   (This is not as surprisin
as it might seem, since the further to the right one is
on this selectivity pattern, the more the selectivity is
expected to reflect predominantly the work of dehydratin
the cation and the less the particular interactions with
the solvent or site.)   The data for nitrobenzene (after
(19), Fig.7) were obtained in the presence of 5% oleic
acid.   The data for the polyether (XXXII) are from
potential selectivity data of Table IV of (31).   The dat

for valinomycin are those of (68) (the comparable data of Lev (cf. Table I) are not presented here in order not to clutter the figure), together with the NH_4 salt extraction data of Table I.

ligand groups in the interior of these two molecules. The most notable feature of the valinomycin selectivity pattern is the extremely favorable selectivity for K^+ over Na^+ (and Li^+), a characteristic which makes this the most attractive molecule for applications in which extremely high K^+ selectivities are needed. Notice also that the selectivity for NH_4^+ is relatively low compared to that of the macrotetralides so that valinomycin is a much more favorable molecule for measuring K^+ in the presence of NH_4^+ than is any macrotetralide.

(3) <u>Cyclic polyethers</u>. Selectivity data for the cyclic polyether XXXII (31,66) are also plotted in Fig.6, for comparison with the glass isotherms and the selectivity for the solvent nitrobenzene. It can be seen that this molecule offers a selectivity sequence Cs > Rb > K > NH_4 > Li. Indeed, the Cs^+ selectivity is substantial. However, it should be noted that because the stoichiometry of the complex depends on concentration, reflecting a series of successive equilibria (31,66), the selectivity sequence for this typical polyether also should depend on its concentration in the electrode in a complicated manner. Thus it is not surprising that at lower polyether concentrations other selectivity sequences can be observed (29,95, 78), but these cannot be usefully discussed here.

(4) <u>Conclusions</u>. Despite the large differences seen in the quantitative selectivity patterns for each of these types of molecules, it is clear that the qualitative aspects of selectivity are comparable at least as far as sequences are concerned. (The macrotetralides are characterized by selectivity sequence IV, valinomycin by sequence III, the polyether by sequence I, the solvent nitrobenzene by sequence I.) The quantitative dif-

ferences from one molecular structure to another reflect the consequences of detailed differences in types (e.g. ester vs. ether) and in the spatial array of the ligand oxygens, as well as the configurational deformability of the molecules. The similarities among the diverse types of molecular structures reflect the underlying feature of ion-water vs. ion-dipole interactions common to all.

This class of molecules lends itself particularly well to "tailoring" molecules so as to take advantage of ionic "shape" as well as size and charge through the particular spatial arrangement of multidentate ligand atoms around the cation. For example, an unusually high ammonium selectivity can be attained through tetrahedrally arrayed ligand oxygens to provide favorable hydrogen-bonding interactions with the ammonium ion.

D. *"Solid-state" crystal electrodes*

A summary of the ions to which solid-state crystal electrodes respond is given in Table III after Ross (84). As mentioned above, the selectivity here depends on a favorable lattice energy for the preferred ion and, apparently less importantly, on mobility in the lattice, thus resembling, at least formally, the situation in solid ion exchangers such as glass. However, whereas for the crystal electrodes, the insolubility of the lattice depends, through solubility-product considerations, upon the species being detected (e.g. the Cl^- ion of the insoluble AgCl crystal), the insolubility of the ion exchanger lattice is essentially independent of the counterion species. The selectivity patterns for "solid-state" crystals can therefore be inferred from considerations of the solubilities of the crystals, often reflecting directly the differences of lattice energies vs. hydration energies, as these enter into the ratios of the solubility product constants of the competing

TABLE III   (after Ross (84))

SOLID-STATE MEMBRANE ELECTRODES

Ion determined	Membrane	Principal interferences
F^-, La^{+++}	LaF_3	OH^-
Cl^-	$AgCl/Ag_2S$	Br^-, I^-, $S^=$, NH_3, CN^-
Br^-	$AgBr/Ag_2S$	I^-, $S^=$, NH_3, CN^-
I^-	AgI/Ag_2S	$S^=$, CN^-
SCN^-	$AgSCN/Ag_2S$	Br^-, I^-, $S^=$, NH_3, CN^-
$S^=$, Ag^+	Ag_2S	Hg^{++}
CN^-	AgI/Ag_2S	I^-, $S^=$
Cu^{++}	CuS/Ag_2S	Hg^{++}, Ag^+
Pb^{++}	PbS/Ag_2S	Hg^{++}, Ag^+, Cu^{++}
Cd^{++}	CdS/Ag_2S	Hg^{++}, Ag^+, Cu^{++}

ions (e.g. the greater selectivity of Ag electrodes to I^- than Cl^- reflects the lesser solubility of the AgI crystal). Indeed, I have observed that the potential differences of metallic silver electrodes in equimolar solutions of NaCl, NaBr, and NaI very closely reflect those expected from the ratio of solubility product constants.

E.   Concluding remarks on selectivity patterns

Two principal conclusions can be reached from the above data.   First, in all of the above ion selective systems, a single qualitative selectivity pattern (described in detail in the next section) suffices to account for the sequences of cation selectivity observed to date.   All sequences being one of

the eleven deduced elsewhere (16) and listed on p.269.   The
reasons for this will be explained in Section V but may be usefully anticipated here by stating that all systems share the
common features of ion-site vs. ion-water interactions as central to their selectivity.   It will be shown in Section V that,
whether the sites are monopolar as in ion exchangers or dipolar
as in neutral ligands, there are significant ranges of conditions in which the same selectivity patterns will be expected
to occur.   These reflect the fundamental asymmetry in the interactions of ions with water, which are highly multipolar in
origin, and ions with sites which are less so.

Second, despite the existence of a single set of selectivity sequences for all of these systems, each system differs
in important quantitative details from the others.   These differences, not surprisingly, are characteristic of the detailed
molecular arrangements in each type of detecting system.   From
this point of view, all glassy systems may be viewed as being
of one type since the differences in molecular arrangements are
insufficiently different from each other to lead to important
differences in selectivity pattern.   By contrast, each type of
carrier molecule having a significantly different geometrical
structure is a prototype for a series of selectivities comparable, in principle, to glass.   It is in this regard that we
have just begun to scratch the surface of the types of selectivities to be expected from such molecules since each molecular
type can form the basis for systematic variation of dipole
moments (e.g. through the introduction of electron donating or
withdrawing groups) and other molecular variables.

V   MOLECULAR BASIS FOR THE EQUILIBRIUM CONTRIBUTIONS TO ELECTRODE SELECTIVITY

This section will demonstrate that the same competition between the free energy of an ion in water vs. the free energy of an ion in the membrane underlies the selectivity to be expected for ion exchangers and neutral lipophilic ion-binding molecules. The differences in selectivity patterns between these systems reflect, on the one hand, the differences in ion migration mechanisms (as outlined in Table II). The differences also result from differences in the numbers, types and detailed orientations of the ligand groups in these systems.

From Table II it is seen that for a solid ion exchanger an equilibrium contribution to the selectivity appears through K_j/K_i, the ion exchanger equilibrium constant. For a liquid ion exchanger, the equilibrium parameters are both K_j/K_i and k_j/k_i, reflecting both the ion exchanger properties and those of the solvent itself. For the neutral carriers, as indicated in Table II, the selectivity reflects solely equilibrium properties, particularly when the configuration of the complexes are essentially the same around all cations. Lastly, for the crystalline electrodes, the equilibrium lattice energies are crucial to the selectivity. In view of the importance of these equilibrium factors for all classes of electrodes, it is worthwhile analyzing the molecular basis of equilibrium selectivity. This is now done, first for the case of negatively charged (monopolar) ion exchange sites and then for neutral dipolar ligands. Although the subject matter of this section has been presented elsewhere (25,31), a brief discussion is given here to make this chapter self-contained.

The equilibrium selectivity of a given reaction (e.g. a typical ion-exchange reaction such as (6)) is directly related to the free-energy change of the reaction through

$$\Delta F_{ij} = -RT \, ln \, K_{ij} \, , \qquad (19)$$

where for neutral carriers it will be recalled that K_{ij} is equal to K_j/K_i as defined by the exchange reaction deduced by subtracting Eq. (17b) from Eq. (17a).

This free energy change can be expressed as the difference between ion-water interactions and ion-membrane interactions by

$$\Delta F_{ij} = (\bar{F}_{i+}^{hyd} - \bar{F}_{j+}^{hyd}) + (\bar{F}_{j+}^{*} - \bar{F}_{i+}^{*}) \qquad (20)$$

where $(\bar{F}_{i+}^{hyd} - \bar{F}_{j+}^{hyd})$ represents the difference in partial molal free energies of hydration of the ions I^+ and J^+, and $(\bar{F}_{j+}^{*} - \bar{F}_{i+}^{*})$ represents the difference of their free energies in the membrane phase. The hydration energy differences are accurately known experimentally for each pair of cations so can be taken as empirical constants whenever we are dealing with aqueous solutions (although, of course, the theory of these energies is of considerable interest *per se*, and is useful in understanding why certain patterns arise). By contrast, the free energies of interaction with the membrane phase must be inferred theoretically. The problem of analyzing the origin of equilibrium specificity thus simplifies to the problem of characterizing the affinities of the various cations for the appropriate set of ligands as a function of molecular structure (i.e. the monopole, dipole, and higher terms for each ligand group, their detailed spatial arrangement, and energies and entropies of molecular rearrangement and strain).

A. *Ion exchangers*

Needless to say, the rigorous calculation of the energies of cation interaction with molecular anionic groups for a solid ion exchanger such as glass, or for a liquid ion exchanger molecule dissolved in a given solvent, has yet to be accomplished; although an approximate monopolar representation of the multipolar silicate and aluminosilicate ion exchange sites of minerals and glasses has successfully accounted for their principal selectivity properties ((17), pp.288-292). I will deal only with the monopolar model here.

Consider first the simplest case in which we compare the selectivity expected for the competition for a given cation between such a site and a single water molecule as outlined diagrammatically in Fig.7 (after (16)). The energies of monopolar interaction (referred to the ions at rest in a vacuum) of a monovalent cation I^+ of radius r_+ with a singly charged anionic site of radius r_- are given by Coulomb's law as:

$$\bar{F}^*_{i+} = \frac{-332}{r_+ + r_-} \tag{21}$$

While its energy of interaction with a single water molecule represented by the tripolar model of Fig.7 is given by:

$$\bar{F}^{H_2O}_{i+} = -332 \left(\frac{q^-}{r_4} - \frac{2q^+}{r_3} \right) \tag{22}$$

representing the net attractions between the cation and the effective charge of the oxygen ($q^- = -0.64$) at a distance r_4 from the center of the cation and the repulsions between the cation and the effective charge ($q^+ = +0.32$) of the two protons at the greater distance r_3 from the center of the cation. The energies are in kcal/mol for distances in angstroms, and charges are expressed as fractions of the electronic charge.

Calculating the values of these energies for the various naked (Goldschmitt) radii of the cations and inserting these in Eq. (20), one obtains the selectivity isotherms, referred to Cs^+,

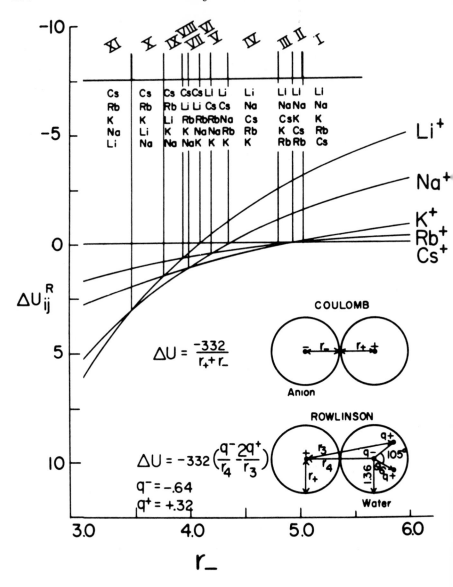

Fig.7 : Selectivity pattern for the hypothetical cation exchange
between a monopolar anion and a single multipolar water
molecule (after (16), Fig.1).   The cation most strongly
selected from water by the anionic "site" is the lowest
on the chart.   Above the graph are tabulated the cationic
sequences (increasing selectivity downwards), defining

eleven rank orders as a function of decreasing site
radius r_-.   Units in kcal/mol and angstroms.

plotted on the ordinate in Fig.7.   This simple procedure leads
to a pattern of selectivity in which, for an anion site of large
largest radius (i.e., lowest electrostatic field strength), the
cations are preferred in the sequence Cs > Rb > K > Na > Li;
while for an anionic site of sufficiently small radius the sequence of preference is reversed, being Li > Na > K > Rb > Cs.
Between these extremes, the cations are seen to pass through
eleven selectivity sequences, corresponding to the 11 Roman
numerals designating the regions between the intersections of
the selectivity isotherms of Fig.7, and corresponding to the sequences previously described on p.269.   Comparing these results
with the data of Fig.2, it is clear that this simple model is
sufficient to account for the observed selectivities of glass
electrodes.

   Turning now to more realistic considerations of selectivity
expected for a monopolar ion exchange site vs. a true aqueous
solution, using the (experimentally known) differences of free
energies of hydration[†] of the cations in water $(\bar{F}^{hyd}_{i+} - \bar{F}^{hyd}_{j+})$
instead of the above calculation.[††]

   If the sites of the exchanger are assumed in one limit to
be widely separated and the water molecules are assumed to be
excluded from their vicinity, then the free energies of inter-
action between cation and site $(\bar{F}^{*}_{j+} - \bar{F}^{*}_{i})$ needed to solve
Eq. (20) are given in first approximation by the energies as cal-

---

[†] Values for these free energy differences referred to CS^+ according to Latimer (54), are: 53.8 ± 1.4 kcal/mol for Li^+, $28.9 \pm$
0.9 for Na^+, 12.7 ± 0.9 for K^+, and 6.7 ± 1.1 for Rb^+.

[††] Interestingly, the differences of hydration energies correspond
closely to 4.8 times the interaction energies with the model
water molecule of Fig.7 (cf. (16), Fig.5C).

culated by Coulomb's law in Eq. (21).   The selectivity expected
in this case is plotted in Fig.8.

On the other hand, if the ion exchanger sites are assumed
to be very closely spaced (for example, with 6 sites coordinated
around each cation and 6 cations around each site), then the free
energies will be given by

$$\bar{F}^*_{i+} = 1.56 \ \frac{-332}{r_+ + r_-} \tag{23}$$

where the factor 1.56 appears as a consequence of the Madelung
constant (1.75) for this coordination state and a factor of
about 8/9 due to the Born repulsion energy.   The expected selectivities in this case are plotted in Fig.9.

Inspection of Figs 8 and 9 yields the following conclusions.
First, the selectivity among cations in each of these states depends upon the radius of the anion r_- (i.e., upon anionic field
strength).   Second, a particular pattern is seen for the selectivity among group Ia cations in that essentially only 11 sequences of cation effectiveness are predicted out of a possible
120.   These sequences are indicated by the Roman numerals I to
XI above Figs 7-9, with only the minor variations indicated by
subscript "a".   This model has recently been critically reviewed for ion exchangers by Reichenberg (79), for zeolites by
Sherry (94), and for biological membranes by Diamond and Wright
(10).   Since Eq. (23) will be recognized as the classical Born-
Landé equation for the energy of an alkali halide crystal lattice, a general similarity between the present considerations
and those that would apply to the lattice energies underlying the
solubility and selectivity of crystal electrodes should be apparent.

(1)   Dipolar ligand groups.   An initial understanding of the
elementary factors underlying the specificity of neutral molecules which bind cations can be gained by extending the above

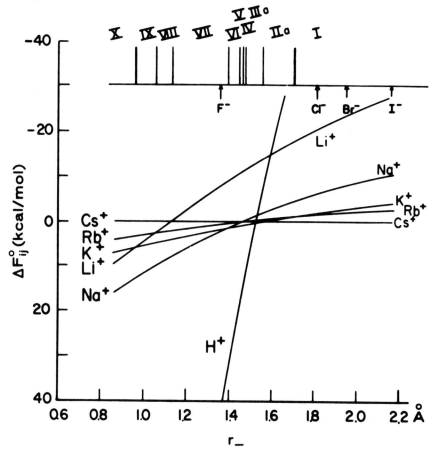

Fig.8 : Selectivity isotherms for widely separated monopolar
ion exchange sites (after (17), Fig.16). Negative
values of free energy change are plotted as a function
of r_- for the case of widely separated sites. The
Roman numerals above the figure indicate the sequences
of ionic selectivity pertaining in the regions separated
by the vertical lines drawn to the intersections of the
various cation isotherms. The more strongly an ion is
preferred the lower its position on the figure.

reasoning to include the consequences for ion binding of the interactions with the essentially dipolar carbonyl and ether
ligands in typical neutral carrier molecules. Theoretical analyses have been carried out for dipoles of finite size (25,31) and

George Eisenman

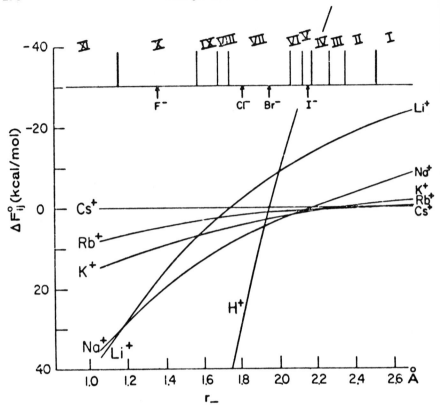

Fig.9 : Selectivity isotherms for closely spaced monopolar sites
(after (17), Fig.17).   Plotted in the same manner as
Fig.8.

point dipoles (67).   The simple relationship between a dipolar
ligand and the previously considered monopolar ligand can be
clearly seen for the finite dipole illustrated in Fig.10, for
which the electrostatic free energy is given simply by:

$$\Delta F = \left[\frac{-332}{r_+ + r_n} + \frac{332}{r_+ + r_p} \right] (q.N), \tag{24}$$

where q is the fractional value of electronic charge, N is

NEGATIVELY CHARGED SITE

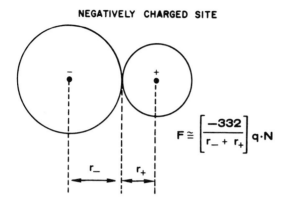

$$F \cong \left[\frac{-332}{r_- + r_+} \right] q \cdot N$$

NEUTRAL DIPOLAR SITE

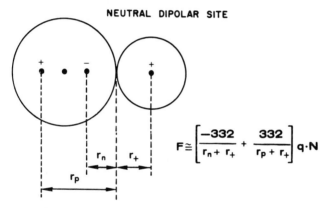

$$F \cong \left[\frac{-332}{r_n + r_+} + \frac{332}{r_p + r_+} \right] q \cdot N$$

Fig.10 : Models for a negatively charged ion exchange site vs. a neutral dipolar ligand (after (25), Fig.13). Described in text.

the coordination number of the ligands, r_+ is the (Goldschmitt) cationic radius, and r_n and r_p are the distances from the surface of the dipole of the negative and positive charges, respectively. (Eq. (24) effectively lumps all repulsions between ligands, as well as the Born repulsion with the cation, through a diminution of the effective charge.) When the dipole separation is sufficiently large $(r_p - r_n > r_n)$, the selectivities must approach those for the monopolar model, since in this situation the energetic contribution due to the positive pole becomes

negligible.   It is therefore not surprising to find a pattern
of selectivity sequences generated in this case (or even over a
considerable range of dipolar charge distribution) which are
qualitatively similar to those of the monopole model (25).
Indeed, all selectivity sequences observed for the macrocylic
molecules to date correspond to one of the eleven selectivity
sequences proposed here (see summaries by Eisenman (25) and
Diamond and Wright (10)).   I will not go beyond these primitive
considerations here since a definitive analysis of the selectivity of neutral carriers must take into account not only the
energies and entropies of the conformational differences of the
ion-binding molecules but also the details of ligand orientation
and coordination, as well as restrictions on cavity size due to
molecular deformability.   More detailed considerations will be
found elsewhere (25,15,31,32,67).

It should therefore be apparent that the energies of interaction of cations with the dipolar ligand groups, in competition
with the hydration energies of the ions, can account for the
salient features of the selectivity characteristic of neutral
sequestering molecules.   Indeed, the principal variable is
likely to be the dipole moment within a series of molecules whose
complexes should have reasonably constant geometry, such as with
increasing methylation of the series, Non-, Mon-, Din-, and Trinaction illustrated in Fig.5.   However, when considering the
differences in selectivity between dissimilar molecules such as
nonactin and valinomycin, such variables as coordination number
and cage size are likely to be more crucial.   Attention has been
called to these factors by Eigen and Winkler (15) and encouraging
progress has recently been made in taking them into account by
Morf and Simon (67).

(2)   Concluding remarks.   Although a real understanding of the
selectivity of these molecules must await the development of
theories having sufficient molecular detail, recognition of the

underlying competition between hydration energies and sequestration energies noted here is crucial in interpreting selectivity data correctly. For example, Prestegard and Chan (76), in order to account for the lack of selectivity in the binding manifested by nonactin in acetone solutions, suggested that the molecule had "little inherent propensity for selective binding, at least in acetone solution". This gives an erroneous picture since it is more likely that the apparent lack of selectivity is due to similar differences in the interactions from ion to ion with the carbonyls of nonactin and the carbonyls of acetone.

Lastly, the unusually high NH_4^+ selectivity of the macrotetralide actins (which have four carbonyl oxygens conveniently arrayed in tetrahedral coordination) as compared to the considerably lower NH_4^+ selectivity of valinomycin (which has six carbonyl oxygens arrayed in octahedral symmetry) is such a clear example of the role that ionic "shape" can play in the selectivity of appropriately oriented ligands, that it should be mentioned.

ACKNOWLEDGEMENTS

This work has been made possible by the continued support of the National Science Foundation (Grants GB 4039, 6685, 16194 and 30835) and the U.S. Public Health Service (Grants GM 17279 and NS 09931). I thank Dr Sally Krasne, Dr Gabor Szabo and Dr Robert Eisenberg for reading this chapter and for their valuable comments and gratefully acknowledge the hospitality of Prof Peter Läuger and the University of Konstanz, where much of this chapter was written.

REFERENCES

1. Baum, G. *Analytical Letters 3*: 105-111 (1970).
2. Bonner, O.D. and Lunney, J. *J. Phys. Chem. 70*: 1140 (1966).

3. Caldwell, P.C.   *J. Physiol. (London) 126*:   169 (1954).

4. Ciani, S., Eisenman, G. and Szabo, G.   *J. Membrane Biol. 1*: 1 (1969).

5. Ciani, S.M., Eisenman, G., Laprade, R. and Szabo, G. "Theoretical Analysis of Carrier-Mediated Electrical Properties of Lipid Bilayers", *In:* Membranes -- A Series of Advances, Volume 2, G. Eisenman, Ed.,Marcel Dekker, New York.(1972).

6. Conti, F. and Eisenman, G.   *Biophys. J. 5*:   247 (1965).

7. Conti, F. and Eisenman, G.   *Biophys. J. 5*:   511 (1965).

8. Conti, F. and Eisenman, G.   *Biophys. J. 6*:   227 (1966).

9. Cremer, M.   *Z. Biol. 47*:   562 (1906).

10. Diamond, J.M. and Wright, E.   *Ann. Rev. Physiol. 31*:   581 (1969).

11. Dole, M.   "The Glass Electrode".   Wiley, N.Y. (1941).

12. Donnan, F.G. and Garner, W.E.   *J. Chem. Soc. (London) 115*: 1313 (1919).

13. Doremus, R.H.   "Diffusion Potentials in Glass".   *In:* Glass Electrodes for Hydrogen and Other Cations, G. Eisenman, Ed. Marcel Dekker, New York.(1967).

14. Dupeyrat, M.   *J. Chim. Phys. 61*:   306 (1964).

15. Eigen, M. and Winkler, R.   "Alkali-Ion Carriers: Dynamics and Selectivity".   *In:* The Neurosciences, F.O. Schmitt, Ed., Rockefeller Univ. Press, N.Y. (1970).

16. Eisenman, G.   "On the Elementary Atomic Origin of Equilibrium Ionic Specificity".   *In:* Symposium on Membrane Transport and Metabolism, A. Kleinzeller and A. Kotyk, Eds. Academic Press, New York. (1961).

17. Eisenman, G.   *Biophys. J. 2*:   part 2, 259 (1962).

18. Eisenman, G.   "The Electrochemistry of Cation Sensitive Glass Electrodes".   *In:* Advances in Analytical Chemistry and Instrumentation.   IV.   C.N. Reilley, Ed., Interscience. (1965).

19. Eisenman, G.   "The Origin of the Glass Electrode Potential" *In:* Glass Electrodes of Hydrogen and Other Cations, G. Eisenman, Ed., Marcel Dekker, New York (1967).

20. Eisenman, G.   "The Physical Basis for the Ionic Specificit of the Glass Electrode".   *In:* Glass Electrodes of Hydrogen and Other Cations, G. Eisenman, Ed., Marcel Dekker, New Yor (1967).

21. Eisenman, G.   "Particular Properties of Cation-Selective Glass Electrodes Containing Al$_2$O$_3$".   *In:* Glass Electrodes of Hydrogen and Other Cations, G. Eisenman, Ed., Marcel Dekker, New York (1967).

22. Eisenman, G.   *Anal. Chem. 40:* 310 (1968).

23. Eisenman, G.   *Annals of the New York Academy of Sciences 148:* Article 1, 5-35 (1968).

24. Eisenman, G.   "The Ion-Exchange Characteristic of the Hydrated Surface of Na$^+$ Selective Glass Electrodes". *In:* Glass Microelectrodes, M. Lavallee *et al.*, Eds., John Wiley & Sons, New York (1968).

25. Eisenman, G.   "Theory of Membrane Electrode Potentials". *In:* Ion-Selective Electrodes, R.A. Durst, Ed., National Bureau of Standards Special Publication 314 (1969).

26. Eisenman, G., Rudin, D.O. and Casby, J.U.   *Science 126:* 831 (1957).

27. Eisenman, G., Bates, R., Mattock, G. and Friedman, S.M. The Glass Electrode, Wiley - Interscience Paperback, New York (1966).

28. Eisenman, G., Sandblom, J.P. and Walker, J.L. Jr.   *Science 155:* 965 (1967).

29. Eisenman, G., Ciani, S. and Szabo, G.   *Fed. Proc. 27:* 1289 (1968).

30. Eisenman, G., Ciani, S. and Szabo, G.   *J. Membrane Biol. 1:* 294 (1969).

31. Eisenman, G., Szabo, G., McLaughlin, S.G.A. and Ciani, S.M.   "Molecular Basis for the Action of Macrocyclic Antibiotics on Membranes". *In:* Symposium on Molecular Mechanisms of Antibiotic Action on Protein Biosynthesis and Membranes, D. Vasquez, Ed., Springer-Verlag.   In Press.

32. Eisenman, G., Szabo, G. and Ciani, S.M. "Electrical Properties of Lipid Bilayer Membranes Exposed to Lipophilic Ion-Binding Molecules: Equilibrium Domain". *In:* Membranes -- A Series of Advances, Volume 2, G. Eisenman, Ed., Marcel Dekker, New York (1972).

33. Eyal, E. and Rechnitz, G.A.   *Anal. Chem. 43:* No.8 (1971).

34. Frant, M. and Ross, J.W.   *Science 154:* 1553, Orion Research Bulletin 94-09 and 94-16 (1966).

35. Frant, M. and Ross, J.W.   *Science 167* (1970).

36. Frensdorff, H.K.   *J. Amer. Chem. Soc. 93:* 600 (1971).

37. Garfinkel, H.M.   *J. Phys. Chem. 72:* 4175 (1968).

38. Garfinkel, H.M. *J. Phys. Chem. 74*: 1764 (1970).

39. Garfinkel, H.M. "Cation--Exchange Properties of Dry Silicate Membranes". *In:* Membranes--A Series of Advances, Volume 1, G. Eisenman, Ed., Marcel Dekker, New York (1971).

40. Guilbault, G.G. and Montalvo, J.G. *Anal. Letters 2*: 283 (1969).

41. Guilbault, G.G. and Montalvo, J.G. *J. Am. Chem. Sco. 92*: 2533 (1970).

42. Guilbault, G.G. and Hrabankova, E. *Anal. Chem. 42*: 1779 (1970).

43. Guilbault, G.G. and Hrabankova, E. *Anal. Chim. Acta* (1971)

44. Haber, F. and Klemensiewicz, Z. *Z. Physik. Chem. (Leipzig) 67*: 385 (1909).

45. Haynes, D.H., Kowalsky, A. and Pressman, B.C. *J. Biol. Chem. 244*: 502 (1969).

46. Hinke, J.A.M. *Nature 184*: 1257 (1959).

47. Horovitz, K. *Z. Physik 15*: 369 (1923).

48. Hughes, W.S. *J. Am. Chem. Soc. 44*: 2860 (1922).

49. Isard, J.O. "The Dependence of Glass--Electrode Properties on Composition". *In:* Glass Electrodes for Hydrogen and Other Cations, G. Eisenman, Ed., Marcel Dekker, New York (1967).

50. Kahlweit, M. *Pflügers Archiv. 271*: 139 (1960).

51. Karreman, G. and Eisenman, G. *Bull. Math. Biophys. 24*: 413 (1962).

52. Kilbourn, B.T., Dunitz, J.D., Pioda, L.A.R. and Simon, W. *J. Mol. Biol. 30*: 559 (1967).

53. Kossiohoff, A. and Harker, D. *J. Am. Chem. Soc. 60*: 2047 (1938).

54. Latimer, W.M. The Oxidation States of the Elements and Their Potentials in Aqueous Solutions, 2nd Ed., Prentice-Hall, New York (1952).

55. Läuger, P. and Stark, G. *Biochim. Biophys. Acta 211*: 458 (1970).

56. Läuger, P. and Neumcke, B. "Theoretical Analysis of Ion Conductance in Lipid Bilayer Membranes". *In:* Membranes--A Series of Advances, Volume 2, G. Eisenman, Ed., Marcel Dekker, New York (1972).

57. Lengyel, B. and Blum E. *Transactions of the Faraday*

Society 157: Vol. XXX, Part 6 (1934).

58. Lev, A.A. and Buzhinsky, E.P. *Cytology (USSR)* *3*: 614 (1961).

59. Lev, A.A. and Buzhinsky, E.P. *Cytologia (USSR)* *9*: 106 (1967).

60. Lev, A.A., Buzhinsky, E.P. and Grenfeldt, A.E. "Electrochemistry of Bimolecular Phospholipid Membranes with Valinomycin at Nonzero Current". Proceedings XXIV International Congress of Physiological Science, Excerpta Medica, Netherlands, p.39 (1968).

61. Lev, A.A., Malev, V.V. and Osipov, V.V. "Electrochemical Properties of Thick Membranes with Macrocylic Antibiotics". *In:* Membranes--A Series of Advances, Volume 2, G. Eisenman, Ed., Marcel Dekker, New York (1972).

62. Levins, R.J. *Anal. Chem. 43*: 1045 (1971).

63. Ling, G.N. "Anion-Specific and Cation-Specific Properties of the Collodion-Coated Glass Electrodes". *In:* Glass Electrodes for Hydrogen and Other Cations, G. Eisenman, Ed., Marcel Dekker, New York (1967).

64. Ling, G.N. and Kushner, L. Quoted in Eisenman, G. (17) (1961).

65. McLaughlin, S.G.A., Szabo, G., Eisenman, G. and Ciani, S.M. *Proc. Nat. Acad. Sci. 67*: 1268-1275 (1970).

66. McLaughlin, S.G.A., Szabo, G., Eisenman, G. and Ciani, S. *J. Memb. Biol.* (1971).

67. Morf, W.E. and Simon, W. *Helv. Chim. Acta 54*: 794 (1971).

68. Mueller, P. and Rudin, D.O. *Biophys. Biochem. Res. Comm. 26*: 398 (1967).

69. Nicolskii, B.P. and Tolmacheva, T.A. *Zh. fiz. Khim. 10*: 504 (1937).

70. Orme, F.W. "Liquid Ion-Exchanger Microelectrodes". *In:* Glass Microelectrodes, M. Lavallee *et al.*, Eds., John Wiley, New York, p.376 (1968).

71. Osterhout, W.J.V. Cold Spring Harbor Symposium *8*: 5 (1940).

72. Ovchinnikov, Y.A., Ivanov, V.T. and Shkrob, A.M. "Chemistry and Membrane Activity of Peptide Ionophores". *In:* Symposium on Molecular Mechanisms of Antibiotic Action on Protein Biosynthesis and Membranes, D. Vasquez, Ed., Springer-Verlag (1971).

73. Pioda, L.A.R., Wachter, H.A., Dohner, R.E. and Simon, W.

Helv. Chim. Acta 50: 1373 (1967).

74. Pioda, L.A.R. and Simon, W. *Chimia (Switzerland) 23*: 72 (1969).

75. Pressman, B.C., Harris, E.J., Jagger, W.S. and Johnson, J.H. *Proc. Nat. Acad. Sci. 58*: 1949 (1967).

76. Prestegard, J. and Chan, S.I. P.M.R. Studies of the cation-binding properties of nonactin. II. Comparison of the Na^+, K^+ and Cs^+ complexes. In Press (1970).

77. Pungor, E. *Anal. Chem. 39*: (13) 28A (1967).

78. Rechnitz, G.A. Personal Communication. (1971).

79. Reichenberg, D. "Ion-Exchange Selectivity". *In:* Ion Exchange, J.A. Marinsky, Ed., Volume 1, Marcel Dekker, New York, p.277 (1966).

80. Ricci, J.E. *J. Amer. Chem. Soc. 70*: 109 (1948).

81. Rosano, H.L., Duby, P. and Schulman, J.H. *J. Phys. Chem. 65*: 1704 (1961).

82. Ross, J.W. Jr. *Science 155*: 1378 (1967).

83. Ross, J.W. Jr. Orion Research Inc., Bulletin 92-81 (1968).

84. Ross, J.W. Jr. "Solid-State and Liquid Membrane Ion Selective Electrodes." *In:* Ion-Selective Electrodes, R.A.Durst, Ed., Nat. Bur. Stand. Spec. Publ. 314, 1-56 (1969)

85. Ross, J.W. Jr. and Frant, M.S. "The Mechanism of Interference at Solid-State Ion-Selective Electrodes", Presented at Pittsburgh Conference on Analytical Chemistry (1968).

86. Ross, J.W. Jr. and Frant, M.S. *Science 167*: 987 (1970).

87. Sandblom, J.P., Eisenman, G. and Walker, J.L. Jr. *J. Phys. Chem. 71*: 3862 (1967).

88. Sandblom, J.P., Eisenman, G. and Walker, J.L. Jr. *J. Phys. Chem. 71*: 3971 (1967).

89. Sandblom, J.P. and Orme, F. "Liquid-Ion Exchange Membranes as Electrodes and Biological Models". *In:* Membranes--A Series of Advances, Volume 1, G. Eisenman, Ed., Marcel Dekker, New York (1971).

90. Schiller, H. *Ann. Physik 74*: 105 (1924).

91. Scholer, R.P. and Simon, W. *Chimia 24*: 372 (1970).

92. Shean, G.M. and Sollner, K. *Ann. N.Y. Acad. Sci. 137*: 759 (1966).

93. Shemyakin, M.M., Ovchinnikov, Y.A., Ivanov, V.I., Antonov,

V.K., Vinogradova, E.I., Shkrob, A.M., Malenkov, G.G., Evstratov, A.V., Laine, I.A., Melnik, E.I. and Ryabova, I.D. *J. Memb. Biol. 1:* 402-430 (1970).

94. Sherry, H.S. "The Ion-Exchange Properties of Zeolites". *In:* Ion Exchange, Volume II, J. Marinsky, Ed., Marcel Dekker, New York (1969).

95. Simon, W. "Alkali Cation Specificity of Antibiotics, Their Behavior in Bulk Membranes and Useful Sensors Based on These". *In:* Symposium on Molecular Mechanisms of Antibiotic Action on Protein Biosynthesis and Membranes, D. Vasquez, Ed., Springer-Verlag (1971).

96. Simon, W. and Morp, W.E. "Alkali Cation Specificity of Carrier Antibiotics and Their Behavior in Bulk Membranes". *In:* Membranes--A Series of Advances, Volume 2, G. Eisenman, Ed., Marcel Dekker, New York (1972).

97. Sollner, K. *Ann. N.Y. Acad. Sci. 57:* 177 (1953).

98. Stark, G. and Benz, R. *J. Memb. Biol. 5:* 133 (1971).

99. Stefanac, Z. and Simon, W. *Chimia (Switzerland) 20:* 436 (1966).

100. Stefanac, Z. and Simon, W. *Microchem. J. 12:* 125 (1967).

101. Szabo, G., Eisenman, G. and Ciani, S.M. *J. Memb. Biol. 3:* 346-382 (1969).

102. Truesdell, A.H. and Christ, C.L. "Glass Electrodes for Calcium and Other Divalent Cations". *In:* Glass Electrodes for Hydrogen and Other Cations, G. Eisenman, Ed., Marcel Dekker, New York (1967).

103. Truter, M.R. and Bright, D. *Nature 225:* 176 (1970).

104. Walker, J.R. Jr. "Liquid Ion-Exchanger Microelectrodes for Ca++, Cl⁻ and K+". *In:* Ion Specific Microelectrodes, N.C. Hebert and R.N. Khuri, Eds., Marcel Dekker, New York. In Press (1971).

105. Walker, J.L. Jr. and Brown, A.M. *Science 167:* 1502 (1970).

106. Wipf, H.K. and Simon, W. *Helv. Chim. Acta 53:* 1732 (1970).

107. Wise, W.M., Kurey, M.J. and Baum, G. *Clin. Chem. 16:* 103 (1970).

ACCURACY AND ERRORS OF THE pO_2 MEASUREMENT BY MEANS OF THE PLATINUM ELECTRODE AND ITS CALIBRATION *IN VIVO*

W. GRUNEWALD

Max-Planck-Institut für Arbeitsphysiologie, Dortmund, Germany

The accuracy of the pO_2 measurement by means of the platinum electrode depends primarily on two factors:

(1) on systematic errors of the platinum electrode

(2) on the calibration method of the platinum electrode.

Systematic errors include the diffusion error as well as errors caused by consumption and response time of the electrode.

The calibration of the platinum electrode influences the measuring accuracy, because generally calibration medium and measuring medium differ with respect to their diffusion properties. Furthermore, an intermediate calibration is related to a change of the measuring point.

The calculation base of these errors is the diffusion field in front of the electrode that is formed during the pO_2 measurement. It overlaps the pO_2 field in the measuring medium and is conditioned by the loss of CO_2 molecules that are reduced on the platinum surface of the electrode. Diffusion field and diffusion flow of the O_2 molecules towards the platinum electrode, the pro-

per measuring signal, are combined. The calculation of the
diffusion field demands first an assumption as to the value of
the O_2 partial pressure on the platinum surface. Together with
Baumgärtl and Lübbers we could show that the pO_2 on the platinum
surface during measurement in the steady state equals zero. In
an agar layer the diffusion field of a macro-electrode (diameter
1.5 mm) was measured by means of a micro-electrode (diameter
1-3 μ). Figure 1 shows the course of the pO_2 over the path of
the micro-electrode towards the periphery (at the top) and the
center (at the bottom) of the macro-electrode. The result of
the consistency between measured and calculated pO_2 field and
of the extrapolation of the measured curve is: pO_2 = zero on
the platinum surface. This fundamental result serves as boundary condition for the further observations on the measuring
accuracy.

     First we consider the diffusion error. We can prove that
the measuring signal of a membrane-covered platinum electrode
is proportional to the pO_2 on the membrane surface. Because of
the "disturbance" of the pO_2 field in the measuring medium
caused by the diffusion field of the electrode, this value is
not identical with the pO_2 value at this point before the beginning of the measurement. This deviation is called the diffusion
error. Figure 2 shows the pO_2 field in a tissue with a defined
O_2 consumption before measurement and the same pO_2 field with the
disturbance induced by the diffusion field of the electrode
(r_0 : radius, r_1 : membrane surface of the electrode). The dif
diffusion error is the difference between the pO_2 values at the
membrane surface ($r = r_1$). This error depends on the nature
of the electrode as well as on the diffusion properties of the
measuring medium and its O_2 consumption. The choice of an
appropriate membrane can largely limit the diffusion field of the
electrode to the membrane, so that the diffusion error is small.
Figure 2 shows the diffusion field of an electrode (r_0 = 2.5 μ,

$$p(z,r) = p_c \left(1 - \frac{2}{\pi} \int_0^\infty \frac{\sin(\lambda R)}{\lambda} \, J_0(r\lambda) \exp(-\lambda z) \, d\lambda \right)$$

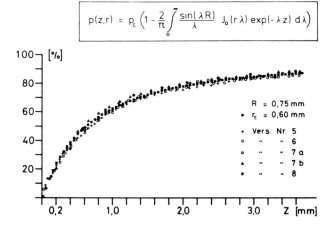

R = 0,75 mm
● r_E = 0,60 mm

+ Vers. Nr. 5
○ " " 6
□ " " 7 a
▲ " " 7 b
● " " 8

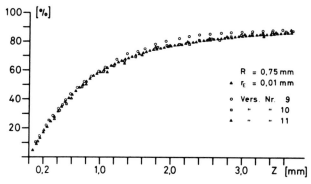

R = 0,75 mm
▲ r_E = 0,01 mm

○ Vers. Nr. 9
□ " " 10
▲ " " 11

Fig.1 : Calculated and measured diffusion field of a macro-electrode (diameter 1.5 mm), i.e. course of the relative pO₂ over the path (z-direction) of the micro-electrode (at the top) towards the periphery (r_E = 0.60 mm) and (at the bottom) towards the center (r_E = 0.01 mm) of the macro-electrode.

■ ■ ■ and ▲ ▲ ▲ respectively, indicate the calculated

pO₂ course $\dfrac{p(z,r_E)}{p_C}$; p_C indicates the pO₂ before the

beginning of measurement.

d_m = 2.0 μ) (a) for the same diffusion conductivity in medium and membrane: K/K_m = 1 and (b) for the quotient of the dif-fusion conductivity of measuring medium (skeletal muscle tissue)

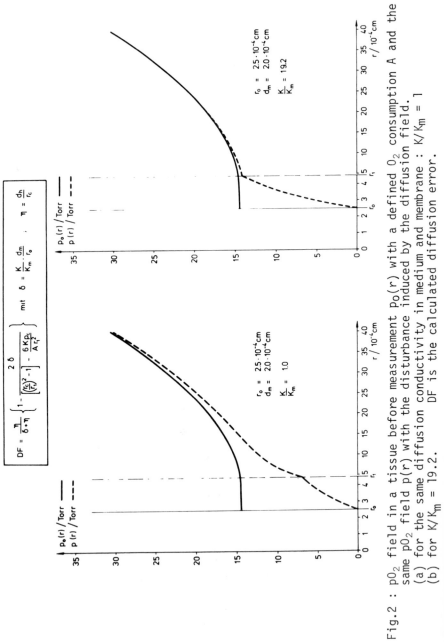

Fig.2 : pO_2 field in a tissue before measurement $p_0(r)$ with a defined O_2 consumption A and the same pO_2 field $p(r)$ with the disturbance induced by the diffusion field.
(a) for the same diffusion conductivity in medium and membrane : $K/K_m = 1$
(b) for $K/K_m = 19.2$.   DF is the calculated diffusion error.

and membrane (polyethylene) K/K_m = 19.2.   The diffusion error
amounts to (a) 52.4%;   (b) 3.3%.

The second systematic error is caused by the consumption of
the electrode.   In addition to the mitochondria, the electrode
itself consumes O_2 in tissue.   Its O_2 requirement must also be
covered by the intracapillary O_2.   Figure 3 shows at the left
side the surface of a platinum electrode inserted into a capillary mesh.   During measurement the O_2 consumption of the electrode induces a steeper pO_2 decrease along the capillary.
Figure 3 shows at the right the pO_2 slope along a capillary of
the observed mesh (a) without electrode, (b) and (c) with
electrodes of different thickness.   In the figure at the top a
very thin membrane is chosen, at the bottom the thickness of
the membrane is chosen in the order of magnitude of the platinum
wire.   The steeper pO_2 decrease along the capillaries effects
a decrease of the pO_2 values in the intercapillary pO_2 field,
particularly at the measuring point.   The resulting measuring
errors are calculated by a diffusion model for brain and myocardium.   Because of the smaller capillary distance in the myocardium, the influence of the O_2 consumption of the electrode on
the measuring accuracy is greater than that in the brain.   (For
instance, when r_o = 2.5 μ at membrane thickness 5 μ, the error
is 1% in the brain, 5% in the myocardium;   without membrane the
error is about 4% and 19%, respectively.)

The accuracy of unsteady pO_2 measurements depends on the
response time of the electrode.   It shows particularly in
measurements of biological media as compared with measurements
in gases.   Figure 4 shows the measuring signal of a membrane-
covered electrode in gaseous media.   With the jump to higher pO_2-
values the signal increases monotonously; T_{90} and T_{95} are the
response times respectively.   Figure 5 shows the measuring signal with the same pO_2 jump to higher pO_2 values in a biological
measuring medium.   The signal becomes greater than the steady

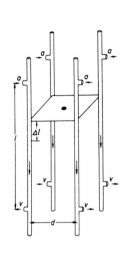

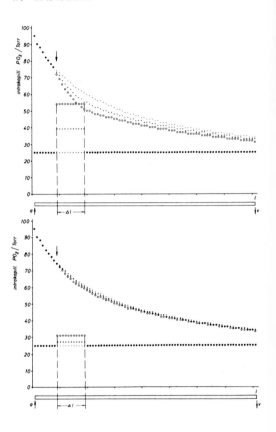

Fig.3 : Left: surface of a platinum electrode inserted into a
capillary mesh (d = capillary distance, l = capillary
length, a = arterial ends, b = venous ends of capillaries, Δl = capillary section where the electrode
effects an additional saturation slope).
Right: pO$_2$ slope along a capillary
• • • without electrode in tissue
+ + + with electrode of 6 μ diameter in tissue
○ ○ ○ with electrode of 12 μ diameter in tissue
In the upper part of the figure the thickness of the
membrane was supposed to be very thin as compared with
the diameter of the platinum wire; in the lower part
it equals the radius of the electrode. The pO$_2$ courses
show an additional decrease from the localization of the
electrode in tissue (arrow). The capillary supply area
seems to be augmented along Δl due to the O$_2$ consumption
of the electrode in tissue.

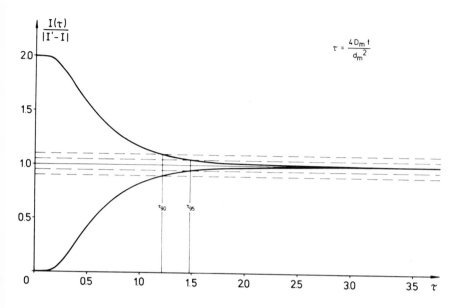

Fig.4 : Measuring signals of a membrane-covered electrode in a gaseous medium after a jump to higher pO_2 values (lower curve). The measuring signal I(t) is related to the signal difference between initial signal I and final signal I'. τ represents a dimensionless parameter connected with time t by the given formula (D_m = diffusion coefficient of the membrane, d_m = its thickness). From τ_{90} and τ_{95}, respectively, the response times T_{90} and T_{95}, respectively, are calculated.

final signal and then approaches it from the higher values. The intensity of the overshoot depends on the quotient of the membrane thickness and the radius of the electrode and, furthermore, on the quotient of the diffusion constant in membrane and medium. This figure shows measured and calculated signal curves. In these experiments a platinum electrode touched an agar layer equilibrated by air at time t = 0. Previously the electrode had measured nitrogen. The experiments were carried out with

W. *Grunewald*

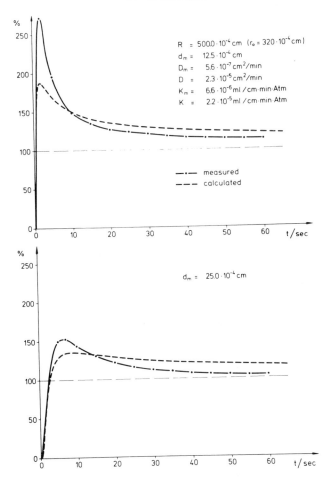

Fig.5 : Calculated and measured signal of a membrane-covered
electrode after a jump to higher pO_2 values corresponding to Fig.4, but measured in a biological medium. The
measuring signal shows an overshoot which depends,
among others, as shown by the figure on the thickness
of the membrane d_m.

(R = radius of the electrode; D_m, D and K_m, K =
diffusion coefficient and diffusion conductivity of
membrane and measuring medium, respectively.)

different quotients between membrane thickness and ratio of the
electrode. Due to insufficient knowledge about diffusion pro-

perties, particularly about the medium, and because of simplifying assumptions, the correlation between measured and calculated signal is still unsatisfactory. Table I shows the response times for diffusion conditions as they exist in the tissue and for electrodes covered with Teflon membrane:

If the quotient $\frac{d_m}{r_o} \approx 0.1$, then we have to face the possibility of an overshoot, $T_{95}*$ amounts to about 3 sec. Under the same conditions the response time for the quotient $\frac{d_m}{r_o} \approx 0.1$ about 0.4 sec. The overshoot and the augmentation of the response time so caused are a phenomenon of *thin* membrane.

In addition to the systematic errors due to the platinum electrode, the calibration method influences the accuracy of the pO_2 measurement. Generally the electrodes are calibrated in liquids, that is, in media showing diffusion properties other than the measuring medium. For calibration the electrode has to be moved from the measuring point. Errors result from this that can be avoided by calibration *in vivo*. We developed, together with Lübbers, a calibration method for the pO_2 measurement in tissue. First, the tissue is saturated with O_2 by respiration with high-percent O_2 mixtures. The pO_2 must be so high that the hemoglobin in the capillary blood is fully oxygenated. Then the blood flow is stopped. Because of the O_2 consumption in the tissue the pO_2 in the blood decreases linearly with time (Fig.6). If the pO_2 falls below the hemoglobin saturation limit, the pO_2 decrease changes its behavior corresponding to the hemoglobin binding curve. The moment at which the pO_2 decrease leaves the linear course, is called the "deviation point". By the hemoglobin binding curve a defined pO_2 can be assigned to that moment. Now we can prove mathematically that the measuring signal of the electrode is proportional to a pO_2 course (Ψ) which is parallel to the pO_2 decrease in blood (Ψ_o). Only at the beginning the two decreases are different. The proportionality constant between measuring signal and pO_2

TABLE I

RESPONSE TIMES (T_{90}, T_{95} AND $T_{90}^\star$, $T_{95}^\star$, RESPECTIVELY) OF TWO ELECTRODES (RADIUS r_0) COVERED WITH A TEFLON MEMBRANE (THICKNESS d_m) IN TISSUE

When the measuring signal $I(t)$ (see Fig.5) gets above the final value I' by more than 10% and 5%, respectively, then the response times $T_{90}^\star$ and $T_{95}^\star$, respectively, take the place of T_{90} and T_{95}. The calculation bases on the following values for membrane and medium:

$D_m = 5.5 \cdot 10^{-7}$ cm²/sec; $K_m = 6.6 \cdot 10^{-6}$ ml O_2/cm·min·atm
$D = 1.5 \cdot 10^{-5}$ cm²/sec; $K = 2.2 \cdot 10^{-5}$ ml O_2/cm·min·atm

$r_0/10^{-4}$ cm	$d_m/10^{-4}$ cm	T_{90}/sec	$T_{90}^\star$/sec	T_{95}/sec	$T_{95}^\star$/sec	t_{max}/sec	$\dfrac{I(t_{max})}{I'}$ (%)
7.5	1.00	-	0.75	-	3.0	0.013	59.5
	6.25	0.32	-	0.39	-	1.00	4.2
	12.50	1.40	-	1.75	-	5.75	1.5
	25.00	5.75	-	7.40	-	30.0	0.5
2.0	0.5	-	0.04	-	0.15	0.004	25.1
	2.0	0.03	-	0.04	-	0.10	3.1
	4.0	0.14	-	0.18	-	0.75	1.1
	10.0	0.95	-	1.20	-	6.50	0.3

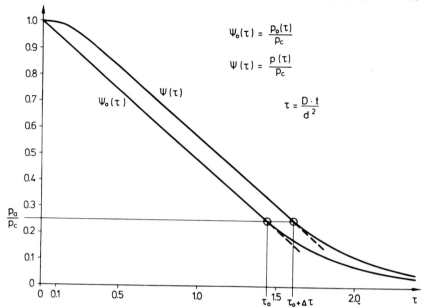

Fig.6 : pO$_2$ decrease after saturation of the tissue with O$_2$ by
respiration with high-percent O$_2$ mixture.
$\Psi_0(\tau)$ = pO$_2$ decrease in blood: $\Psi(\tau)$ = pO$_2$ decrease
measured by Pt electrode: p_c = pO$_2$ at the beginning of
measurement (stop of blood flow): p_a = pO$_2$ at "deviation point" (τ_0) in blood; the electrode reaches this
value delayed by $\Delta\tau$: τ = dimensionless parameter correlated by the given formula to time t (D, d are diffusion coefficient and mean thickness of the sheet
between platinum surface of the electrode and blood vessels; the latter are taken to be distributed over a
hemisphere).

course is the calibration constant in quest.   The "deviation
point" appears in this pO$_2$ course shifted in time compared with
the deviation point of the pO$_2$ decrease in blood.   The pO$_2$
value at the "deviation point" remains constant in both cases.
Thus the pO$_2$ value found by the binding curve can be assigned to
the measuring signal at the "deviation point".   The calibration
constant is the quotient of both values; pO$_2$ measurements with
this calibration method were carried out on the human skin by
R. Huch and Lübbers.

W. Grunewald

REFERENCES

1. Grunewald, W.   Diffusion error and O_2 consumption of the Pt electrode during pO_2 measurements in the steady state. *Pflügers Arch. ges. Physiol. 320:* 24-44 (1970).

2. Grunewald, W.   Response time of the Pt electrode with measurements of non-stationary oxygen partial pressures. *Pflügers Arch. ges. Physiol. 322:* 109-130 (1971).

3. Huch, R.   Quantitative, kontinuierliche Sauerstoffpartial-druckmessung (pO_2-Messung) auf der hyperämisierten Erwachsenenhaut.   Inaugural-Dissertation Marburg (1971).

4. Schuler, R. and Kreuzer, F.   Rapid polarographic *in vivo* catheter electrodes.   *Respir. Physiol. 3:* 90-110 (1967).

THE ELECTRON MICROPROBE

T.A. HALL and H.J. HÖHLING

Cavendish Laboratory, Cambridge, England
Institut für Medizinische Physik, Universität Münster, Germany

FUNDAMENTALS

Electron-probe X-ray microanalysis is a method for the
analysis of chemical elements *in situ* within microvolumes of a
specimen. The elemental composition is determined by the
spectroscopic analysis of the X-rays which are generated when a
finely focused beam of electrons is directed onto the selected
microvolume (Fig.1). Most often the X-ray spectral analysis
is carried out by means of a diffracting crystal, which is
oriented to pass on to the detector only X-rays of a particular
wavelength characteristic of the assayed element. (This alternative technique of "energy dispersive" spectroscopy is discussed below.) While the beam is stationary on the analyzed
microvolume, the diffractor may be rotated to pass successively
through the orientations associated with different elements, to
give a spectrum identifying the constituent elements, as in
Fig.2.

The electron beam may be scanned over an area of the specimen, and any one of several locally-varying signals may be used
to modulate the brightness and thus form an image on a synchron-
ously scanned cathode-ray display tube. The signals for scanning electron images can be based on the backscattering of the
beam electrons, on the ejection of "secondary" electrons from
the specimen, on the electron current travelling from the speci-

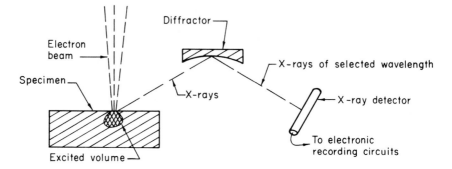

Fig.1 : Basic principles of electron-probe X-ray microanalysis.
(From Hall (10).   Courtesy of Academic Press.)

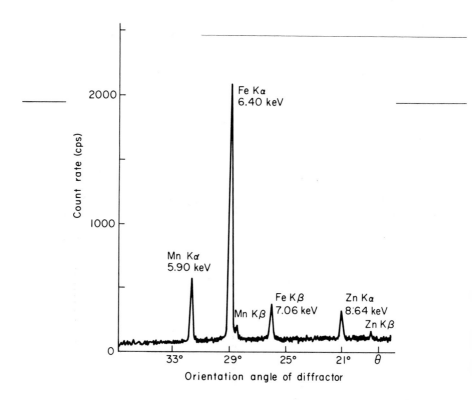

Fig.2 : Spectral peaks recorded as the diffractor is rotated,
with the beam stationary on the specimen.   (From Hall
et al. (12).   Courtesy of Charles C. Thomas,
Publishers.)

men to earth, or in the case of thin specimens, on the transmission of electrons through the specimen. These modes of image formation are the same as in the scanning electron microscope and indeed, with recent improvements in the imaging performance of microprobe analyzers and the recent availability of effective X-ray attachments for scanning electron microscopes, there is no longer a definite dividing line between these instruments.

By setting the X-ray spectrometer to a characteristic line of a particular element and modulating the display tube according to the *X-ray* intensity, one can get an X-ray image, which gives a map of the distribution of the selected element in the scanned area. Figure 3 shows scanning images of a 5 μm section of sclerotic aorta, based on backscattered electrons and on calcium X-rays.

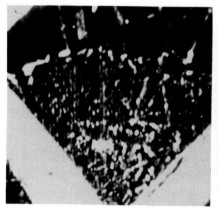

Fig.3 : Scanning images of a section of sclerotic aorta.
Left: Backscattered electron image (with strong background from the bars of the supporting grid).
Right: Calcium X-ray image. (From Hall *et al.* (1966).
Courtesy of John Wiley and Sons.)

However in biological material most often the X-ray intensities from the elements of interest are too low, and the peak-to-background ratios are too poor, to allow the production of

useful scanning X-ray images.   Commonly an electron scanning
image is formed on a display tube with a persistent phosphor,
and X-ray data are then obtained either along a linear scan or
with a stationary beam.   The location of the line of scan or
the position of the static probe may be set precisely as
desired by reference to the features seen in the scanning
image.   The most common mode of operation in biological work
is static-probe analysis in microareas selected within scanning
images.   Several elements can be assayed simultaneously, either
by using several diffracting crystals or by "energy dispersive"
X-ray spectroscopy.

The main features of the technique are these: Chemical
elements are analyzed regardless of their state of chemical
binding;   no staining or other chemical pre-treatment is required for the sake of the analysis;   all elements of atomic number above 5 may be routinely assayed;   the method is "non-
destructive" (with qualifications, discussed below);   the
spatial resolution of the analysis is of the order of 1 μm;
the minimum measurable amount of an element is usually in the
range 10^{-16} - 10^{-17} g;   and the minimum measurable *local* mass
fraction is generally of the order of 10^{-4} (100 ppm).

TECHNICAL PROBLEMS AND LIMITATIONS

Microprobes are used most in metallurgy and mineralogy.
In biology special problems arise, mainly with regard to the
maintenance of good spatial resolution in the analysis, histological correlation (i.e. identification of the microvolumes to
be analyzed), the achievement of quantitative assays, the pre-
servation of the natural elemental distributions within the
specimen during preparation, and specimen stability under the
electron beam.

Spatial resolution

The electron beam is scattered within the specimen. Because soft tissue has a relatively low stopping power for electrons, this scattering can lead to the excitation of X-rays throughout an undesirably large volume, with correspondingly poor spatial resolution. Good resolution can be preserved either by the use of low-energy, short-range electrons, or by the use of high-energy electrons for the study of thin sections mounted on thin supporting films. If a low-energy beam is used, the electron energy must be no more than a few kV above the excitation threshold of the assayed element; the situation is then as sketched in Fig.1 and the excited volume has a mean dimension of the order of 1 μm or less. Alternatively, for specimens of dried soft tissue less than 5 μm thick, if one uses electrons of 30 kV or more, the situation is as depicted in Fig.4, and the mean lateral spread of the beam is restricted to the order of 1 μm or less.

The scattering of the electron beams within biological specimens has been discussed especially by Andersen (1,2) and also by Hall (10).

Histological correlation

Optical microscopes are fitted to most microprobe instruments and to some scanning electron microscopes. However optical microscopy is often an inadequate guide for positioning the electron beam within a specimen, especially in unstained preparations, and staining is undesirable because of the risk of removing or displacing tissue constituents. For viewing uneven specimen surfaces, most types of electron scanning image are usually satisfactory. When one has to see cytological structure in flat tissue sections, the problem is to get enough contrast, and most types of scanning image often prove unsatisfactory. In scanning electron microscopes the best contrast is

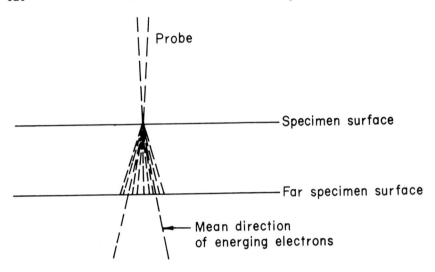

Fig.4 : The nature of the X-ray emitting volume in thin specimens.   (From Hall (10).   Courtesy of Academic Press.)

afforded by the *transmission* scanning image, which can be obtained for sections up to 1 μm thick or somewhat thicker (24).
Similar good contrast is combined with the best spatial resolution in the conventional EM images of ultra thin sections produced in combination electron microscope-microanalyzers (see
below).

Quantitation

In thick (i.e. electron-opaque) specimens, the local mass
fraction of an element can be deduced directly by comparing
characteristic X-ray intensities from the specimen and from a
standard.   The experimental procedure is very simple but the
theory is elaborate and there are many types of biological
specimen to which it cannot be accurately applied.   (The correction for self-absorption of X-rays within the specimen is
often large and impossible to calculate accurately.)

If the beam voltage and specimen thickness are such that most of the electrons pass through the specimen with the loss of only a small fraction of their energy, a different method of quantitation is used. In these circumstances the intensity of a characteristic X-ray line is no longer a direct measure of elemental mass fraction; the characteristic intensity is now a measure of the *amount* of element in the analyzed microvolume, and in order to determine mass fractions, one needs a separate measure of the total mass in the same microvolume. In most biological specimens the intensity of the X-ray bremsstrahlung or continuum radiation (observed simultaneously with the characteristic line) can be used as the requisite measure of total mass.

For the theory of quantitative analysis in thick specimens the reader may refer to Andersen (1), Philibert (20) or Birks (3). Quantitation in thin specimens has been expounded by Höhling and Hall (13) and by Hall (10), the latter work containing an extensive critique of the problems of quantitation both in thick and thin specimens.

Specimen preparation

Preparative techniques involving liquids are likely to remove or displace the elements one wishes to study. For the production of tissue sections, an alternative standard method is to freeze quickly a block of tissue, cut frozen sections and remove the ice by sublimation. This avoids the danger of washing away elements, and the method is quite suitable for the study of bound elements, especially for measuring the elemental amount per unit mass of dry tissue.

There are very serious problems in the preparation of tissue sections for the analysis of *electrolytes in situ*. Sublimated sections are far from ideal for this purpose since the elements which were in aqueous solution are inevitably displaced during drying, and furthermore measurements of mass fractions in dry tissues are not at all the same as measurements of the original electrolyte concentrations.

A different procedure for the study of electrolytes is to cut sections from blocks which have been rapidly frozen and then dried and embedded in vacuum (cf. (15); for a modified and improved technique, (16)). In this method also, displacement of the solute elements undoubtedly occurs, and it is difficult to judge how the electrolytes are finally redistributed in the embedded blocks. It must be noted that when the specimen is embedded, the measured elemental mass fractions refer to the specimen *including* any embedding medium which may be present in the assayed microvolume. Such data may be less cogent than mass fractions when referred to the dry weight of the tissue proper. However, in the case of electrolytes, the mass fraction in embedding medium may conceivably be closer to the original electrolyte concentration than is the elemental mass fraction in a dried, unembedded preparation.

There is hope that with the aid of a cold stage, microprobe studies can be carried out on frozen sections in which the water is retained in the form of ice (6). In such preparations the electrolyte distributions should be better, though still not ideally, preserved; and the measured mass fractions should correspond more closely to the electrolyte concentrations *in vivo*.

The microprobe has also been used for the assay of electrolytes and other elements in microvolumes of fluid extracted from tissues by micropuncture (14,18).

Specimen stability under the electron beam

Under suitable conditions, which are now fairly well established, tissue sections appear to be morphologically stable under the beam. When a conducting coating has been evaporated onto the specimen surface, it is quite unlikely that morphological

damage will be seen with beam currents under 10 nA;   50 nA
may well be withstood without apparent damage;   and much higher
currents can be tolerated when heavier coatings are evaporated
onto thin specimens (10,11).   However, even without apparent
damage, sections may suffer considerable losses in total dry
mass and in some minor elements.

Relatively volatile elements, such as Na, K, Se, and
S, may be partially removed because of a rise in the temperature of the specimen.   Such a rise is not very rapid, so that
one can check for this effect simply by observing the steadiness
of the X-ray line intensities.

Two other effects must be considered: loss of elements,
prominently C, N, O and H, in the gaseous products of
radiation-induced reactions (23), and displacement of ionic
elements due to local electrostatic fields built up by the beam
in the specimen (4,17,25,26).   These effects are likely to
occur at very low radiation doses, before enough X-rays have
been generated for analysis, even with highly efficient energy-
dispersive X-ray spectrometers, and before ordinary detection
methods can reveal that changes have taken place.   Special experiments will be necessary to establish the size of these
effects and the conditions for avoiding them.

RECENT DEVELOPMENTS

Three recent developments are greatly extending the effectiveness of the electron microprobe in biology:

Combined conventional electron microscopy and X-ray microanalysis

In "EMMA" instruments, X-ray microanalysis is combined with
conventional electron microscopy, making it possible to carry out
elemental analysis with reference to features which can be seen
only in the conventional transmission electron microscope. Since
the spread of the beam due to scatter in conventional 1000 Å sec-

tions is slight, it has been possible to achieve a spatial resolution of the X-ray analysis in the neighbourhood of 1000 Å
(19).

Transmission scanning microscopy

This mode of scanning electron microscopy provides images
with the same contrast as conventional electron micrographs,
for specimens of thickness up to the order of a few μm, with
image spatial resolution of the order of 0.1 μm. When a scanning electron microscope is fitted with attachments for transmission electron images and for X-ray spectroscopy, one may
have the benefit of the contrast available in transmission electron images combined with an X-ray intensity which is higher
than available from ultra thin specimens.

"Energy-dispersive" spectroscopy with cooled silicon detectors

One can analyze the X-rays from a specimen without using
any diffractor, by the so-called "energy-dispersive" technique.
In this method the entire spectrum of X-rays impinges on a
detector, which responds to each incident X-ray quantum by producing an electrical pulse proportional in height to the quantum
energy. The pulse-height distribution is then recorded by a
multichannel pulse-height analyzer, giving simultaneously a record of the entire X-ray spectrum. Energy-dispersive analysis
has been practised for some time with gas-filled X-ray counters
(5,9), but the method has recently become much more convenient
and powerful with the advent of greatly improved, cooled, semiconducting silicon detectors.

In order to illustrate the technique, Fig.5 shows the
pulse-height distribution photographed from the display tube of
a multichannel pulse-height analyzer connected to a silicon detector exposed to X-rays from the two elements nickel (atomic
number 28) and copper (atomic number 29). The main peaks cor-

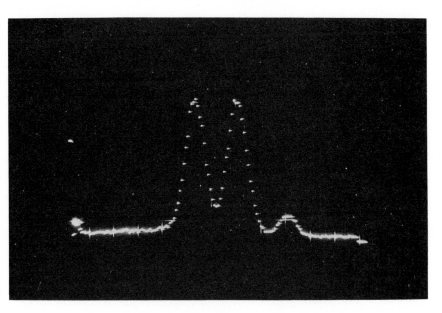

Fig.5 : A pulse-height spectrum generated in an energy-
dispersive silicon detector by equal intensities of X
radiation from nickel (atomic number 28) and copper
(atomic number 29).   The K-alpha lines (the two main
peaks) are easily resolved.   The copper K-beta line is
seen at the right, but the nickel K-beta line is obscured by the copper K-alpha peak.   (From Hall *et al*.
(12).   Courtesy of Charles C. Thomas, Publishers.)

respond to the K-alpha radiation of the two elements, and the
small peak is due to copper K-beta radiation.   Excellent
examples of spectra obtained in biological applications of
energy-dispersive analysis are given in other chapters in this
volume (Gehring *et al*.;   Gullasch).

The pulses generated in a silicon detector by mono-
energetic quanta are not all exactly equal;   there is a distribution of pulse heights about the average.   Hence there is an
overlap between close spectral lines.   The wavelength or quan-
tum-energy resolution of energy-dispersive spectroscopy is inferior to the resolution of diffracting crystals, so that dis-

crimination between close X-ray lines or against continuum
background is poor. On the other hand, when the detector is
placed close to the specimen, one can detect a much higher fraction of the X-rays leaving the specimen. The evaluation of the
relative roles of diffractive and energy-dispersive spectroscopy
is quite complicated, but in practical terms the following gen-
eralizations seem valid:

　　　1.   When a silicon detector is located 10 cm from the
specimen, the detection efficiency for a given particular X-ray
line is similar to the efficiency of the most sensitive diffractive spectrometers. Hence one cannot much improve the sensitivity of a microprobe analyzer by fitting a silicon detector at
a distance of 10 cm or more, although one does gain the advantage of collecting the entire X-ray spectrum simultaneously.

　　　2.   In scanning electron microscopes, silicon detectors
can readily be mounted at a specimen distance of the order of
1 cm.   In this case there is a great improvement in the efficiency of X-ray detection, permiting the analysis of smaller
volumes with lower beam currents, thus improving spatial resolution and reducing the danger of specimen damage.   However, the
minimum detectable mass fractions for energy-dispersive spectrometers are generally not as good as for diffractive spectrometers because of the poorer peak-to-background ratios and the
greater difficulty in excluding extraneous backgrounds.   For
elemental mass fractions of the order of 1%, as encountered in
electrolyte studies, the closely mounted silicon detectors perform beautifully in quickly identifying elemental constituents
and in roughly indicating their amounts.   For quantitative
analysis, it remains to be seen how well the relatively high
backgrounds can be taken into account.

　　　The reader may want to refer to a recently published, extensive discussion of energy-dispersive X-ray spectroscopy (22), or
to a still more recent analysis of performance capabilities (21).

REFERENCES

1. Andersen, C.A. *In:* Methods of Biochemical Analysis 15, pp.147-270. (D. Glick, Ed.) Interscience Publishers, New York (1967).

2. Andersen, C.A. *Brit. J. Appl. Physics 18*: 1033-1043 (1967).

3. Birks, L.S. "Electron Probe Microanalysis". Interscience Publishers, New York (1963).

4. Borom, M.P. and Hanneman, R.E. *J. Appl. Physics 38*: 2406-2407 (1967).

5. Dolby, R.M. *J. Sci. Instr. 40*: 345-351 (1963).

6. Echlin, P. *In:* 4th Annual Scanning Electron Microscope Symposium, Part I, pp.225-232. (O. Johari, Ed.) Ill. Inst. Tech. Research Institute, Chicago (1971).

7. Gehring, K., Dörge, A., Nagel, W. and Thurau, K. (this volume).

8. Gullasch, J. (this volume).

9. Hall, T.A. *In:* Encyclopedia of X-Rays and Gamma Rays, pp.653-655. (G.L. Clark, Ed.) Reinhold Publishers, New York (1963).

10. Hall, T.A. *In:* Physical Techniques in Biological Research, 2nd Ed., Vol. IA, pp.157-275. (G. Oster, Ed.) Academic Press, New York (1971).

11. Hall, T.A., Hale, A.J. and Switsur, V.R. *In:* The Electron Microprobe", pp.805-833. (T.D. McKinley, K.F.J. Heinrich and D.B. Wittry, Eds.) John Wiley and Sons, New York (1966).

12. Hall, T.A., Röckert, H.O.E. and Saunders, R.L. de C.H. "X-Ray Microscopy in Clinical and Experimental Medicine". In press. Charles C. Thomas Publishers, Fort Lauderdale, Florida.

13. Höhling, H.J. and Hall, T.A. *Die Naturwissenschaften 56*: 622-629 (1969).

14. Ingram, M.J. and Hogben, C.A.M. *Analyt. Biochem. 18*: 54-57 (1967).

15. Ingram, M.J. and Hogben, C.A.M. *In:* Developments in Applied Spectroscopy *6*, pp.43-54. Plenum Press, New York (1968).

16. Läuchli, A., Spurr, A.R. and Wittkopp, R.W. *Planta (Berlin)*

95: 341-350 (1970).

17. Lineweaver, J.L.   *J. Appl. Physics 34*:   1786 (1963).

18. Morel, F. and Roinel, N.   *J. de Chimie Physique 66*: 1084-1091 (1969).   (Cf. also Morel, Roinel and Le Grimellec, *Nephron 6*:   350-364 (1969).)

19. Nottingham.   Proc. Nottingham Conference on X-Ray Microanalysis in Biology, July 23, 1971.   *Micron*.   In press.

20. Philibert, J.   *In:* "X-Ray Optics and Microanalysis", pp.114-131.   (G. Möllenstedt and K.H. Gaukler, Eds.) Springer-Verlag, Berlin, Heidelberg and New York (1969).

21. Russ, J.C.   *In:* Proc. 4th Annual Scanning Electron Microscope Symposium, Part I, pp.65-72.   (O. Johari, Ed.) Ill. Inst. Tech. Research Institute, Chicago (1971).

22. Russ, J.C. (editor) *et al.*   "Energy-Dispersion X-Ray Analysis: X-Ray and Electron Probe Analysis".   Special Technical Publication 485, American Society for Testing and Materials, Philadelphia (1971).

23. Stenn, K. and Bahr, G.F.   *J. Ultrastruc. Res. 31*:   526-550 (1970).

24. Swift, J.A. and Brown, A.C.   *In:* Proc. 3rd Annual Scanning Electron Microscope Symposium, pp.113-120.   (O. Johari, Ed.)   Ill. Inst. Tech. Research Institute, Chicago (1970).

25. Varshneya, A.K., Cooper, A.R. and Cable M.   *J. Appl. Physics 37*:   2199 (1966).

26. Vassamillet, L.F. and Caldwell, V.E.   Paper 40 in Proc. 3rd Nat. Conf. on Elec. Microprobe Analysis, Elec. Probe Anal. Soc. Amer., Chicago (1968).

LOSS OF MASS

IN BIOLOGICAL SPECIMENS

DURING ELECTRONPROBE X-RAY

MICROANALYSIS

H. J. HÖHLING*, T.A. HALL**, W. KRIZ***,
A.P.v. ROSENSTIEL****, J. SCHNERMANN***** and U. ZESSACK*

*Institut für Medizinische Physik, Münster, Germany
**Cavendish Laboratory, Cambridge, England
***Anatomisches Institut, Münster, Germany
****Metaalinstituut TNO, Delft, The Netherlands
*****Physiologisches Institut, München, Germany

Applying the "relative quantitive" method of Hall *et al*. (1) to our probe measurements of intracellular electrolytes (Na, K, Cl) in kidney sections we obtained several new results concerning the distribution of these elements in the different portions of the nephron (2,34). After this first step of probe analysis we used the theory of full quantitative elemental determination (5,6) in order to get real elemental weight fractions. The results of our first measurements were disappointing as we got far too high values for the intracellular content of Na, K and Cl.

To clarify this discrepancy we tested the experimental conditions of the probe analysis for the Marshall-Hall theory by analyzing the elemental content in thin sections of known mineral standards. Secondly we tested the method on known organic standards, i.e. kidney-homogenate sections and Agar-Agar sections.

PREPARATION AND MEASUREMENTS

Mineral standards

Powders of "Sanidin" (K-, Na-Feldspar), "Albit" (Na-Feldspar) and NaCl were embedded in methacrylate. Ultra thin sections of these minerals were transferred to the nylon foil of microprobe-holders. The sections were covered on the top-side with an Al-layer of approximately 450 A thickness.

Organic standards

Rat kidneys were homogenized mechanically and by ultrasound. The homogenates were used in the original composition, after reducing and after increasing the concentrations of the electrolytes. The final concentrations were determined by conventional chemical analysis. In addition to the kidney-homogenates Agar-Agar solutions were used to which defined amounts of electrolytes were added. From these organic materials 6 μm sections were cut in a cryostat, transferred to a microprobe-holder, freeze-dried and covered with Al.

In the mineral sections we performed static probe measurements, i.e. we placed the beam onto individual mineral grains, the probe remained fixed on each grain while X-ray data were counted. In the organic materials we scanned microareas of 80 × 80 μm. We used a high voltage of 30-35 kV, and a specimen current of 59 nA (if not otherwise noted).

EXPERIMENTS, RESULTS AND DISCUSSIONS

Tests of the mineral standards

The quality of the experimental conditions for the quantitation, using the minerals, was checked in a way in which the full equation for determining elemental weight-fractions, i.e. equation 4 of Hall and Höhling (7), need not be applied. Using the symbols

n_x = counts for chosen characteristic radiation of element x,

n_w = continuum counts,

N_x = number of atoms of element x in the analyzed microvolume,

ΣN = number of atoms of all constituent elements in the analyzed microvolume,

Z = atomic number

the following equation results:

$$\frac{(n_x/n_w)_1}{(n_x/n_w)_2} = \frac{\left[N_x/(\Sigma NZ^2)\right]_1}{\left[N_x/(\Sigma NZ^2)\right]_2}$$

The subscripts 1 and 2 identify different materials, in our case different standards. The values $N_x \Sigma NZ^2$ are the G-values of equation 4 of Hall and Höhling (7). So the ratio of the corresponding counting rates of standard 1 and 2 (on the left side) must correspond to the ratio of their G-values (on the right side of the equation). As Fig.1 shows, there was a good agreement between the measured values and the predicted values indicating that we can use the minerals as standards according to the Marshall-Hall theory.

Tests using a known kidney-homogenate (or Agar-Agar) as specimen and a known mineral as standard

In this series of probe analysis kidney homogenates were used as specimens, the minerals as standards. We applied the full equation of quantitation (7) and compared the probe values with the values obtained by conventional chemical analysis. In all cases the probe values were far too high. In an original kidney homogenate, for example, the mass fractions of the elec-

Pair of Materials (1)/(2)	Probe kV	Continuum Energy Band, keV	Takeoff Angle	R(1)/R(2) Mean of Measured Values	R(1)/R(2) Predicted
CaO/Wollastonite	40	20-25	20°	1.72±.05	1.70
Bytownite/Wollastonite	40	20-25	20°	0.385±.02	0.437
Wollastonite/CaWO$_4$	35	12-22	75°	6.9±2.	7.7
NaCl/Sanidine	30	at 15	52$\frac{1}{2}$°	26.2±.6	25.8
Albite/Sanidine	30	at 15	52$\frac{1}{2}$°	7.85±.27	7.65

Fig.1 : Comparison of measured (column 4) and predicted values (column 5). The results of the first three pairs of materials have been obtained by Hall and Werba (1971), the last two pairs by our group.   See text for further explanation.

trolytes which we got by conventional chemical analysis were: Na 0.55%, Cl 0.53% and K 0.86%. The corresponding probe values were: Na 0.93%, Cl 0.98% and K 1.47%.

The most probable reason for this discrepancy is a loss of mass during the measurements. A loss of organic material during electron bombardment in the electron-microscope has been described by Stenn and Bahr (8). In order to answer the question whether there is a loss of mass we started some additional experiments.

We used kidney-homogenate sections as well as Agar-Agar sections. One group of sections was coated on the top-side with a single layer of Al of approximately 450 Å thickness, a second group was coated on both sides. Both groups were analyzed with a "detail-scan" of 80×80 μm using in one run a specimen current of 59 nA, in a second run, 10 nA.

Repeated measurements of the same microarea

In these experiments we put the detail-scan onto a "fresh" microarea and immediately started to count the "white" X-ray photons (as information on the total local mass) as well as, in the same run, the characteristic counts of Na, K and Cl. We scanned such an area for a period of about 2 minutes, counting the pulses for any 12 seconds. Whereas the counting rate of the characteristic radiation remained fairly constant, the counting rate of the continuum radiation decreased in the beginning of a run (Figs 2 and 3). These data indicate a loss of mass, which apparently was not due to a loss of the elements K, Na and Cl. Figures 2 and 3 demonstrate that the loss of mass is most drastic in the single coated sections analyzed with 59 nA specimen current, less prominent in the double coated sections analyzed with 59 nA specimen current as well as in the single coated sections analyzed with 10 nA specimen current. No decrease in the counting rate of the "white" pulses is visible in

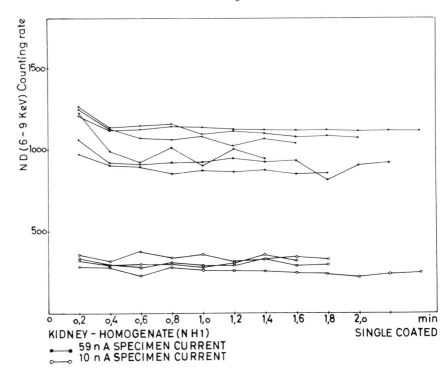

Fig.2 : Repeated measurements of the same microarea in a single
coated section. The counting rate of the white pulses
(ND) shows a decrease in the beginning of the measurements. The decrease is more prominent in the analysis
with 59 nA than with 10 nA specimen current.

the double coated sections analyzed with 10 nA specimen current. It is, however, unknown if there was a decrease during
the first seconds or perhaps even milliseconds of electron-
bombardment.

Comparison of mean values from single and double coated sections
In this series of analyses the mean values of several
(approx. 10) measurements of different areas of single and double
coated sections of the same homogenate (as well as the same Agar-
Agar) were compared. It is apparent from Fig.4 that there is no
difference in the counting rate of the characteristic radiation

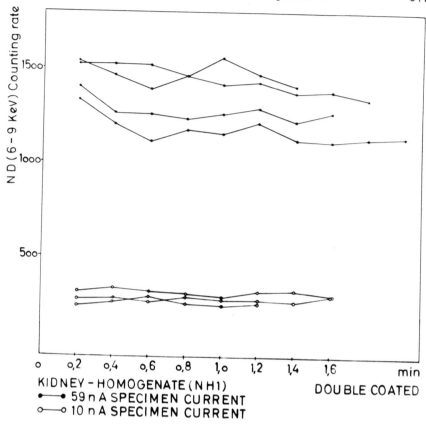

KIDNEY-HOMOGENATE(NH1)   DOUBLE COATED
●——● 59 n A SPECIMEN CURRENT
○——○ 10 n A SPECIMEN CURRENT

Fig.3 : Repeated measurements of the same microarea in a double
coated section.   A decrease of the counting rate of the
white pulses (ND) is only visible in the analysis with
59 nA specimen current.   There is no decrease with
10 nA specimen current.   A possible decrease in the
first seconds remains obscure.

of the elements Na, K and Cl in single and double coated
sections.   There is, however, a considerable difference in the
counting rate of the "white" pulses probably indicating a higher
loss of mass for the single coated section than for the double
coated one.   This is the reason why the relative mass fractions
of the elements are higher in the single coated than in the
double coated sections.   Very similar results have been obtained

H. J. Höhling et al.

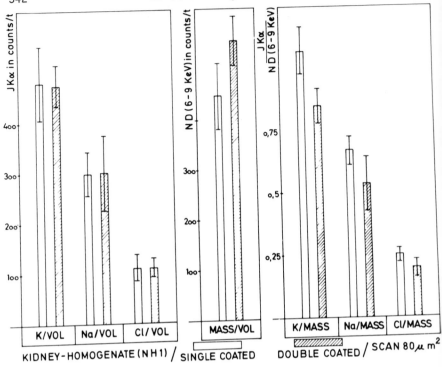

Fig.4 : Comparison of counting rates of single and double
coated sections of the same kidney homogenate.
The columns and brackets represent the mean ± standard
deviation. There are no differences in the counting
rates of the characteristic radiation of K, Na and
Cl (IKα in counts/t; left side). There is a considerable difference in the counting rates of the continuum radiation (ND in counts/t; middle). Therefore
there are differences in the elemental weight fractions
between single and double coated sections (IKα/ND;
right side).

in the Agar-Agar specimens.

It is apparent from these data that a loss of mass occurs
during electron bombardment. There are, however, ways to avoid
or at least to decrease this loss of mass. One possibility is
a thicker coating, another possibility is to use lower probe
currents. It will be necessary to determine the exact amount
of the loss of mass under defined conditions and in different

biological soft tissues.   It is possible that there may be differences in behavior under the electron beam between sections of
kidney-homogenates and those of kidneys.   Nevertheless, the
problems seem to be solvable so that in future Hall's method of
quantitation may be applicable to biological soft tissues.

REFERENCES

1. Hall, T.A., Hale, A.J. and Switsur, V.R.   *In:*   The Electron-microprobe, p.805.   (McKinley, Heinrich and Wittry,
   Eds.)   John Wiley and Sons, New York (1966).

2. Schnermann, J., von Rosenstiel, A.P., Kriz, W. and Höhling,
   H.J.   *Pflügers Arch. ges. Physiol. 319:*   R 80 (1970).

3. Kriz, W., Höhling, H.J., Schnermann, J. and von Rosenstiel,
   A.P.   *Anat. Anz. Suppl. ad 128   (Verh. Anat. Ges. 65):*
   217 (1971).

4. Kriz, W., Schnermann, J., Höhling, H.J., von Rosenstiel,
   A.P. and Hall, T.A.   Electron probe microanalysis of electrolytes in kidney cells: problems and results.   Proc. of
   the Symp. on Renal Handling of Sodium, Brestenberg 1971.
   Karger, Basel.   In press.

5. Marshall, D.J. and Hall, T.A.   *In:*   X-ray Optics and
   Microanalysis, p.374.   (Castaing, Deschamps and Philibert,
   Eds.)   Hermann, Paris (1966).

6. Hall, T.A.   *In:*   Physical Techniques in Biological Research
   2nd Ed., vol. IA, p.157.   (G. Oster, Ed.)   Academic Press,
   New York (1971).

7. Hall, T.A. and Hohling, H.J.   *In:*   Vth International Con-
   gress on X-ray Optics and Microanalysis, p.581.
   (Möllenstedt and Gaukler, Eds.)   Springer, Berlin-
   Heidelberg-New York (1969).

8. Stenn, K. and Bahr, G.F.   *J. Ultrastruct. Res. 31:*   526
   (1970).

ENERGY DISPERSIVE X-RAY

MICROANALYSIS

IN RED BLOOD CELLS AND

SPERMATOZOA

J. GULLASCH

*Siemens AG, Bereich Mess- und Prozesstechnik,
Karlsruhe, Germany*

The only way to detect the small amounts of physiological electrolytes present in thin biological specimens by means of X-ray microanalysis seems to be the use of solid state detectors. Such a detector has been used and is attached to the microprobe Elmisonde* in order to investigate the inorganic constituents of human red blood cells and bull spermatozoa. Both these cells appear to be particularly suitable because of the ease with which air-dried specimens can be prepared without the danger of heavy ionic redistribution during preparation. In addition, their ionic compositions are fairly well known and the X-ray analytical findings can therefore be easily related to this detail. In red blood cells K, Cl, P and Fe can be detected and quantitatively distinguished from the amounts present in the surrounding material. In spermatozoa a characteristic distribution of K and Ca between the head and the tail could

---

* Trademark

be demonstrated.

There are three main difficulties in the investigation of elemental compositions of thin biological specimens by means of electron microprobes or scanning electron microscopes.

Firstly, the weight fractions of some interesting inorganic constituents to be analyzed are very small. This requires a high sensitive X-ray detector in order to detect the low intensities of the characteristic X-rays.

Secondly, biological samples contain up to 80% water which must be either removed by conventional drying techniques or transformed into the frozen state. In the latter case the sample must be maintained at very low temperatures during the analysis in order to prevent water-sublimation in the vacuum.

However, both methods create some particular problems: drying techniques inevitably produce ionic redistribution giving rise to misleading results. On the other hand the analysis of water-containing specimens requires beam currents small enough to prevent water-sublimation by heating up the analyzed area during electron bombardment.

Thirdly, the quantitative determination of ionic concentrations in transparent specimens is difficult because of the high statistical error due to the low X-ray intensities. In addition, matrix corrections are not as accurate as bulk specimens. Finally there is a loss of mass during analysis as stated in the paper of Dr. Höhling. Nevertheless at the present time the electron probe microanalysis seems to be one of the most attractive and promising methods for the elemental analysis of biological specimens. Most of the problems listed above may be overcome in the near future by technical improvements particularly designed for the needs of biological applications.

In order to make some steps towards the biological applicability of electron probe microanalysis some preliminary experiments were performed in red blood cells and spermatozoa on the Elmisonde. Both preparations appear to be particularly suitable because of the ease with which air-dried specimens can be prepared without the danger of intolerable ionic redistribution during preparation. In order to overcome the problem of sensitivity a special lithium-drifted silicon detector has been used with an active area of 40 mm^2. This detector has been connected with an extremely low noise preamplifier and a computer-calculated shaping amplifier. With this device it is possible to detect all elements down to fluorine and to discriminate adjacent elements from silicon upwards provided that the concentration differences between adjacent elements are not too large.

In Fig.1 the results of a microanalysis obtained in a human red blood cell is shown. In the upper panel specimen current pictures with negative and positive signal polarity indicate the analyzed area being either inside or outside one single red blood cell. In the corresponding X-ray spectra obtained with a multi-channel analyzer the characteristic peaks represent the elements P, Cl, K and Fe. The presence of the Cu lines L, $K\alpha$ and $K\beta$ is due to the fact that the specimens are mounted on usual nylon-coated copper grids, which are hit by scattered electrons exciting their characteristic radiation. An interesting result is the detection of iron as a constituent of hemoglobin.

This result indicates the sensitivity of energy dispersive X-ray detection. The weight fraction of iron in air-dried red blood cells roughly calculated is in the range of 0.002%. By comparison of both spectra obtained inside and outside of the red blood cell characteristic qualitative and quantitative differences can be seen, particularly with respect to potassium which is accumulated inside the cell as one would expect from the physiological situation.

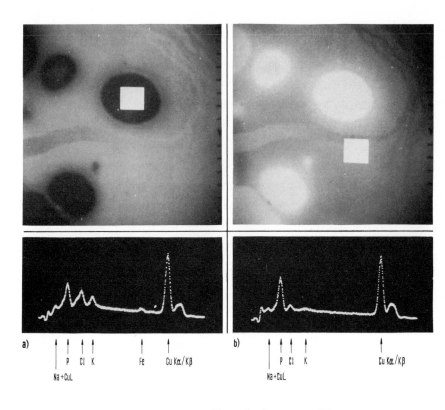

a) red cell   **Chemical composition**

b) surrounding material   **of a human red blood cell**

Fig.1 : Energy dispersive X-ray spectra obtained from (a) inside
and (b) outside a human red blood cell.   See text for
further explanation.

Figure 2 demonstrates results obtained in a single bull
spermatozoon.   Measurements were made in the head area, the tail
and the surrounding material respectively.   Again remarkable
and characteristic differences were found.   One of them might
be of physiological relevance.   That is the high amount of calcium present in the tail region as compared with the head area
or the surrounding electrolytes.

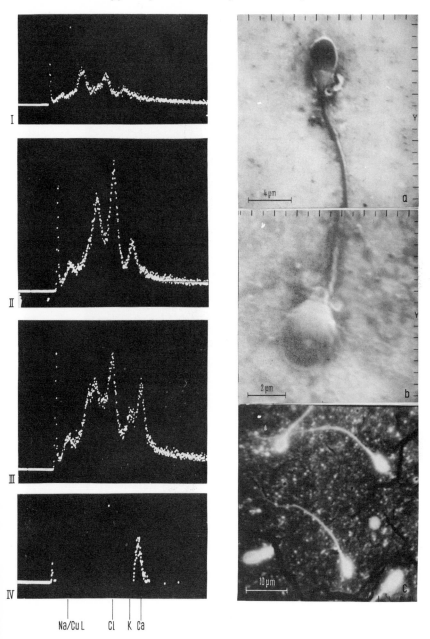

Na/CuL   Cl   K Ca

Fig.2 : Energy dispersive X-ray spectra obtained in the head
(a), the tail (b) and the surrounding material (c) of a
single air-dried bull spermatozoon.   See text for
further explanation.

Could it be that calcium is of some importance in the control mechanism of sperm motility as it is in muscular function?

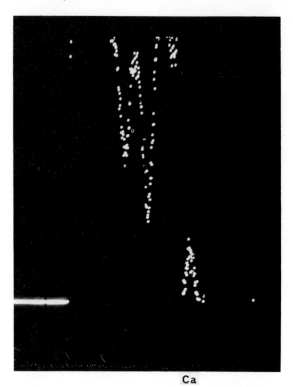

Ca

Fig.3 : Energy dispersive X-ray spectra obtained in the surrounding material (a), the head (b) and the tail region (c) of a single air-dried bull spermatozoon. (d) represents the difference spectrum between (c) and (b) indicating that the tail contains much more calcium than the head.

In Fig.3 the same facts as before in other spermatozoa are shown. We have again the accumulation of Ca in the tail region. The difference-signal between tail and head region given on the right show that all elements disappeared except Ca.

Summarizing our preliminary results it can be pointed out that the small amounts of inorganic constituents present in thin layers of soft biological materials can be qualitatively detected within the spatial resolution limits of the method. The use of a solid state detector in combination with a multi-channel analyzer permits the simultaneous registration of all constituents cutting short the measuring time to acceptable values.

ACKNOWLEDGEMENT

We would like to thank Dr. Kaufmann from the Institute of Physiology, University of Freiburg for selection and preparation of the specimens and for many helpful discussions.

THE USE OF THE SCANNING ELECTRON MICROSCOPE IN CONNECTION WITH A SOLID STATE DETECTOR FOR ANALYSIS AND LOCALIZATION OF ELECTROLYTES IN BIOLOGICAL TISSUE*

K. GEHRING**, A. DÖRGE, W. NAGEL and K. THURAU

Institute of Physiology, University of Munich, Germany

The X-ray analysis of elements in biological tissue with the electron probe permits the determination of the intra- and extracellular concentrations of electrolytes. In order to achieve this, two main conditions must be fulfilled:

Firstly, the tissue must be prepared in such a way that dislocation of the electrolytes is kept to a minimum, and secondly, it is essential that relatively low electrolyte concentrations in a small volume of the tissue can be measured accurately.

To fulfil the first condition, tissue was shock-frozen in isopentane at $-150^{\circ}C$ and then cryosectioned at $-50^{\circ}C$ with a

---

* Supported by Deutsche Forschungsgemeinschaft.
** Recipient of a scholarship from the Stiftung Volkswagenwerk.

steel knife.   After cutting, a silver specimen grid, covered
with a collodium foil was pressed against the knife to mount the
section.   For freeze-drying, the grid containing the section
was placed in a special holder, and a Teflon plate was placed
over this holder so that the tissue section lay between the
Teflon and the collodium foil.   This ensured that only water
escaped from the section.   After freeze-drying, the Teflon
plate was removed and another grid, coated with collodium foil
was placed over the section.   In this manner, frog skin and
Albumin-Agar were cut into sections of 2.5 μ thickness.

The determination of electrolyte concentrations in the tissue section was performed by scanning small areas to allow
measurements at the cellular level.   Since the excited volume
is very small, and because the electrolyte concentrations are
low, the X-ray emission produced is also low.   Increasing the
beam current to increase the X-ray emission is not a suitable
solution to this problem because it damages the tissue.   Therefore, we attempted to increase the sensitivity of the X-ray
detector system to achieve sufficient count rates at the lowest
possible beam current.   For this purpose, the Cambridge S4
scanning microscope was used in connection with an energy-
dispersive X-ray detector (Nuclear Diodes, Chicago, Ill.).   By
placing the detector crystal close to the specimen, a beam current of only 0.4 nA was sufficient for the quantification of
the X-ray spectra.   As indicated in Fig.1, the distance between
the specimen and the detector crystal was less than 10 mm.
A 7.5 μ thick beryllium window between the specimens and the
detector crystal absorbs only 40% of the Na radiation.   The
modified specimen holder used is shown in Fig.2.   In order to
obtain a transmission electron image, a scintillation crystal
associated with a light pipe is placed under the specimen.   The
best acceleration voltage for Na excitation for these tissue
slices was found to be between 10 and 15 kV.   The beam cur-

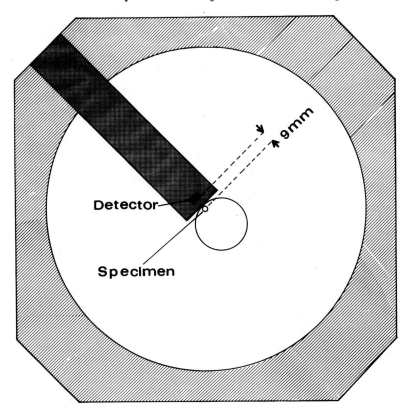

Fig.1 : Cross-section of the specimen chamber of the scanning
electron microscope, showing the position of the specimen and the solid state detector.

rent, which was set at 0.4 nA, measured by use of a Faraday
cage, remained stable during the analysis. Eighty percent of
this current is absorbed as specimen current, i.e. most of the
electrons are slowed to zero velocity in the specimen. This
means that for quantification of our data, the theory of Hall
(2), which requires a small energy loss of the probe electrons
when passing through the specimen, may not be applied. In
order to circumvent this problem, the following steps were
carried out: The ratio of the characteristic counts for each
element and the counts of the white radiation found in the tis-

K. Gehring et al.

Fig.2 : Specimen holder of the scanning electron microscope with 7 carbon tubes. Three tubes contain silver grids with sections sandwiched between collodium foils.

sue section was compared with the ratio determined from the Albumin-Agar standard sections. These standards had the same thickness and approximately the same weight fraction of dry matter as the sections of the frog skin. The electrolyte concentrations of the standards were in the same range as expected in the frog skin. A calibration curve for Na obtained from these standards is shown in Fig.3.

During the first 10 to 20 sec of analysis, a decrease in the count rate was observed. There is evidence that this

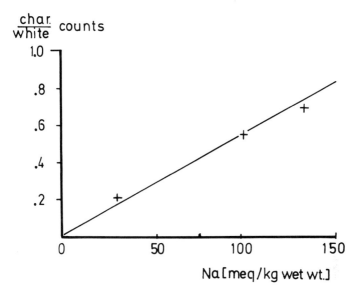

Fig.3 : Calibration curve obtained from 2.5 μ thick Albumin-
Agar sections with different Na concentrations. The
ratio of the count rate of the characteristic emission
of Na and the white radiation of the Albumin-Agar
sections is plotted against the Na concentrations
measured by flame photometry.

decrease is mainly due to the loss of matter from the covering
collodium foil. In separate experiments, we measured the decrease of count rate observed during the bombardment of collodium membranes only. The reduction in the count rate was practically the same as that observed in the sandwiched tissue sections. Since in no case was the covering collodium foil des-
troyed by the bombardment, it is unlikely that a considerable
loss of matter had occurred from the tissue sections.

In preliminary experiments, we determined electrolyte concentrations in the frog skin before and after the action of
ouabain. These experiments were chosen because this tissue had
been previously analyzed under similar conditions by chemical
methods. We were thus able to compare our data with those re-

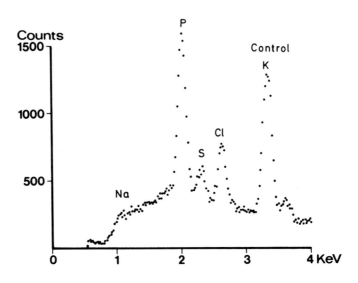

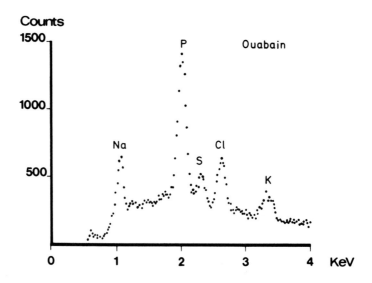

Fig.4 : X-ray spectra between 0.5 and 4 keV obtained in epithelial cells of frog skin sections in control and after ouabain.

ported by Aceves and Erlij (1).

For X-ray analysis, an area of $3\mu^2$ was scanned in order to eliminate the scatter caused by the inhomogeneity of the freeze-dried structures. The counting time was 100 sec, the count rates were approximately 700 cps. Figure 4 shows the spectra obtained from the intracellular measurements in two pieces of the same frog skin, one serving as control, the other after its exposure to ouabain. Both pieces had been incubated under short circuit conditions with Ringer solution, using an Ussing type chamber. After ouabain, the short circuit current decreased to less than 15% of the control. Inspection of Fig. 4 shows clearly that the ouabain-treated frog skin has a higher Na peak and a lower K peak than the control. A quantitative

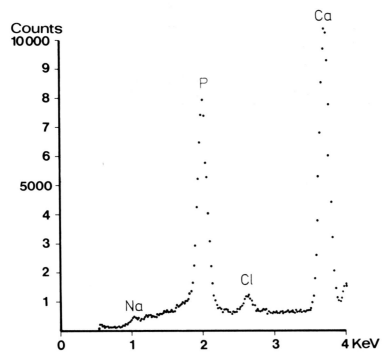

Fig.5 : X-ray spectrum between 0.5 and 4 keV of the electron dense layer between the loose and compact connective tissue of the corium.

analysis indicated that the intracellular Na concentration increases from 35 to 120 meq/1 and the K concentration decreases from 120 to 20 meq/1. Quantitatively similar results were obtained from all epithelial cells analyzed, which is consistent with the inhibitory action of ouabain on membrane transport systems. The absolute values of the intracellular Na concentrations, as measured by this method, are in agreement with the results of Aceves and Erlij (1), who chemically analyzed the isolated epithelium of the frog skin.

The electron dense layer between the loose and compact connective tissue in the corium was found to have a high content of Ca and P, and a low content of Na and Cl (Fig.5).

REFERENCES

1. Aceves, J. and Erlij, D.   Sodium transport across the isolated epithelium of the frog skin.   *J. Physiol. 212:* 195 (1971).

2. Hall, T.A.   The microprobe assay of chemical elements. *In:* Physical Techniques in Biological Research. (G. Oster, Ed.)   Academic Press, New York and London (1971).

THE ION MICROPROBE*

H. LIEBL

Max-Planck-Institute for Plasma Physics, Munich, Germany

INTRODUCTION

A researcher working with microscopic material will sooner
or later wish to see more than just the structures of the sample
in his microscope; he will want to know in addition the chemical composition of what he sees. Important advances have been
made with autoradiographic methods, especially after artificial
radioactive isotopes became available. For many problems,
however, autoradiography is not applicable, either because of
lack of suitable isotopes or because of undesirable consequences
of the radioactive decay. About 20 years ago the electron
microprobe X-ray analyzer was developed by Castaing (1) and has
proved to be a very successful tool for microanalysis in the
years since, as has been demonstrated in other papers. The
method has an intrinsic limitation to low concentrations due to
the continuous bremsstrahlen background. Another limitation is
poor sensitivity of elements with low atomic numbers and the
exclusion of the first few elements in the periodic table.
Furthermore, the inherent limit of the spatial resolution is a
volume of a few cubic microns, due to the scattering of the
electrons in the sample.

When a research method has been developed to its limits, a
different method is sought. So a new approach based on a dif-

* Part of this work was performed under a contract of association with EURATOM.

ferent physical concept was made 9 years ago by Castaing and
Slodzian (2) with the development of the Ion Microanalyzer and
4 years ago by the author with the Ion Microprobe Mass Analyzer.
The method used in these two instruments is mass spectrometric
analysis of material sputtered from the sample, combined with
the necessary ion optics to achieve a lateral resolution comparable to that of an optical microscope. Mass spectroscopy
is an analytical method with high sensitivity, capable of identifying single ions.   Ions of all elements can be analyzed
with the same sensitivity using the same mass spectrometer. A
completely new aspect is the possibility of isotopic analysis,
which has a considerable range of applications, epsecially in
the biological sciences.

TWO WAYS OF MICROANALYSIS

Figure 1 shows schematically the two possibilities of
microscopy combined with analysis: the imaging method (a), and
the microprobe method (b).   In both cases the sample is induced
by some kind of primary radiation to emit some kind of secondary
radiation through which the atoms or molecules that make up the
surface can be identified.   In case (a) the secondary radiation
is used to image the surface directly on a screen by means of
appropriate optics.   On its way it is analyzed by a filter so
that the image is formed by radiation coming from one atomic or
molecular species only.   The magnified image thus shows the distribution of that particular species over the sample area under
investigation.   A spot analysis can be made if a detector is
placed behind a small hole in the screen and the sample is moved
sideways until the image of the spot to be analyzed coincides
with the hole in the screen.   Now if the secondary filter is
made to scan the radiation, a spectrum is plotted representing
the composition of that spot.   In case (b) the primary radiation is focused on a fine beam probe that can be deflected across

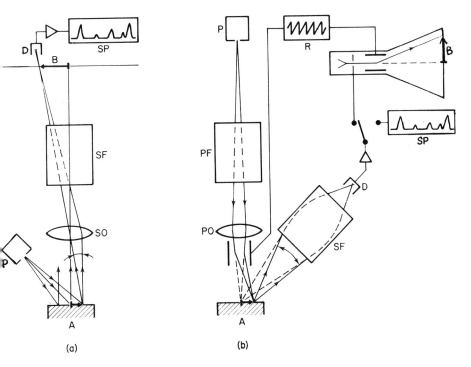

Fig.1 : Two ways of microanalysis: (a) Direct imaging combined
with analysis; (b) Microprobe technique. A = investigated area, B = image of A, P = primary source,
SO = secondary optics, SF = secondary filter, D = detector, SP = spectrum, PF = primary filter, PO = primary
optics, R = scanning.

the sample. If an image is wanted, the output signal from the
detector behind the secondary filter is used to modulate the
spot brightness of an oscilloscope, the primary beam is made to
scan the sample line by line in a TV-like manner and the spot
deflection of the oscilloscope is synchronized with the primary
beam deflection. Thus, if the secondary filter is tuned to
pass the radiation from one atomic or molecular species only,
the raster image appearing on the oscilloscope screen shows, on
a magnified scale, the distribution of that species over the
scanned sample area. Spot analyses are made by directing the

primary beam to the spot of interest and plotting the spectrum
of the secondary radiation.   Case (b), of course, is the principle underlying the electron microprobe X-ray analyzer, too.

MASS SPECTROMETRIC ANALYSIS

Mass spectrometric analysis is based on the phenomenon that
a beam of ions of equal speed or energy, passing in vacuum
through a transverse magnetic field (a magnetic prism) is split
according to the mass, or atomic weight, of the ions, similar to
a light beam being split up by a glass prism according to wavelength (Fig.2).   A magnetic prism not only disperses the ion
beam into its components, but also focuses, so that additional
lenses are unnecessary, as in the case of the optical spectrograph.   Another convenience is that the magnetic field strength
can be varied simply by turning a knob.   Therefore, the mass
spectrum is usually not recorded on a photographic plate like an
optical spectrum, but a slit is placed on the focal line with an
ion collector behind it.   As the magnetic field strength is increased from zero, the different mass components of the ion beam
will successively reach the collector.   An X-Y recorder will
thus plot the mass spectrum, if the collector current is made to
drive the vertical Y-deflection and the field strength the horizontal X-deflection.   The X-axis can be calibrated directly in
mass numbers (atomic weight) from which elements or compounds
are easily identified.

The sample has to be ionized first in the ion source (Fig.
3).   Gaseous samples can be ionized directly by passing a beam
of electrons through the gas.   Liquid or solid samples can also
be ionized by electron impact, but they have to be vaporized
first.   The ions formed are accelerated by an electric field.
Another way of getting ions from a solid sample is by sputtering,
as the physicists call it.   The sample is bombarded by a beam
of energetic primary ions such as argon or oxygen ions that them-

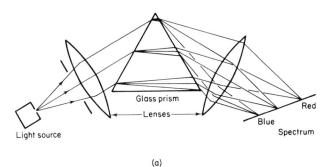

(a)

(b)

Fig.2 : Analogy between (a) optical spectrograph and (b) mass spectrometer.

selves have been formed in an ion source and accelerated by some 10 kV. Each of these primary ions knock several atoms out of the surface. This so-called sputtering is similar, on an atomic scale, to a billiard ball being hurled into a box filled with other billiard balls with such great force that it dives under the surface. The balls around the point of impact will be knocked outward, and some of them will jump up from the surface. A fraction of the sputtered atoms come off already ionized, so they can be directly accelerated into the mass spectro-

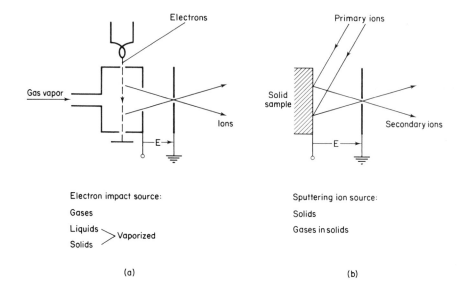

Electron impact source:

Gases

Liquids
Solids } Vaporized

(a)

Sputtering ion source:

Solids

Gases in solids

(b)

Fig.3 : Ion sources (E = acceleration voltage).

meter.   The sputtering ion-source type is obviously the one most suited for microanalysis.

ION PROBE MICROANALYZERS

Mass spectrometry applied to microanalysis according to the two ways sketched in Fig.1 means that the boxes indexed P are the primary ion sources, and the secondary filters (SF) are mass spectrometers, in addition to the primary filter (PF) indicated in case (b).   The secondary optics (SO) of case (a) and the primary optics (PO) of case (b) are made up of electrostatic lenses (Fig.4) that focus ion beams, just as electron beams are focused by magnetic lenses in electron microscopes or light beams by glass lenses in optical microscopes.   Ion lenses are inherently of poor optical quality and there is no way to correct for it, as can be done with glass lenses in microscope objectives.   Therefore, only very narrow ion beams can be used

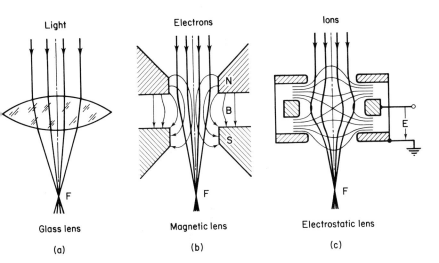

Fig.4 : Lenses (N = north pole, S = south pole, B = magnetic field, E = lens voltage, F = focus).

in imaging systems to keep the image aberrations small. For this reason, the scanning method (Fig.1b) has an advantage over the imaging method (Fig.1a) as far as absolute sensitivity is concerned: the spatial resolution depends in case (a) on the quality of the imaging by the secondary ions, whereas in case (b) it depends only on the fineness of the primary beam. Therefore a mass spectrometer with high transmission can be used as a secondary filter. Since analysis by sputtering is a process consuming the sample, the absolute sensitivity aspect is of importance.

The ion microanalyzer of Castaing and Slodzian (2) works according to case (a) of Fig.1. It will be discussed in detail in the following chapters.

The Ion Microprobe Mass Analyzer (3) (Fig.5) works according to case (b) of Fig.1. For reasons explained below, the primary ion beam is mass analyzed also and the sample can be

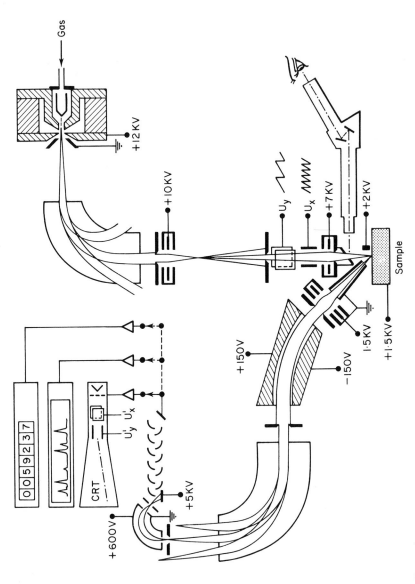

Fig. 5 : Ion microprobe mass analyzer (IMMA) (3).

TEST GRID IMAGES

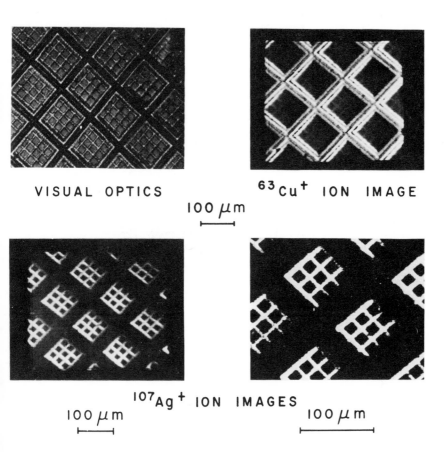

Fig.6 : Test pictures taken with IMMA (coarse mesh = copper, fine mesh = silver) (4).

viewed with an optical microscope during the analysis. The mass spectrometer analyzing the secondary ions is designed for very high transmission. This is achieved through a special combination of the magnetic field with an electric sector field and an ion lens. Figure 6 shows test pictures taken with this instrument (4). Development of ion microprobe instruments is

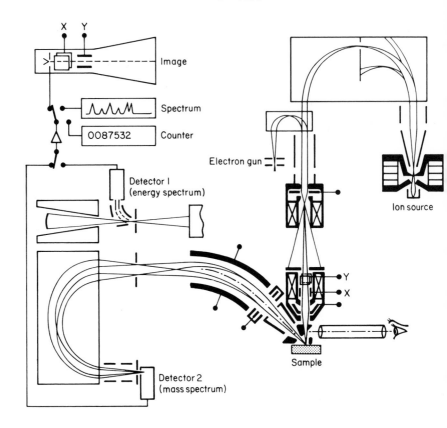

Fig.7 : Universal microprobe analyzer (UMPA) (7).

going on in several other laboratories (5,6).   A new microprobe apparatus for surface research in ultra high vacuum that
combines an ion microprobe with an electron microprobe (Fig.7)
has been developed by the author during the past few years.
The primary ion beam is mass analyzed by a 180° magnet and
focused by two special lenses that can also focus the electron
beam coming onto the optical axis through another small 180°
magnet.   Microscopic sample viewing is again provided.   The
electron probe can be used to release loosely bound molecules
or atoms from the sample, or simply for scanning electron
microscopy.   This instrument is now in the testing stage.   The

analyzer for the secondaries is a combined mass and energy
spectrometer that takes not only mass spectra but also energy
spectra of secondary ions or electrons.   This gives additional
analytical possibilities which to elucidate would lead us too
far afield at this time.   Although not provided in the present
laboratory instrument it would be a simple matter to attach an
X-ray spectrometer to it and thus have an Ion Microprobe Mass
Analyzer combined with an Electron Microprobe X-ray Analyzer.
Further, by attaching a laser to the viewing microscope, it
could also be used as a laser microprobe mass analyzer.

APPLICATIONS

The Ion Microprobes have been built and used by people who
did not have biological applications primarily in mind.   Most
applications made so far have been in the fields of metallurgy,
semiconductor physics and mineralogy.   Figure 8 shows as
example ion micrographs of a mineral, taken by Andersen (8) with
the Ion Microprobe Mass Analyzer.   Figure 9, which shows ion
micrographs of some unknown organic whisker on a steel surface
(4), is an admittedly poor example of an application to organic
materials.

It is one thing to produce impressive micrographs, but
quite another to make quantitative analyses.   The problem is
much more difficult than with electron-probe X-ray analysis.
In the latter case the inner shells of the atoms, which have
nothing to do with the chemistry of the atom, are involved in
the X-ray production.   Ionization, however, occurs in the outer
shell, where the atomic chemistry occurs.   Therefore, the fraction of the sputtered atoms which comes off as ions, depends
heavily on matrix effects, surface effects, and the primary ion
species.   The dependence on the primary ion species is the main
reason for the mass analysis of the primary beam, which is done
in the instruments shown in Figs 5 and 7, since it is virtually

HELVITE (Mn, Fe)₄ Be₃ Si₃ O₁₂ S

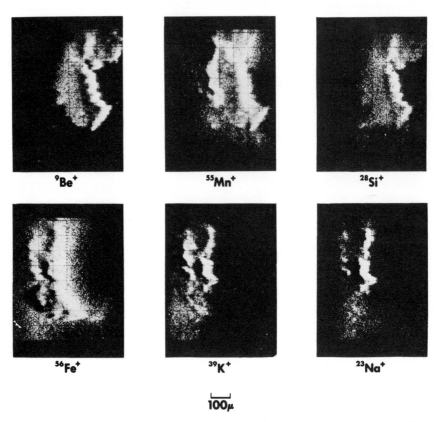

Fig.8 : Secondary ion micrographs of mineral sample (taken with
        IMMA (8)).

impossible to extract from an ion source an ion beam consisting
of a single ionic species.   Further, the problem of possible
interference of sample constituents with primary beam impurities
is avoided.

As complex as the problem of quantitative analysis looked
at the beginning, the progress made during the past few years
has been very encouraging.   Notably, Andersen has worked out a
method for quantitative analysis that has been applied, for instance, very successfully to microanalysis of rock samples

ORGANIC WHISKER ON STEEL SURFACE

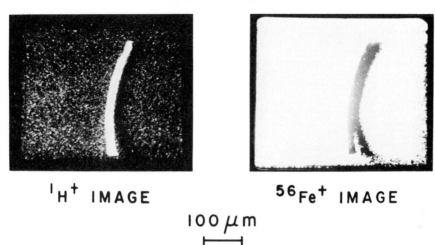

¹H⁺ IMAGE   ⁵⁶Fe⁺ IMAGE

100 μm

Fig.9 : Secondary ion micrographs of organic whisker (taken with IMMA (4)).

brought back from the moon by Apollo 11 (9). It has turned out that using reactive gas ions, like oxygen ions, as the primary species, offers great advantages in many cases over using noble gas ions, such as argon ions, which had generally been used, since it tends to enhance and to stabilize the secondary ion yield and thus make quantitative work much easier.

The Ion Microprobe will be most useful in applications that cannot be handled by electron microprobes; the two methods should therefore be considered as complementary, not competitive. These applications are:

Low concentration work

Due to the virtually non-existent background in a mass spectrum, trace elements can be detected in very low concentrations, down to ∿10 ppb for most elements; for others, such as the alkalis, which have an exceptionally high secondary ion yield,

even lower.   For comparison, the concentration limit for electron-microprobe X-ray analysis is   100 ppm   in favorable cases.

Low absolute detection limit

Low concentration detectability is not the whole story.
One part per billion out of 1 kilogram is still one microgram,
which is a lot of material in terms of microanalysis.   The detection limit with the Ion Microprobe goes down $\sim 10^{-18}$ g for
many elements (electron microprobe $\sim 10^{-16}$ g), for the alkalis
again even lower.   For example, as few as   100 K   atoms can be
detected, which is less than 10^{-20} g.   Typically, for non-
alkalis the sample consumption to produce one ion micrograph of
a major constituent is about 1% of an atomic layer with the
scanning type instrument (about one atomic layer with the
imaging instrument).

Depth analysis

During sputtering analysis the surface is being eroded.
The atoms sputtered away originate from the topmost five or so
atomic layers of the sample at any instant.   The depth resolution, therefore, can be   10   to   20 Å,   if care is taken that
the surface is eroded evenly.   Depth profiles of concentrations
can be measured, and ion micrographs of successive layers can be
taken.

Isotopic analysis

Since the isotopes of an element are chemically identical,
they are ionized with exactly the same probability.   Therefore,
isotopic ratios can be measured with very high accuracy.   This
opens up a whole new field, namely to do tracer work with stable
isotopes on a microscale.   Isotope dilution techniques could be
applied and the isotopic ratio measurements done *in situ* with
very high sensitivity and accuracy. About 70% of all elements have

stable isotopes that can be enriched and used as tracers. The
mass resolution is sufficient to separate the isotopes of all
elements on the periodic chart, unless they happen to fall on
the same mass number. There is a lot to be said in favor of
the use of stable isotopes as tracers as compared to radio-
tracers. An excellent survey of potential applications of
stable isotopes in biology and medicine has been made by White
(10).

Precise, isotopic ratio measurements can be performed conveniently with the instruments shown in Figs 5 and 7 applying
the isotope switching technique (11) (Fig.10) to the ion microprobe mass analyzer, which was introduced by the author mainly
with geological applications in mind, where isotopic ratio
measurements can be used for age determinations of minerals.

The two isotope beams leave the mass spectrometer magnet
on separated trajectories. Between the magnet and the detector
two pairs of deflector plates are mounted. The voltages
applied to them are switched in rapid sequence, about 100 times
per second, so that the two isotope beams reach the detector
alternatively. The output signal is fed into a dual scaler
which is gated in synchronism with the deflection switching.
Due to the rapid switching any variations of the secondary ion
production are averaged out. With this technique, not only
isotopic ratios can be measured, but also ratios of different
elements, if their masses are not too far apart.

FUTURE ASPECTS

The lateral resolution reached so far is about 1 μm for
both types of instruments. With the imaging type this could be
improved only by sacrificing sensitivity or by increasing the
secondary ion acceleration voltage which would meet technological
difficulties. With the scanning-type instrument the lateral
resolution can be improved without sacrificing sensitivity, since

ION DETECTOR ASSEMBLY

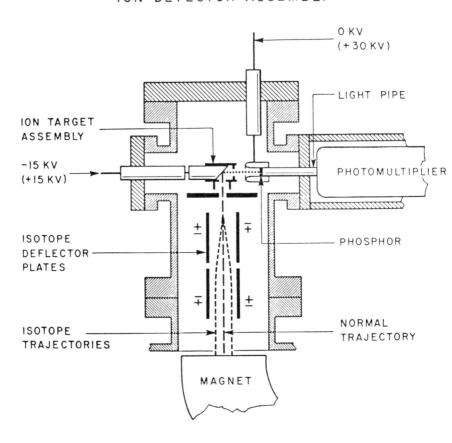

Fig.10 : Peak switching for isotopic ratio measurements (11).

it is given by the probe diameter.   Several researchers (12,13)
have already succeeded in focusing ion beams down to diameters
of 0.2 or 0.3 μm using conventional ion sources.   There is
justified hope that much finer ion probes can be made with sufficient intensities with the development of finer and brighter
ion sources (14).

On the application side, it seems that elemental analyses
can be made with sufficient accuracy.   However, the identifi-

cation and analysis of compounds from their fragmentation spectra, as is being done successfully in organic chemistry with mass spectrometers using electron impact ionization, is a problem barely touched in sputtering mass analysis. Promising contributions have been made by Benninghoven (15) who investigated various inorganic and organic compounds attached to metallic surfaces.

In conclusion it can be stated that the Ion Microprobe has great potential for applications in the life sciences. However, in view of the rather complex interdependence of the technology and methodology involved and their applicability to biological research, it appears that a real pioneering effort of cooperation between physicists and biologists is necessary if fast progress is to be made.

REFERENCES

1. Castaing, R. Thesis, Univ. of Paris, Publ. ONERA No.55 (1951).

2. Castaing, R. and Slodzian, G. *J. de Microscopie 1*: 395 (1962).

3. Liebl, H. *J. Appl. Phys. 38:* 5277 (1967).

4. Courtesy of ARL, Sunland, California.

5. Long, J.V.P. *Brit. J. Appl. Phys. 16*: 1277 (1965).

6. Tamura, H., Kondo, T., Doi, H., Omura, I. and Taya, S. *In:* Recent Developments in Mass Spectroscopy, 205. University of Tokyo Press (1970).

7. Liebl, H. *Int. J. Mass Spectrom. Ion. Phys. 6* (1971).

8. Courtesy of C.A. Andersen, Hasler Res. Ctr. ARL, Goleta, California.

9. Andersen, C.A., Hinthorne, J.R. and Fredriksson, K. Proc. Apollo 11 Lunar Science Conf. 1, 159.

10. White, F.A. Mass Spectrometry in Science and Technology, Chapter 14, 317. John Wiley & Sons, New York (1968).

11. Robinson, C.F., Liebl, H.J. and Andersen, C.A. 3rd Nat. Conf. Electron Microprobe Analysis, Chicago (1968).

12. Drummond, I.W. and Long, J.V.P.   Proc. 1st Int. Conf. Ion Sources, Saclay (1969).

13. Hill, A.R.   *Nature 218*: 202 (1968).

14. Heil, H.   Private communication.

15. Benninghoven, A.   *Z. Phys. 230*: 403 (1970).

IMAGING MASS SPECTROMETER

K.H. GAUKLER

Institute of Applied Physics, University of Tübingen, Germany

There are two different methods of microanalysis by secondary ions:
1. The ion microprobe
2. An apparatus which is a combination of an ion emission
    microscope and a mass spectrometer.
The ion microanalyzer was developed by Castaing and Slodzian
(1,2,3) and improved by Rouberol *et al.* (4) and is now commercially available (IMS 300 CAMECA, Paris).

The principle of this apparatus is shown in Fig.1.   The
primary ion source is a duoplasmatron.   The primary beam is
focused by a double condenser system which permits obtaining a
spot size of 25 μm to 400 μm on the specimen.   The acceleration potential is adjustable up to about 12 kV.   The maximum density of the bombarding ion beam is of the order of some
mA/cm^2.   This corresponds to a sputtering rate of several ten
nm/sec of the specimen.   The electrostatic immersion lens ac-
celerates the secondary ions and forms an image of the specimen
by all secondary ions which are emitted from the specimen.   The
dimensions of the specimen are 25 mm in diameter and 8 mm
in height.   An optical microscope with a magnification of 70
permits observation of the specimen when it is placed in view
position.   The mass spectrometer is a homogeneous magnetic
field which separates the masses so that after a deflection of
90° the images of the different masses are separated and one

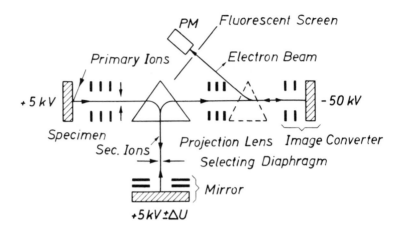

Fig.1 : Scheme of the ion microanalyzer IMS 300.

of them with a given mass may be selected by a diaphragm.   By
a proper choice of the magnetic field it is possible to select
all masses from hydrogen to approximately mass 300.

The mass resolution without using the electrostatic mirror
is rather poor because of the different initial energy of the
secondary ions.   The mirror is adjusted to reflect only ions
with an energy lower than a certain value, thus eliminating ions
of higher energy and improving the mass resolution.   Under normal conditions the mass resolution M/ΔM is 300.

The ion beam of the selected mass is reflected through the
magnetic field in such a way that the ion image is preserved.
By a further electrostatic immersion system the ion image is
converted to an equivalent electron image which represents the
distribution map of a selected mass of the surface of the specimen.

There are different modes of operation.   The electron
image may be displayed on a fluorescent screen for direct visual
viewing.   The image observed on a 1"-diameter screen corresponds
to a sample diameter of approximately 150 to 300 μm. By speci

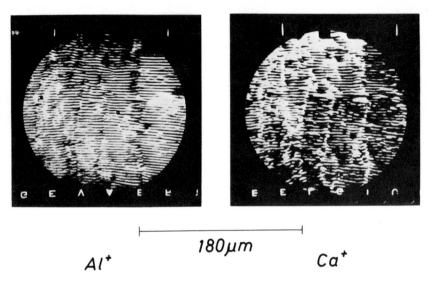

Al^+ *180μm* Ca^+

Fig.2 : Test of spatial resolution, Al-Ca alloy (4).

men motion and proper choice of magnification it is possible to choose the desired sample area. This image may be photographed by a vacuum camera with 30 exposures on a 35 mm film.

Figure 2 shows a resolution test of an Al-Ca alloy. The distance of the stripes is approximately 1 μm. The resolution is of the order of the light microscope. The usual time of exposure is between a few milliseconds and several seconds.

It is also possible to analyze insulators by first depositing a conductive grid upon the surface of the sample to carry away the electrostatic charge. Figure 3 shows pictures of the distribution of hydrogen, boron and sodium in a mineralogical sample. By secondary ions it is possible to detect the light elements, here for example hydrogen, contrary to X-ray microanalysis. Another method of analyzing thin foils of insulators - especially suitable for biological specimens - will be discussed in the next paper.

K.H. Gaukler

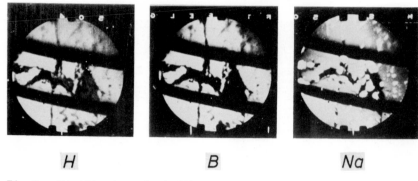

H   *B*   *Na*

Fig.3 : Biotite (granite) (4).

     Besides photographing the ion image there is a second mode
of operation.   The screen may be removed and the electron beam
impinges on a scintillator which is coupled to a photomultiplier
tube.   A diaphragm placed in front of the scintillator permits
the spatial selection of a small area of the electron image.
This small area corresponds to a small area of the specimen.
By continuous variation of the magnetic field of the mass spectrometer we obtain a mass spectrum of this selected small sample
area.   As with other mass spectrometers the sensitivity of this
instrument is high for most elements.

$$I_{el} \sim j_p \cdot F \cdot \gamma$$

where I_{el} is the measured electron current, j_p the primary
ion density, F the size of the analyzed area (which is variable from some μm to 250 μm), and γ the secondary ion yield
that is the number of secondary ions per incident primary ion.
There are many variables which affect the value of γ, for
example the composition of the matrix, the bombarding gas and
especially the chemical bonding of the components of the specimen.

Fig.4 : Yield of various metals. Primary ions: Ar^+ (9 keV).

Figure 4 shows the measured electron current for various metals bombarded by Ar^+-ions. These values correspond to positive secondary ions. The sensitivity for the metals is quite different. The yield of aluminum and tin for example differs by a factor of 1000, so that the sensitivity for aluminum is very high. The yield for lithium and sodium in chlorides and fluorides is higher than that of aluminum metal by a factor of 100, so that the detection limit of alkali metals is of the order of ppb. The limits of sensitivity for this instrument lie in the region of mass concentrations between ppm and ppb

for the different elements. To increase the sensitivity for
fluorine, chlorine, oxygen, sulphur or negative radicals, one
must record the spectrum of the negative secondary ions.

It is also possible to obtain a depth analysis of the
specimen with this instrument by using the erosion effect of
the sputtering primary ion beam. One can study the different
composition of the specimen surface layer by layer with a resolution in depth of some atomic layers. Figure 5 is a simple
example of the results obtained by this method. It shows the
diffusion of boron in a single crystal silicon.

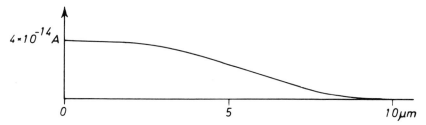

Fig.5 : Diffusion of boron in silicon.   Primary ions: 0^+;
        surface concentration of boron: $4 \cdot 10^{19}$ at/cm^3.

The instrument described has the capability of performing
microprobe analyses with ppb sensitivity. It can produce
spatial images of a particular mass and provide isotopic abun-
dances from an area of a few μm.   Depth analyses can be performed.

REFERENCES

1.   Castaing, R. and Slodzian, G.   *J. Microscopie 1*:   395
     (1962).

2.   Slodzian, G.   *Anal. Phys. 9*:   591 (1964).

3.   Castaing, R.   Optique de rayons X et microanalyse, p.48.
     Hermann, Paris (1966).

4.   Rouberol, J.M. *et al*.   Proc. V. Int. Congress on X-Ray
     Optics and Microanalysis, p.311.   Springer-Verlag,
     Tübingen (1969).

THE APPLICATION

OF THE ION MICROPROBE

TO BIOLOGICAL MATERIALS -

FIRST RESULTS

PIERRE GALLE and JEAN-PIERRE DUMERY

Laboratoire de Biophysique, Université Paris,
Faculté de Médicine, 94 Creteil, France

The ion microprobe offers in biology entirely new possibilities:
- by reason of its very high sensibility
- and by reason of the fact it permits the separate study of different isotopes stable or not of the same element.

The apparatus we have used is the ion microprobe introduced in 1962 by Castaing and Slodzian, and built by the CAMECA Company (France).

The apparatus consists of a primary ionic gun emitting ions: ionized ion gases oxygen (O^+ or O^-) or in our study nitrogen ions N_2^+, which sputter the sample in a uniform field; this image is directly obtained by electrostatic lens, without using any electronic or mechanical scanning. This process has the enormous advantage of giving a very homogeneous secondary ionic emission.

The ions pertaining to the superficial layers of the sample are plucked away, with an ionization rate depending on their

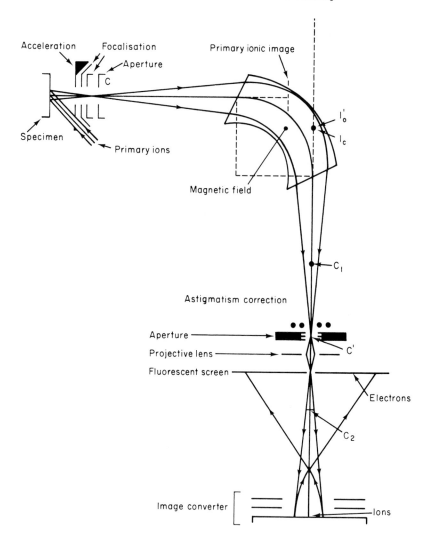

Fig.1 : Principle of the apparatus.

nature, and then are collected by an ionic optic consisting of:
- an acceleratory field and electrostatic lens which give an enlarged image.
- a double electromagnetic prism with an electrostatic mirror working also as a mass spectrometer which first

operates a sorting of the different mass of the isotopes that are present and second, an energy filtration of the ions (Fig.1).

Finally an enlarged image corresponding to the repartition of the different ions on the surface of the sample is obtained.

The secondary ionic beam could also be focused on a photo-multiplier in the way to make a quantitative analysis of the corresponding ions, thus permitting a dosage of very little volumes of tissues.

RED BLOOD CELLS

The first biological applications of this apparatus had been done by Galle, Blaise and Slodzian in the study of the re-partition of diffusible ions in red blood cells.

Figure 2 shows different views of the distribution of sodium $^{23}Na^+$ at the surface of a red blood cell spreading out on a support. These views have been obtained successively between the tenth second and the tenth minute of the sputtering. At the beginning the plasma appears very bright and the red blood cells dark. Then progressively the plasma is volatilized, its image disappears and only the cells remain visible. If the sputtering continues the cells disappear one after another.

RAT THYROID SECTIONS

We then studied the ionic repartition in rat thyroid sec-tions. The sections which have been used are semi-thick from 2 to 4 μ, and have been prepared after coagulation. This thickness was chosen in order that the sample could emit enough ions to enable several different kinds to be observed. When the thickness is greater than 5 μ, the sample becomes a poor conductor and the pictures are bad or non-existent.

The sections are placed on polished silicon supports. Silicon is a good choice because it is possible to obtain it very

Pierre Galle and Jean-Pierre Dumery

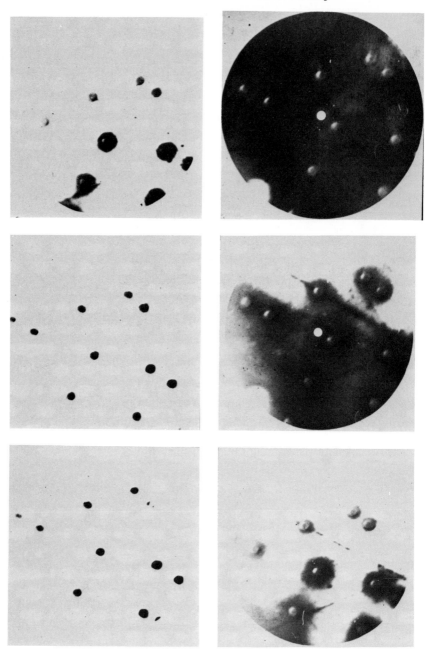

Fig.2 : Distribution of Sodium ^{23}Na$^+$ at the surface of red blood
cells.

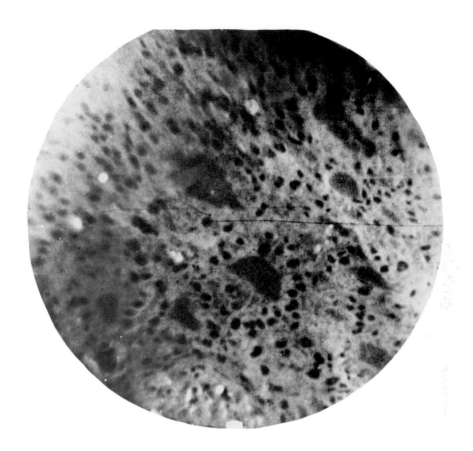

Fig.3 : Distribution of Sodium $^{23}Na^{+}$ in rat thyroid section.

pure and generally, certain special cases excepted, it is free
of biological tissues.

Figure 3 represents a rat thyroid section showing the distribution of the sodium ($^{23}Na^{+}$) (diameter of the zone 200 μm).
We note the existence of thyroid vesicles surrounded by the
nuclei of the cells and between the two the cytoplasm.

If ions which are emitted from such a section are analyzed
with a photomultiplier equipped with a sufficiently small aper-

Fig.4 : Rat thyroid section sputtered during a long time.

ture, it is possible to evaluate the dosage of an element present in a part of the cell; the nucleus, a mitochondria for instance.   Such evaluation is impossible with the electron microprobe.

As we have seen with red blood cells, if thyroid sections
sputtered for a long time, they broke up in an heterogeneous way,
the total ionization being different according to the zone.   The
different views of Fig.4 show that first the nuclei, then the

cytoplasm and finally the vesicles disappear.

With this technique it is also possible to obtain images of extra cellular spaces. This is practically impossible with the electron microprobe due to the sublimation of the ions.

In Fig.5 different views of a same thyroid zone are represented, showing the repartition of negative ions Iodine 127, Chlorine 35, Sulfur 32, and Oxygen 16. The primary positive nitrogen ionic beam sweeps over a 400 µ diameter zone of this preparation and of the image near the border. We see that iodine is essentially located in the vesicles, as is oxygen. On the view showing the repartition of the sulfur the repartition of oxygen O^{16} is also found, although it is not so bright. It corresponds to the ionization O_2^- mass 32 of the oxygen.

We do not show comparative views of the repartition of the positive sodium ions (mass 23) and potassium ions (mass 39) as they are practically identical.

Figure 6 represents two views of the same section showing the repartition of the sodium (23) and the potassium (39), which are practically the same. This, we think, is due to the fact that the cutting had to be done at thirty degrees below zero; the edge of the knife gives rise to such a pressure on the coagulated tissue that it causes the local fusion of the ice and secondly the equi repartition of these ions.

DISCUSSION

We have seen that the thyroid sections break up at different speeds according to the zone. This shows that it is impossible to deduce from the instantaneous brightness the corresponding local concentration; it is necessary to integrate all the information coming from all the thicknesses of the tissue.

Another example is shown in Fig.7. This slide shows the emission of two natural isotopes of potassium: isotope 39

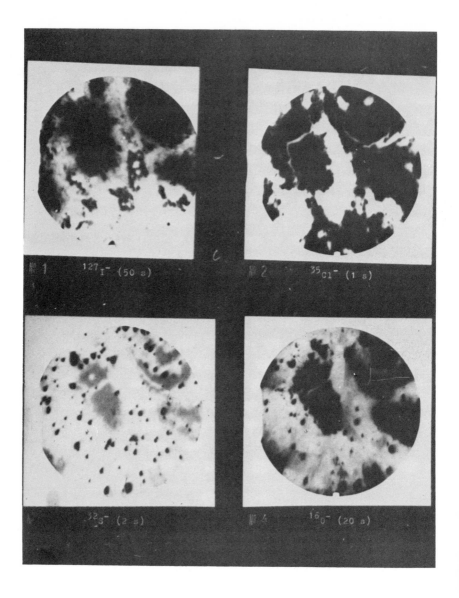

Fig.5 : Different views of a same thyroid section showing the repartition on Iodine $^{127}I^-$, Chlorine $^{35}Cl^-$, Sulfur $^{32}S^{--}$ and Oxygen $^{16}O^{--}$.

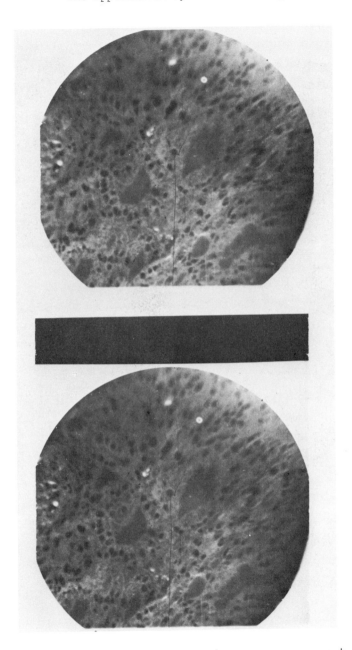

Fig.6 : Repartition of Sodium $^{23}Na^+$ and Potassium $^{39}K^+$ on a same thyroid section.

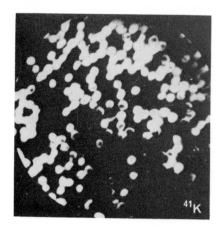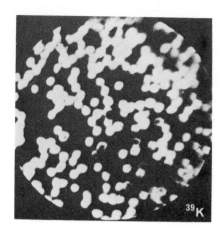

Fig.7 : Repartition of two natural isotopes of Potassium. [39]K
and [41]K on red blood cells smears.

(left) and 41 (right). We observe on these views, images of
red blood cells which have practically the same brightness even
if they are not at the same concentration. If the information
is integrated in all the thicknesses of the sample we could find
that the isotope 39 constitutes 95% of total natural potassium
where the isotope 41 is only 5%.

Such an image of red blood cells obtained with sodium ions
has permitted us to determine the maximum sensibility of the
method knowing that the intracellular concentration of sodium is
approximately 3.10^{-16} g/μ^3 and the quantity of sodium breaks up
in 0.03 sec, which is the time necessary to obtain a photograph
of a surface of 1 μ^2 is 10^{-20} g corresponding to 300 atoms.
This is the minimal quantity necessary now to get a photograph.

The sensibility of the ion microprobe is 10,000 times
greater than the electron microprobe.

The limitations of the biological applications are essentially due to the preparation of the samples, and it is not now
possible to avoid the diffusion of mobile ions.

REFERENCES

1. Castaing, R. and Slodzian, G.   *J. Microscopie 1*: 395-410
    (1962).
2. Galle, P., Blaise, G. and Slodzian, G.   7th Internat.
    Congr. of Electron Microscopy, p. 489. Grenoble (1970).

ADVANCES IN

MINIATURIZED ELECTRONICS FOR

BIOMEDICAL APPLICATIONS*

R. STUART MACKAY

Boston University and Boston University Medical Center, Boston, Massachusetts, U.S.A.

When vacuum tubes could be replaced by semiconductor devices such as transistors, a whole new series of possibilities for biological investigation and medical procedure became possible. In this discussion three aspects will be mentioned. The first will consider the use of small radio transmitters within or on the bodies of animal or human subjects to sense and communicate biological information with minimum disturbance to the normal patterns of the subject. The second will be mention of the possibilities of fabrication of very small computers which can be carried by human subjects to signal significant or critical conditions which should be noted and acted upon. And finally, mention will be made of some techniques developed for the fabrication of modern electronic components which can also contribute to the fabrication of improved physiological sensors.

---

* The recent studies on the fetus are being aided by a grant from the John A. Hartford Foundation, New York, while the satellite oriented studies are aided by NASA Grant NGr 22-004-024.

BIO-MEDICAL TELEMETRY

Radio transmitters swallowed, surgically implanted or carried externally can be used to study animal and human subjects
with minimum disturbance to normal activity patterns.   They
have been used to study fish, porpoises and alligators freely
swimming, birds while flying, animals in burrows in the ground
and fetuses in the uterus of conscious active mothers.   It is
possible to sense and transmit a variety of information.   Humans
can be monitored while recovering from illness or while working
in a hazardous environment, as well as being involved in fundamental studies.   Drugs can be tested with minimum effect by
extraneous influences.   Two extreme examples include tiny
transmitters placed in the eye to record pressure changes produced by unexpected sights (in glaucoma studies) and the following of animal movements over great distances from artificial
earth satellites.

Several examples will be used to illustrate some of these
possibilities.   Figure 1 shows a radiograph of a rhesus monkey
into which has been surgically implanted four radio transmitters
simultaneously active on different frequencies.   The upper two
transmitters sense and transmit temperature and electrocardiogram.   Both are placed subcutaneously, with the leads for the
latter being sutured near the top and bottom of the sternum.
The next lower transmitter contains a simple accelerometer to
indicate motion or activity.   The transducer associated with
the bottom transmitter is in a cuff strapped on the outside of
the abdominal aorta, and it senses instantaneous absolute blood
pressure independent of changes in the elastic properties of the
intervening vessel wall.

Since all known plastics seem permeable to body fluids, the
transmitters are coated with wax (1) to form a moisture barrier,
and then with an outer layer of silicone rubber.   Each of these

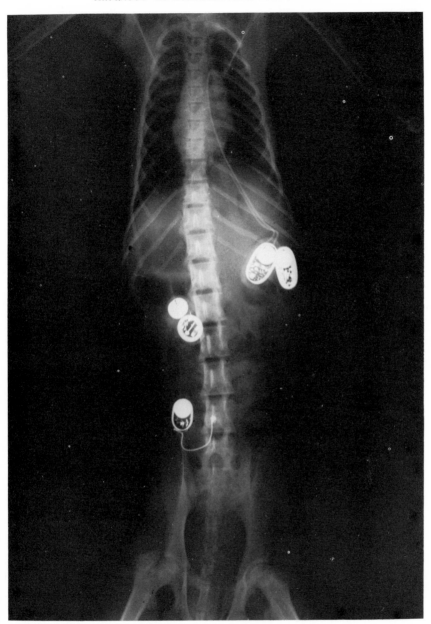

Fig.1 : Four simultaneously active physiological monitoring
transmitters surgically implanted in a rhesus monkey.
The variables transmitted are temperature, electrocardiogram, activity and blood pressure.

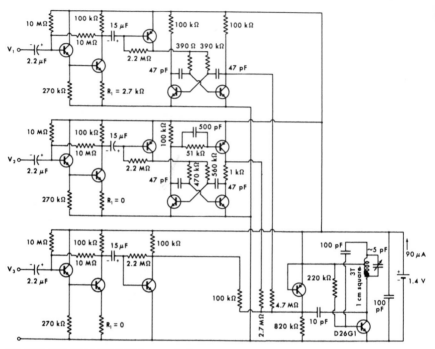

Transistors: NPN, D26E-5; PNP, D30A-3; except final

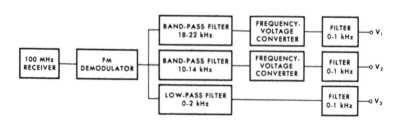

Fig.2 : Single transmitter of three independent voltages used
for transmission of vectorcardiogram, with receiver
system at bottom.   One channel of frequency modulation
is employed and two channels of FM-FM using multivibra-
tors as subcarrier oscillators.

transmitters is less than half the size of a man's thumb, and yet they continuously fill a cage with a reliable signal for over a year as the animal moves about in play or other activity.

In other cases, rather than using multiple transmitters, several variables are transmitted from a suitably placed single transmitter.   As an example, Fig.2 shows the circuit employed for simultaneously transmitting three independent voltages as are required to describe the spatial vectorcardiogram.   The two transistors at the lower right constitute a single channel transmitter which is modulated in frequency by voltage induced variations in capacitance of the left transistor which is connected as a diode.   (This arrangement is slightly more effective than the corresponding circuit given in (1).)   An input voltage V_3 is amplified to produce modulation.   Superimposed upon this is a 12 kHz signal which is varied in frequency in response to changes in V_2.   Also superimposed is a 20 kHz signal which is modulated by input V_1.   The received signal, at a radio frequency after detection, allows the three voltage signals to be recorded separately as shown in the bottom of the figure.

Using standard small commercially available components, the entire transmitter can be connected and embedded in plastic with a size as shown in Fig.3.   This can be done in any laboratory, and great size reduction is not provided by the use of special integrated circuits, as the antenna and battery (the latter is not shown) still tend to limit the overall reduction.

In work with Ben Jackson and George Piasecki such transmitters were surgically implanted in fetal dogs and sheep by surgery through the uterus to allow monitoring before, during, and after birth.   Since the fetal skin seems to function as a partial insulator, it appears impossible to obtain an undistorted vectorcardiogram from electrodes placed on the maternal abdomen, while this is no problem with electrodes placed within the fetus, and indeed, the maternal electrocardiogram is excluded.

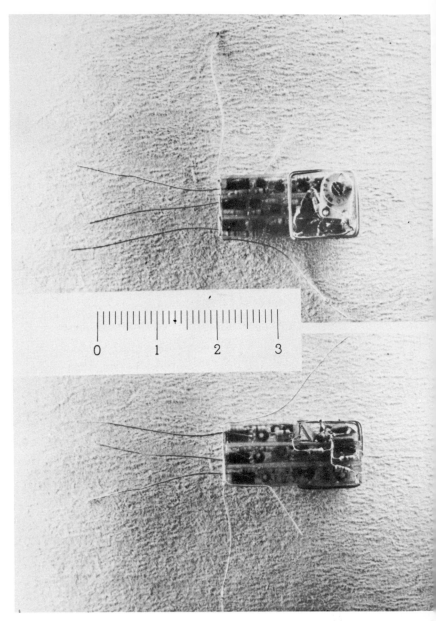

Fig.3 : Front and back views of a transmitter using the circuit
of Fig.2, without the battery.　The three channels are
along the three rows, with the radio frequency portion
being on the right.

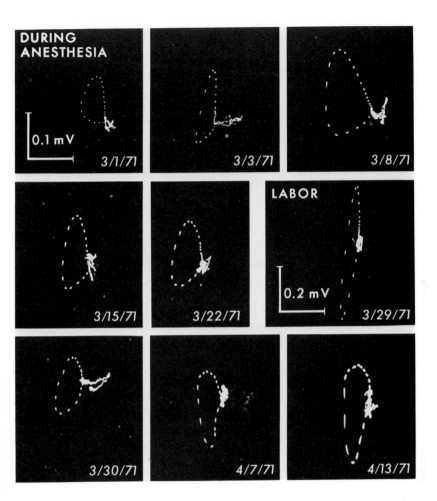

Fig.4 : Transverse vectorcardiogram loops from transmitter in
the fetus of a sheep.   The record extends from before
birth to after, with the indicated voltage scale applying to all loops except about the time of the start of
labor where there was the indicated increase in amplitude.

Of the three voltages, different pairs can be plotted
against each other to display the vector loops in three planes.
In Fig.4 the transverse vectorcardiogram of a sheep is shown,
from approximately four weeks before birth to two weeks after.

In the display oscilloscope the trace is dotted at a 1 kHz
rate, and the dots formed into "tear drops" with the blunt end
leading to show the direction of progression.   In this particular example some variability in loop form was seen before
birth, though there was no dramatic change in the net electrical
balance of the heart.   A slight anterior shift appeared after
birth, becoming clear within two days of the birth.   The
sagittal loops during this same period reflected the same variation, and in addition showed a slight caudal shift exclusively
after birth, with the overall loop assuming a more anterior-
posterior than cephlad-caudad orientation.   The size change in
the animal can put a strain on the transmitter leads and electrodes if precautions are not taken.   These observations were
made from a separate room by closed circuit television.   After
the birth, the sheep could again become pregnant.

Some transmitters are swallowed or introduced through
other normal body openings.   An example is seen in Fig.5 which
shows an ingestible capsule containing a glass pH electrode.
The necessary high input resistance to this circuit is achieved
by using a field-effect transistor, plus feedback, to give an
overall input resistance of approximately 5000 megohms.   Thus
this same input configuration can also be used to accept the
signals from microelectrodes.   In the present case, the amplified signal is used to change the click rate of a transmitter so
that the subject can be approached with a simple receiver and
the rate noted at the loudspeaker to indicate the momentary level
of activity.

Physiological processes can be monitored from animals in
relatively inaccessible places, such as underground or in the
water.   An example from the Master's Thesis of Matthew Weintraub
is the trout shown in Fig.6 with a radio transmitter on its back.
Changes in heart rate in response to changing oxygen content of
the water was monitored, and thus an electrode pair was placed

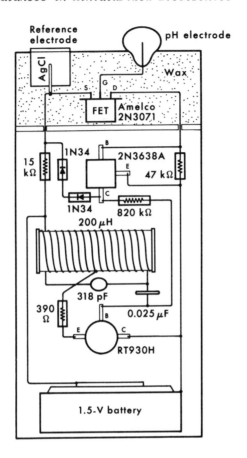

Fig.5 : Capsule containing transmitter circuit with sufficiently
high input resistance to function from a glass electrode.

between the pectoral fins and near the anus.   Representative
recordings are seen in Fig.7.   At the top is seen a recording
showing both heart beat and respiratory movements, the latter
being suppressed in the center recording.   When the fish is
freely swimming, a synchronism between the two motions was often
seen, as in the bottom recording.

In some experiments on animals in the field, a simple transmitter of sounds is extremely valuable.   The signal at the receiver then tells something of environmental conditions, pre-

Fig.6 : Fish with frequency-modulated neutral buoyancy radio
        transmitter on back to allow untethered recording of
        heart rate.

dation, laughing, eating, biological sounds, vocal activity,
locomotion, respiration, etc.   Movement in customary vicinities
can be monitored by attaching a low speed penwriter through a
diode to the speaker connection of a receiver so that radio
field strength changes from the transmitter are recorded.
     Internal transmitters can be turned on and off to conserve
battery power by magnetic switches.   This can also be done by
radio signals, pulses of light, or pulses of sound coupled into
the body.
     Some of these small transmitters need their signal for only
a few centimeters or a few meters in order to be useful in moni-

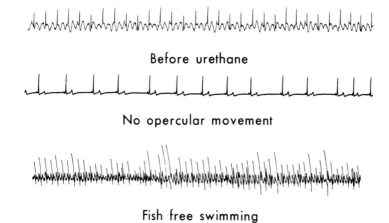

Before urethane

No opercular movement

Fish free swimming
9.0 mg O₂/liter H₂O

Fig.7 : Signal recorded from electrodes of the experiment in
Fig.6.   A pure electrocardiogram is seen if respiratory
movements are stopped.   Synchronism between respiration
and heart beat is sometimes observed.

toring internal activity.   If a signal is then required at a
great distance, for example in studying freely roaming animals
in the field, then a retransmission booster can be employed.
This combination of a small receiver and somewhat larger transmitter can be carried externally by the animal, and it need not
be larger than a cigarette package.   If the subject is not
again to be seen or caught, the transmitter can automatically be
dropped when its life is ended by electroplating away a piece of
metal comprising one link in the harness, as in Fig.8.   Such an
arrangement can also be used to release automatically a recorder
storing information on a dolphin, etc.   In the figure, a silver
wire having an initial breaking strength of about 75 lb is
quickly cut by the small battery when the transistor is activated by depletion of the batteries supplying power to the transmitter.   The depleting process is restricted to a small region
by the epoxy coat, and the ceramic partition minimizes a tendency

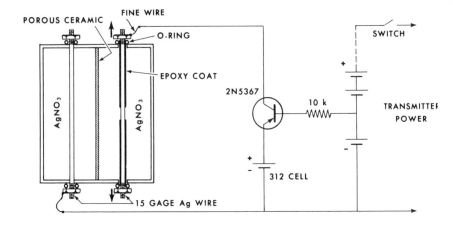

Fig.8 : Mechanical connection which is electroplated away to
        release a transmitter or recorder from an animal when
        the main power source becomes exhausted.

for dendrites to bridge the cell.   Approximately twice the
charge is required to break a copper member having the same
strength.

     To study creatures migrating over very great distances, it
seems desirable to monitor them with the help of artificial
earth satellites.   Experience by many with the Transit navi-
gation system has demonstrated that a satellite can indicate
the location of a transmitter on the earth if the transmitter
frequency is constant and the satellite can record the received
frequency which is Doppler shifted by a changing amount during
passage.   We are presently working on a one pound external
transmitter that should allow us to follow the movements and
some physiological parameters in the Greater Albatrosses for a
month.   For representative conditions, refraction in the iono-
sphere and the effect of the possible velocity of the bird can
introduce an uncertainty in position of about two miles, which
suggests a warranted transmitter stability of about one part in
10^8 during the 5 min time of satellite passage.   Our circuit

employs a crystal oscillator encased in layers of polystyrene,
copper, paraffin (for moisture resistance and adequate specific
heat with medium electrical loss), and styrofoam, feeding a
frequency multiplier and output circuit by fine wires to achieve
adequate temperature compensation.

The above has somewhat emphasized research applications,
but the same methods can be used in the clinical care of human
subjects in hospitals, or the monitoring of human welfare in
hazardous environments such as in the oceans or in outer space.
As an example of clinical utility, it is often found that a
small voltage-monitoring transmitter will yield a "cleaner" and
more artifact-free or reliable signal of heart beat or electro-
encephalogram than does carrying the same signal directly by
wires, because the latter will often introduce ground loop problems and will pick up interfering electrical noise.   Furthermore, this method of monitoring gives the subject greater freedom of movement and less restraint (whether he is ambulatory or
not), while interfering less with attendants, and it essentially
removes the hazard of electrical shock associated with more
direct monitoring.

In many cases the effectiveness of the method will depend
on the input transducer that activates the telemetry system,
and the utility of the methods increases when new sensors of
physiological information are found.   As an example, we are now
better able to study decompression sickness since we have found
that the bubbles associated with the "bends" can be sensed ultra-
sonically.   Figure 9 shows a set of images of the diverse structures in the leg of a guinea pig as the pressure is changed.   In
work with George Rubissow, we produced these images by a scanning
process, but a stationary transducer can monitor the formation
and disappearance of bubbles in any selected region, thus allowing the use of telemetry.   We have seen bubbles form in many
animals, including fish (especially in the white muscle of the

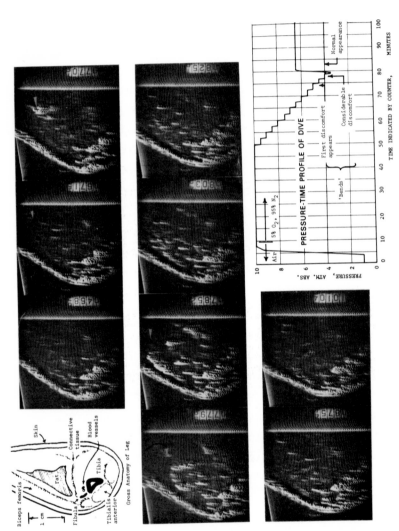

Fig.9 : Cross-sectional image of guinea pig leg using pulsed 12 MHz ultrasound while the pressure is changed to cause bubble formation and the "bends". In addition to intensity modulation in this display, deflection modulation as the scan progressed was also used for easy quantitation of data.

latter), and have successfully decompressed animals by acoustic observation alone.

Standard parameters can sometimes be observed in different ways that are useful. Thus if one wishes to know the temperature of a site without actually contacting it, in many cases one can implant two thermistors a short distance away, with a known separation. The two readings then indicate temperature gradient which, with distance to the point of interest and either temperature, allows computation of the remote temperatures. This assumes the heat flow field is uniform, which can be true if the sensors are not too spread out with respect to the curvature of various structures, and if discontinuities in conductivity are reasonably small. (As a specific example, if a thermistor one millimeter to the left of a thermistor indicating 20° indicates 19°, then one millimeter to the right the temperature should be about 21°.) One can alternatively regard such a procedure as a direct measurement of heat flow, knowing the temperature difference maintained across a given thickness of material of known thermal conductivity; in studying the thermal insulating properties of various tissues, this can be a convenient approach.

The converse process to telemetry of using inwardly induced power, and telestimulation, can supply a controlled disturbance to a system whose response then gives information about the normal interactions of its component parts. Telemetry and telestimulation are often combined for maximum effectiveness.

Much of this material has been covered in a recent book (1), and the interested reader is referred to this for further information and references.

SPECIALIZED PORTABLE COMPUTERS

In monitoring the activities of an individual, various physiological parameters can be telemetered to a central point for analysis and evaluation. If a critical condition in some parameter is noted, then a warning can be transmitted or help sent. But modern developments in miniaturized complex circuits make possible the construction of individualized special purpose computers which can be carried by a subject to form directly an analysis of the state of the person. The general principle is clear enough, and the convenience of the implementation depends largely on the ever progressing technology of integrated circuit design. Two possible example of applications will be given here.

Irregularity in heart rate is a sign for some groups that some medicine needs to be taken or that a prosthesis is on the verge of failing. Small computers can be built to detect such irregularities from either an acoustic or electrical signal from the heart. For those with coronary artery disease, ventricular premature beats appear as a portent of trouble that may be averted. Automatic processing should then include the several factors of prematurity, "compensating following pause", and differing shape from the average.

In a similar way, one might wish to monitor the brain wave pattern of divers or tunnel workers, electroencephalographic parameters being of importance because brain tissue is very sensitive to internal state changes.

Other computers can employ a model of some aspect of the functioning of a human subject. Thus, for example, the equations used to prepare diving tables can be modelled by electrical circuits (2) to form a computer that can predict the onset of the bends in less than real time. Not only can such a computer guide the ascent of a diver (giving the fastest possible safe ascent in minimum time), but by coupling it to a sensor in the air tank, an indication could be given to the diver of the last possible moment at which a safe ascent could be made without running out of breathing air.

Models of human functioning in the form of personalized computers can be made increasingly compact and could be incorporated
in many applications, for example to help stabilize the functioning of prostheses or sensory extending devices.

IMPROVED PHYSIOLOGICAL SENSORS

The techniques that have been developed for fabricating integrated circuits and their tiny components can have application
in the construction of small devices in general, and physiological sensors in particular.   In this case, we refer to the technology of building small circuits rather than the technology of
their use.   Two examples that have been exploited by several
groups will be mentioned.

A force transducer surrounded by an insensitive coplanar
annulus can be used, as in the blood pressure measurement of
Fig.1, to measure the absolute pressure in various body cavities
without penetrating the cavity wall (1) but in some cases it is
required to place a small pressure transducer within the body
cavity.   Such a transducer can be built by placing four strain
gauge elements on a diaphragm, and electrically connecting them
in a bridge circuit as shown in Fig.10.   Temperature changes
will not unbalance the bridge since all gauge elements are similarly affected.   Any asymmetric geometry or even anisotropy
that differently affects different strain sensing elements in
response to a pressure change will give a system in which a
bridge unbalance indicates pressure.   In the figure, assuming
the elements are on the inner surface of the diaphragm, an increase in pressure will cause an increase in resistance in the
center elements and a decrease in those closer to the edge.

If such a unit is to be made smaller, then the gauge elements must be small, perhaps being formed of silicon which gives
good sensitivity.   Evaporation and related techniques can be
necessary for size reduction.   If the diaphragm itself is of

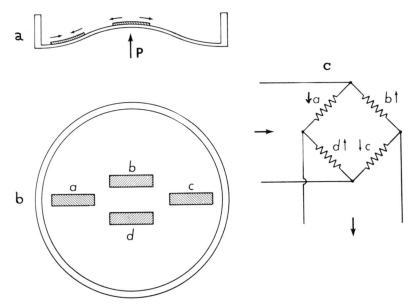

Fig.10 : Deflections of a diaphragm on the end of an air-filled
         compartment can be used to monitor surrounding press-
         ure: (a) side view showing that central strain gauges
         are stretched by a pressure increase while those at
         the edge contract; (b) top view of four bridge elements
         (c) four elements connected in a bridge circuit as shown
         give an output when the pressure changes but are rela-
         tively insensitive to temperature changes.

silicon then etching can be used to form a monolithic unit which
is extremely thin, and hence sensitive though of small diameter.
Groups at Stanford and elsewhere have reduced the size of such
units down to one millimeter.

Similarly, various groups have used these etching and metal
depositing techniques to form microelectrodes having good conducting properties with shielding, in some cases with several in
one bundle. The usual concepts of oxygen electrode measurements
and antimony pH measurements can also be exploited here, while
retaining the customary strengths and weaknesses of the methods.
Since amplifiers, in some cases employing high input impedance
stages, can also be deposited directly at the base of the elec-

trode, one in addition has the possibility of relative freedom
from noise pickup, plus rapidity of response associated with
somewhat reduced lead capacity.

The diversity of application of these fabrication techniques will undoubtedly increase as they become more routine.

REFERENCES

1.　Mackay, R.S.　Bio-Medical Telemetry: Sensing and Trans-
mitting Biological Information from Animals and Man.
Second Edition.　John Wiley and Sons, Inc., New York,
London and Sydney (1970).

2.　Bradner, H. and Mackay, R.S.　Biophysical Limitations on
Deep Diving:　Some Limiting Performance Expectations.
Bull. Math. Biophys. 25:　251-272 (1963).

APPLICATION OF

MASS SPECTROMETRY

IN THE PHYSIOLOGICAL SCIENCES

G. SPITELLER

Institute of Organic Chemistry, Göttingen, Germany

INTRODUCTION

Mass spectrometry was introduced as a method for determination of the structure of natural products about fifteen years ago (1,2). Due to its high sensitivity it has become one of our most powerful tools for the investigation of nearly all types of natural products and metabolites of drugs.

PRINCIPLE

The molecules of the investigated compound are evaporized into the "ion source" of the mass spectrometer where they are bombarded with electrons. The ions thus produced are accelerated in an electrical field and enter a magnetic sector field through a slit, which is arranged in such a manner that ions of the same mass-to-charge ratio are collected in a plane. Either all ions are collected at the same time on a photographic plate, producing a spectrum similar to an optical spectrum (mass spectrograph), or one ion species after another is registered electrically. This is achieved by forcing the ions to flow through a "separation funnel" with continuous alteration of the magnetic field (mass spectrometer). A *mass spectrum* (Fig.1)

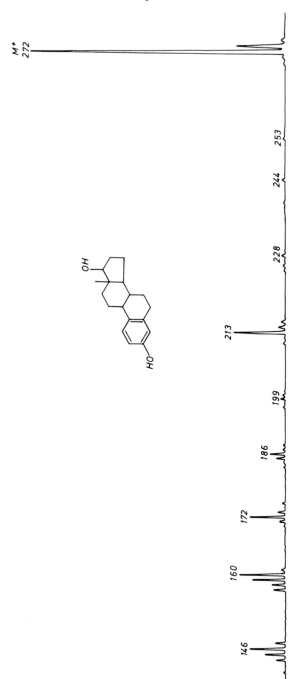

Fig.1 : Part of the original mass spectrum of oestradiol.

indicates the *mass* (usually only ions with the charge one are produced) and the *amount* of produced positively charged ions.

Thus, mass spectrometry differs greatly from the other "spectroscopic" methods (I.R., U.V. and N.M.R. spectroscopy) applied for characterization of organic molecules: in these, excitation energies typical for the presence of functional groups are measured. After the measurement the compound is re-covered unchanged. In contrast, with mass spectrometry the energies transferred from the attacking electron to the molecule vary considerably. The ionized molecules suffer chemical degradation and cannot be recovered unchanged. We do not measure absorption energies but the mass and the amount of the produced positively charged ions.

SCOPE AND LIMITATION

Volatility

We are only able to obtain mass spectra from molecules that can be evaporated without thermal decomposition. This seems to be a serious limitation, but actually the situation is not so bad: the minimum vapor pressure to obtain a mass spectrum is in the order of 10^{-6} Torr. Such a small vapor pressure is pro-duced by many molecules thought to be involatile, e.g. simple amino acids and monosaccharides. Excluded from investigation are compounds of very high polarity, e.g. salts which are not able to dissociate or molecules of large size, e.g. polymers.

Sample amount

The amount of sample necessary to run a mass spectrum varies greatly: usually sample sizes in the 50 microgram scale are sufficient. In special cases, for instance by direct combi-nation of a gas chromatograph with a mass spectrometer, even smaller sample sizes (down to the nanogram range) may be enough.

Compared with other spectroscopic methods mass spectrometry is
of higher sensitivity.

Purity

If a mixture of compounds is investigated, the molecules
of each compound are ionized independently from all others.
The resulting spectrum is therefore a combination of the spectra
of the single compounds (Fig.2). If a mixture of two compounds
of low and high molecular weight is investigated the interpretation of the spectrum of the compound with high molecular
weight is hardly influenced by the spectrum of the compound with
the low molecular weight. Therefore even rather large amounts
of solvents may be present. Small amounts of impurities do not
influence the interpretation very much, because in most cases
they show different molecular weights and key fragments (see
below). Thus valuable conclusions about the structure are
available even from mixtures and impure samples.

Resolution

High resolution mass spectra allow the determination of
molecular formulae in a much better way than elementary analysis.

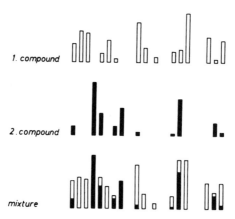

Fig.2 : Schematic representation of a spectrum of a mixture.

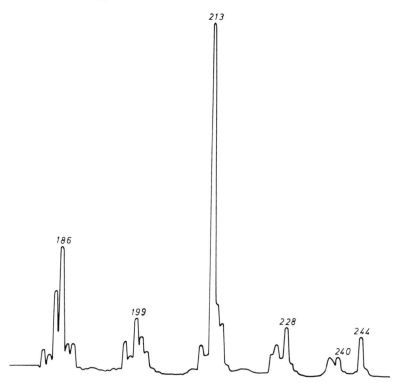

Fig.3 : Part of a mass spectrum showing not fully resolved
        peaks (part of the mass spectrum of oestradiol, shown
        in Fig.1).

The ion beams, separated in the magnetic field of a mass spectrometer, contain ions of slightly differing energies.   Therefore
after passing the magnetic sector field they are not indicated
as a sharp line but rather as a broad peak.   With increasing
mass the separation of the ion beams becomes less efficient and
ions with the highest energy of one mass overlap with those of
the lowest energy of the next higher mass: the ion beams are
not fully resolved (Fig.3).

    With the aid of an additional electrical field it becomes
possible to focus ions of different energies and thus to increase the resolution to such an extent that ions of the same

nominal mass, but different molecular formula, are separated:
Carbon, with a nominal mass of 12, does not have exactly the
same mass as 12 hydrogen atoms:

    C : 12.000   H : 1.008   12 H : 12.096

For instance CO_2 and acetaldehyde (CH_3CHO) have both the
nominal mass 44, but the exact mass of CO_2 is 43.9898 and
of acetaldehyde 44.0261. A mixture of both compounds shows
in its mass spectrum a doublet at mass 44.   (An instrument
having a resolution of 10,000 is theoretically able to separate two ions of mass 9999 and 10,000.)   Actually compounds of such a high molecular weight are not volatile enough.
(The advantage of the instrument therefore lies not in its
ability to investigate molecules of higher molecular weight, but
in enabling exact mass determinations of molecules with lower
molecular weight.)   In practice, molecules with molecular
weights above 1200 could only be investigated in very rare
cases.

REPRESENTATION OF SPECTRA

    As already mentioned, a mass spectrum provides data about
the mass and the amount of the produced positively charged
ions.   To facilitate the recognition of these data they are
best represented in the form of a line diagram (Fig.4) showing
the masses as ordinate and the intensity values as abscissa.

    Not only does the energy transferred by the attacking elec-
tron contribute to the energy content of a molecular ion, but
also the thermal energy which the molecule obtained by colli-
sions before the ionization.   The higher the energy content of
a molecular ion, the higher is its tendency to undergo cleavage
reactions.   The produced fragments can be cleaved further.
Therefore spectra of one and the same compound may look different (Fig.4), depending upon the applied reaction conditions.
Compared with other spectroscopic methods mass spectrometry is

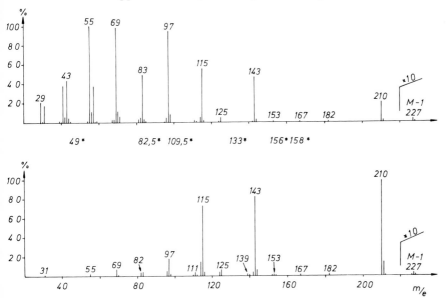

Fig.4 : Line diagram of the mass spectrum of pentadecanol-7
run at an electron energy of 70 and 15 eV.

therefore less suited for quantitative analysis.   But although
the intensity values vary to a great extent, a compound is always cleaved in the same way, so that fragments of the same
mass always occur.   These reactions are highly specific and
usually allow an unequivocal identification.

THE MECHANISM OF FRAGMENTATION IN THE MASS SPECTROMETER

Although we know very little about the processes occurring
in the ion source, we are able to draw a picture which seems to
be not too far from reality: in a molecule the bonding and non-
bonding electrons move in their orbits; this causes an electromagnetic field.   If we introduce a charge - for instance the
attacking electron - into this field, polarization occurs.   This
polarization becomes stronger and stronger the nearer the electron is approached until finally ionization - removal of an
electron from its orbit - occurs.   Usually there are in a mole-

cule different types of orbitals, e.g. sigma orbitals corresponding to single C-C bonds and p-orbitals (in heteroatoms) of
functional groups.   Electrons in these latter orbitals are
much more easily polarized than those in sigma orbitals.   Consequently ionization has the highest probability of occurring
at functional groups.

Therefore if only one functional group is present in a
molecule, predominantly, but not exclusively, molecular ions
with the charge at the functional group are produced.

$$CH_3-(CH_2)_5-\underset{\underset{H}{|\overset{\cdot\cdot}{O}|}}{CH}-(CH_2)_7-CH_3$$

$$\Big\downarrow -e$$

$$CH_3-(CH_2)_5-\underset{\underset{H}{\overset{+}{\underset{\bullet}{|\overset{\cdot\cdot}{O}|}}}}{CH}-(CH_2)_7-CH_3$$

The excess energy of the molecular ion is usually enough to induce cleavage reactions, but it is too low for any bond to be
broken - usually it is just sufficient to ensure that those reactions which need the lowest amount of energy can occur.
Obviously the weakest bonds are those in the neighborhood of the
center; therefore nearly all degradation reactions start at the
ionized group.   The type of degradation reaction depends highly
upon the substituent (3).   If the substituent is able to produce
a stable radical and if it is connected with a rather weak bond
to the carbon skeleton, it is lost preferentially:

$$R \overset{\cdot}{\underset{\curvearrowleft}{\smash{\frown}}} \overset{\bullet+}{X} \longrightarrow R^+ + \overset{\bullet}{\underline{X}}$$

In other molecules the rupture of a C-C bond adjacent to the functional group is favored, because thus stable onium ions are produced:

$$R \text{—} CH_2 \text{—} X \longrightarrow R\bullet + CH_2 = \overset{+}{X}$$

In a third type of reaction a sterically nearby hydrogen atom is abstracted and in a second degradation step the substituent is lost as a small molecule:

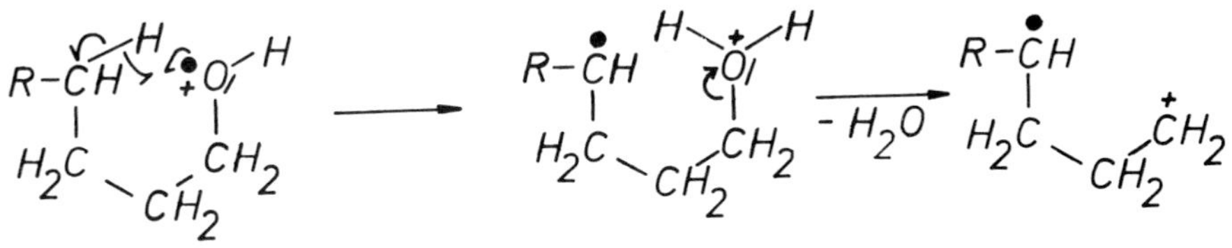

The presence of such fragments allows us to determine which substituents are present and also to locate them in the aliphatic chain. For instance the spectrum of the secondary alcohol *1*, reproduced in Fig.4, contains a prominent ion corresponding to the loss of water and two additional key fragments produced by cleavage of bonds adjacent to the hydroxy group:

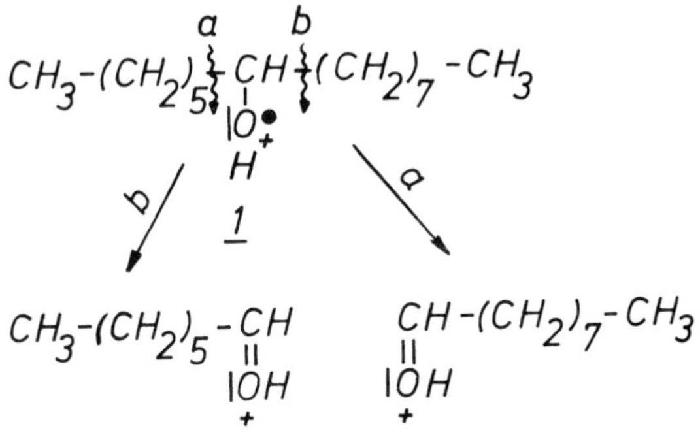

$$CH_3\text{-}(CH_2)_5 \overset{a}{\mid} CH \overset{b}{\mid} (CH_2)_7 \text{-} CH_3$$

$$\underset{1}{\overset{\mid}{\underset{H}{\overset{+}{|0\bullet}}}}$$

$$CH_3\text{-}(CH_2)_5\text{-}CH \qquad CH\text{-}(CH_2)_7\text{-}CH_3$$
$$\overset{\mid\mid}{|OH} \qquad\qquad \overset{\mid\mid}{|OH}$$
$$+ \qquad\qquad\qquad +$$

$$^m/_e\ 115 \qquad\qquad\qquad ^m/_e\ 143$$

The molecular ions of this alcohol are of extremely low stability, demonstrating that molecular ions must not necessarily be found in mass spectra.   Therefore it must always be kept in mind that the ion found at highest mass could already be a fragment ion and that the actual molecular weight is higher.   In such cases it is very useful to prepare a derivative, which is easily ionized but which needs a rather high additional amount of energy for cleavage.   Especially well suited for this purpose are all compounds which contain sulfur.   Unfortunately the primary fragments produced with electrons of 70 eV - the usual energy - still contain a sufficient amount of energy for further degradation reactions, proceeding by stepwise loss of small hydrocarbon fragments or of other small molecules.   Thus, mass spectra show many fragments of low mass.   Such final hydrocarbon degradation products are produced from so many molecules that they actually tell us nothing about the original structure of the molecule.

In order to get more interpretable spectra we have to try to stop the reactions at the primary steps (4).   This can be achieved by lowering the average energy of the molecular ions, as shown in Fig.4.

If two or more functional groups are present in a molecule, ionization may occur at all groups, producing different types of molecular ions.   Because each of the molecular ion species decomposes in a specific way we have to expect that spectra become more and more complicated the more functional groups are present.   Actually the situation is much less serious.

The ionization probability is obviously mainly dependent upon the electron density.   Usually the electron density at each functional group is different;   therefore, ionization at one group dominates.   So even from rather complicated molecules, typical key fragments are produced (Fig.5).   For instance, in testosterone 2 (Fig.5a) ionization occurs mainly at the α, β-

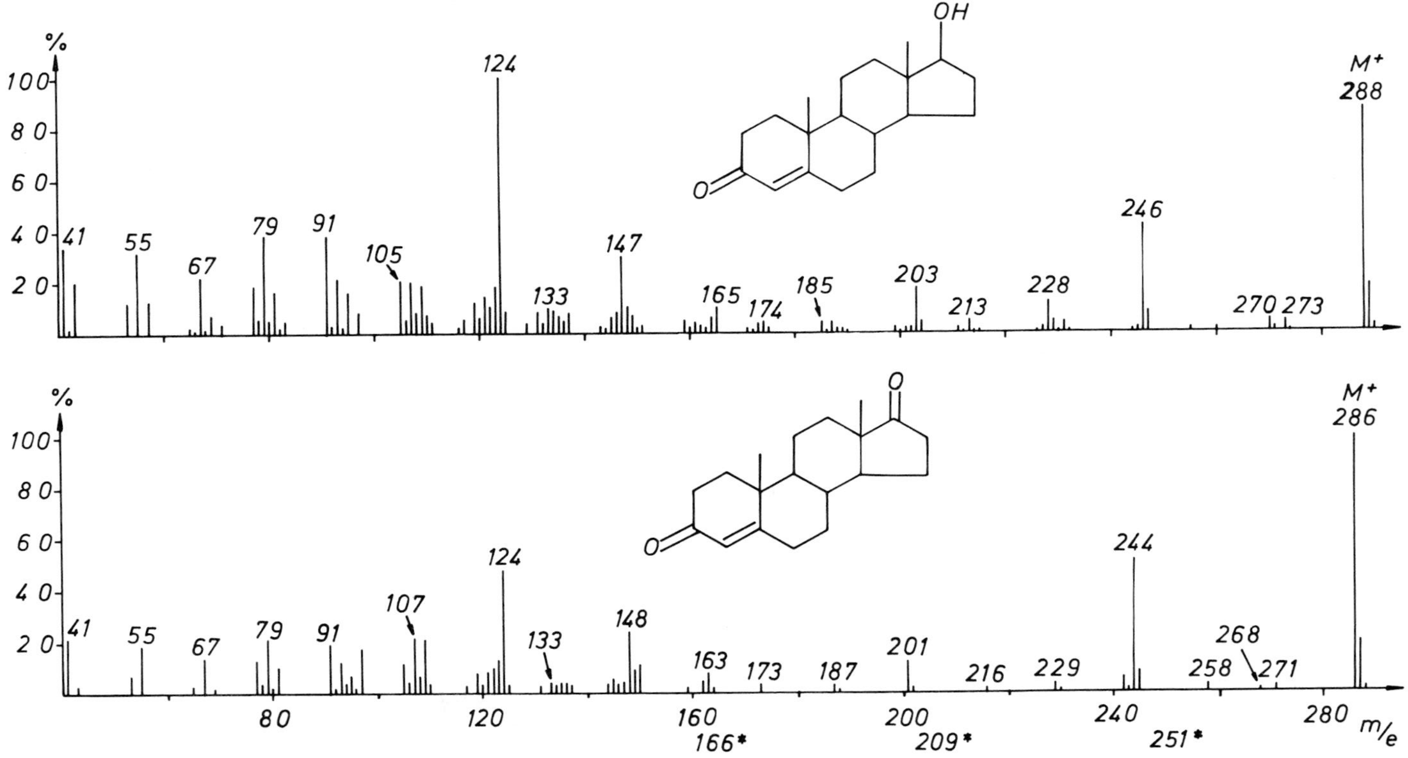

Fig.5 : (a) Mass spectrum of testosterone (upper part); (b) Mass spectrum of androst-4-en-3,17 dion (lower part).

unsaturated carbonyl system.   This results in the production
of key fragments at mass 124 and M-42 ions.   If the hydroxy
function in position 17 is exchanged by a keto group, no alteration in the main cleavage reactions is observed;   again key
fragments of mass 124 and M-42 are produced, *3* (Fig.5b).

The fragments of mass 124 - together with the ions at M-42
- are therefore a strong indication of the presence of a Δ^4-3
keto function in an otherwise unsubstituted A/B ring system.

m/e 124

$M-(CH_2=C=O)$

M-42

2   *3*

However, no information can be gained from the spectra about
the location of the hydroxy resp. carbonyl function in position
17.   These functional groups induce no typical degradation reactions.   This difficulty can be surmounted.   The electron
density in position 17 can be enhanced by the introduction of
additional functional groups, and consequently the course of
the main degradation processes is changed (Fig.6).

In 17 β-methoxy-17α-methyl-androst-4-en-3-on *4* ionization
occurs mainly at the ether oxygen and not at the α,β-unsaturated
carbonyl function (lower electron density).   This causes the
production of the key fragments of mass 72,85 containing the
carbon atoms of the D-ring system.   In addition M-71 ions are

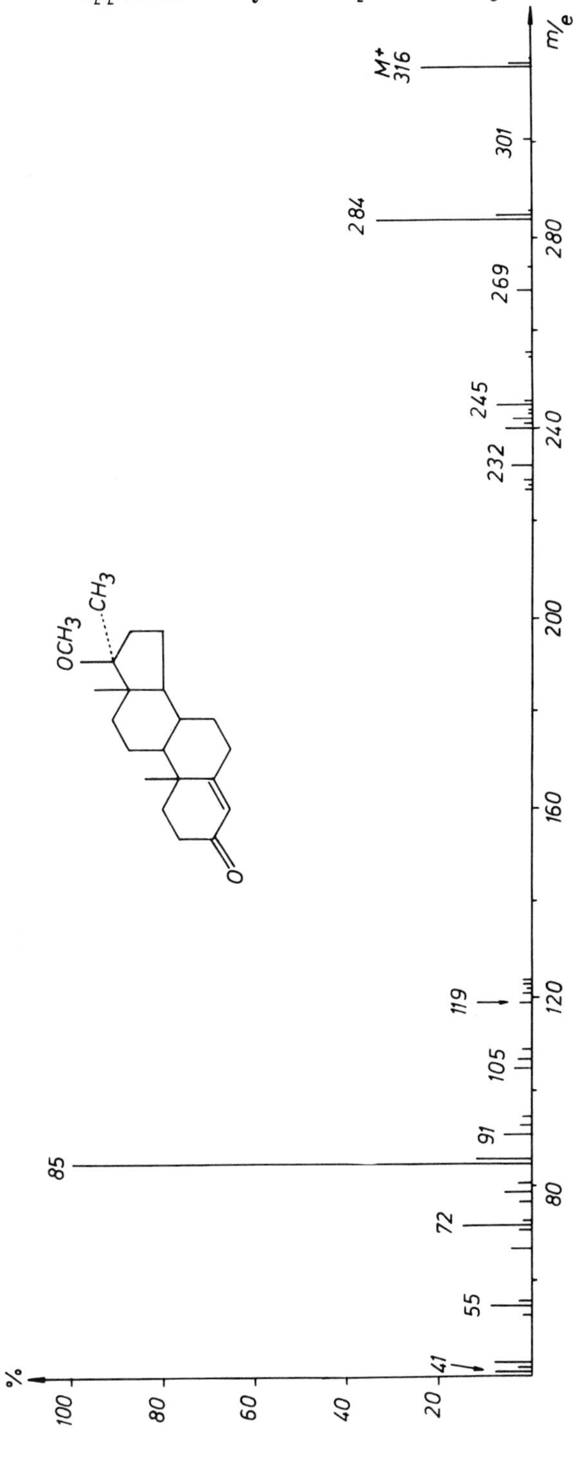

Fig.6 : Mass spectrum of 17β-methoxy-17α-methyl-androst-4-en-3-on.

produced by loss of C-16 and C-17.

Very different structural features may produce fragments of
the same mass.   Therefore one key fragment alone is insufficient
to postulate the presence of a certain structure element.   It
gives us only a hint.   But usually the production of two or more
key ions is induced by a structural element.   Seldom will a
structural element induce the production of the same two or more
key ions as will another.   Therefore with an increasing number
of key ions the hints become more and more in the favor of proof
of a certain structure.

INFORMATION AVAILABLE ON MASS SPECTRA

In most cases molecular weight and molecular formula for
a compound can be determined by mass spectrometry.   If no molecular ions are obtained, preparation of derivatives is advisable.
Molecular ions decompose in a structure-specific manner.   The
presence of functional groups is recognized by typical "key dif-

ferences" corresponding to the loss of the substituent as a small molecule; location of the functional group is possible if key fragments are produced.

The identification of a compound is based mainly on the presence of these key ions and not only or dominantly on intensity values, as they may vary depending upon the average energy of the molecules.

If key fragments of only one part of the molecule are obtained, the introduction of a functional group with high electron density may shift the preferential site of ionization and induce different cleavage reactions in the starting material, leading to key fragments typical for the other part of the molecule.

Due to the fact that mass spectrometry is an extremely sensitive method, with the use of pure compounds not absolutely necessary, the chemical reactions can be performed on the 10 microgram scale. This combination of chemical reactions with mass spectrometry enables structure determinations with minute sample amounts.

COMBINATION OF MASS SPECTROMETRY WITH CHROMATOGRAPHIC METHODS

Impure samples or mixtures can be separated by thin-layer chromatography. The layer containing the separated compound is scratched off and the purified sample is gained by extraction of the silica gel with a solvent. With the aid of an injection needle the solution of the sample is transferred into a small furnace, evaporated, and this furnace is then introduced into the mass spectrometer. Gas chromatography is an extremely useful method for the separation of volatile compounds; mass spectrometry is an equally useful method of characterization for volatile compounds. Both methods supplement each other ideally.

It is not possible to combine both instruments directly because mass spectrometers only work properly at a vapor pressure

below 10^{-5} Torr while the effluents of a gas chromatographic
column have a much higher pressure.   Therefore between the gas
chromatograph and the mass spectrometer a so-called "separator"
must be installed.   Its task is not only to reduce the pressure
but also to enrich the sample in respect of the carrier gas (5,
6).   In the Becker-Ryhage separator, for instance, the effluents flow through a tiny orifice aligned to a second one.   In
the space between the two orifices the molecules of the lighter
carrier gas are pumped off much faster than the heavier molecules of a compound (Fig.7).

This combination instrument is especially useful for the
separation of compounds of relatively high volatility.   Less
volatile compounds need higher temperatures;   these cause higher

To vacuum

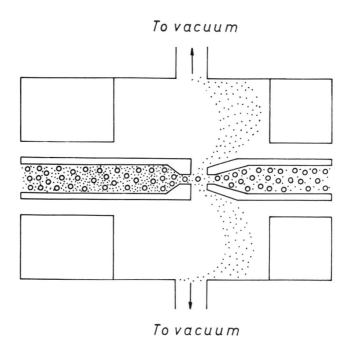

To vacuum

Fig.7 : Schematic representation of a Becker-Ryhage-separator
        (6).

thermal excitation of the samples and therefore a higher tendency for fragmentation reactions, resulting in an accumulation
of fragments of low mass and a diminished intensity of the key
ions.   In addition, if the temperature is raised during the
making of the gas chromatogram (temperature programming), more
and more degradation products are eluted from the column and the
interpretation of the spectra becomes more and more difficult
(7).

During the investigation of compounds in both a gas
chromatograph and a mass spectrometer often hundreds of spectra
are obtained within two hours.   It sometimes takes days to
count all the spectra and to prepare them for an interpretation
by eliminating contributions of "background" (of compounds
eluted in previous fractions or of "column blood").   This can
now be done by connecting the mass spectrometer with a small
computer; it immediately plots the spectra in the form of line
diagrams (8).

The problem of automatic identification of the plotted
spectra is less satisfactorily solved (9).   All identification
systems used so far compare standard spectra with those of the
"unknown compounds".   They are based mainly on the comparison
of intensity values - and these can vary greatly depending upon
the conditions used, as shown before.   Big computers are
necessary to consider all the different factors influencing the
spectra.   The computer is combined with the output of the mass
spectrometer, and this is combined with the gas chromatograph.
The combination of so many instruments makes the whole system
rather sensitive.   A small mistake in one part - for instance
the gas chromatograph - causes a complete shut down.   The dead
time is therefore extremely high, between 30 and 80%.   In addition, to run such a large combination of instruments requires a
team of experts, not only in mass spectrometry but also for the
setting up and improvement of computer programs.   The extremely

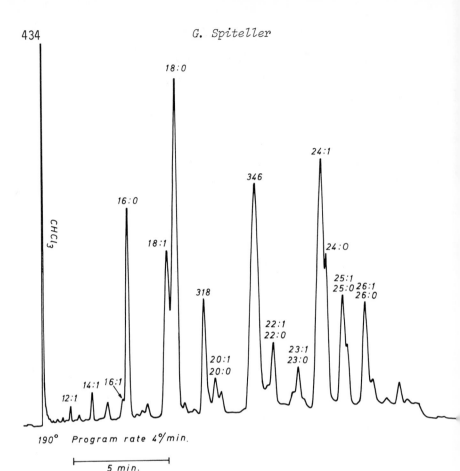

190° Program rate 4°/min.

|———————————|
5 min.

Fig.8 : Gas chromatogram of methylated fatty acids isolated
from the sphingolipid fraction of brain.　The two peaks
of "unknown compounds" are indicated by the mass of the
molecular ions, measured in the mass spectrometer.　All
other peaks correspond to common fatty acids, either
saturated or containing one double bond, which was indicated above the peaks by the carbon number and number of
double bonds.

high cost of buying and maintaining such arrangements will probably make their use the privilege of only a few institutions in
the world, at least in the near future.

Nevertheless, even without computer facilities mass spectrometry is a very useful method in physiological sciences.　This
could be shown with a few examples as it was easier for me to

talk about problems in close connection to our work, although
at many other places work of this type has been done (10).
Dr. Lesch of the medical department of the University of
Hannover investigated the sphingolipid fraction of the brain of
persons where cause of death was complications related to liver
cirrhosis caused by the abuse of alcohol.   After hydrolysis he
methylated the acids and analyzed them by gas chromatography.
He detected peaks in the gas chromatograms, not present in the
same extracts of persons where death was caused by other dis-
eases.   He asked us to help him in the identification of these
compounds with the use of our combination instrument.   A small
sample of his mixture was introduced into the instrument.   The
gas chromatogram was obtained by measuring the total ion current (Fig.9).   The quality of the gas chromatogram is not as
good as with an analytical gas chromatograph, but sufficient to
recognize the different peaks.   The mass spectra of the two unknown compounds showed molecular ions at mass 318 and 346.   The
spectra indicated some similarities.   Both spectra contained
key fragments at mass 150, 164 and 177.   Both compounds showed
key differences at M-98 and M-138.   This, together with the mass
differences between the molecular ions of 28 mass units
(obviously corresponding to two CH_2-groups), allowed the conclusion that the unknown compounds were homologs.

This assumption was confirmed by an exact mass determination
of the molecular ions.   For this purpose another example was
analyzed in a gas chromatograph hooked up with a high resolution
instrument.   The spectra were registered on a photographic plate.
The exact mass determination of important peaks was carried out
under a microscope.

The molecular formula $C_{21}H_{34}O_2$ of the compound with the
lower molecular weight 318 allowed us to deduce that it contained
five double bonds or rings - the corresponding saturated hydrocarbon has the molecular formula $C_{21}H_{44}$.   Similarly the formulae

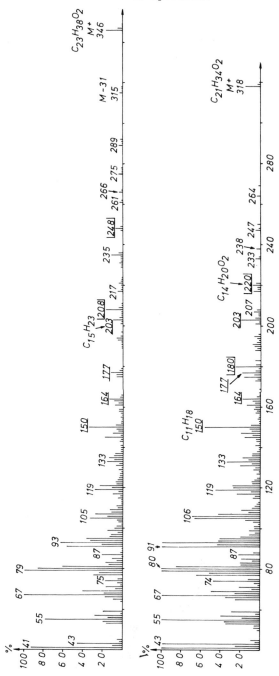

Fig.9 : Mass spectra of the two "unknown compounds" in the gas chromatogram shown in Fig.8. The quality of the spectra is not good, due to rather high thermal excitation; especially the unspecific peaks in the low mass region are of high intensity, nevertheless the key ions can be recognized.

of the ions suspected to be key ions were determined.

Key ions at mass 74 and 87 indicated that we were probably
dealing with aliphatic methyl ester, a quite reasonable assumption because all other compounds in the gas chromatogram were
identified to be esters. Consequently the original compounds
were suspected to be methylated acids with 20 and 22 carbon
atoms respectively containing 4 double bonds.

Comparison spectra were not available, but the assumption
was corroborated when we compared the spectra with those of
methyl-linolenate *5*. Methyl-linolenate has three CH_2-groups
less at the hydrocarbon end. Its spectrum shows a key ion at
mass 108; both of our acid derivatives at mass 150 were to 42
mass units heavier - exactly the mass of three CH_2-groups (Fig.
10). Similarly methyl-linolenate loses C_4H_8 (56 m.u.) and
Dr. Lesch's compounds C_7H_{14} (98 m.u.).

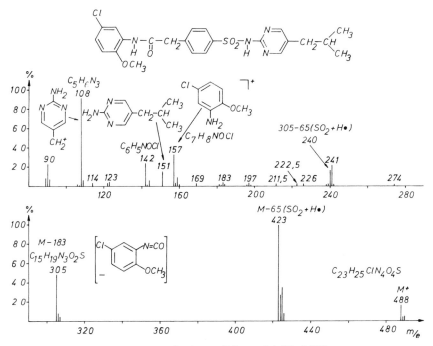

Fig.10a : Mass spectrum of the sulfonamid BS 1051.

$CH_3-CH_2-CH_2-CH_2-CH_2-CH=CH \{ CH_2-CH=CH-CH_2 \} CH=CH-CH_2-CH=CH-CH_2-CH_2-CH_2-CH_2-CH_2-COOCH_3$

          M-98 $m/e 150$

$CH_3-CH_2-CH_2-CH_2-CH_2-CH=CH \{ CH_2-CH=CH-CH_2 \} CH=CH-CH_2-CH=CH-CH_2-CH_2-CH_2-COOCH_3$

          M-98 $m/e 150$

          $CH_3-CH_2-CH=CH \{ CH_2-CH=CH-CH_2 \} CH=CH-CH_2-CH_2-CH_2-CH_2-CH_2-CH_2-CH_2-COOCH_3$

               M-56 $m/e 108$

                                            <u>5</u>

It is remarkable that such high unsaturated fatty acids
have never been found before in sphingolipids, only in glyco-
phosphatides.   The careful preparation of the samples excluded
a contamination of Dr. Lesch's samples by these compounds.

Another example shows the application of mass spectrometry
in the study of metabolites.   Dozent Gerhards at Schering in
Berlin investigated the metabolism of a new sulfonamide.   He
succeeded in isolating a minute amount of metabolite by thin-
layer chromatography and asked us to help him with the identification.

At first we studied the mass spectrum of the starting
material (Fig.10).   It was characterized by a molecular ion of
only low intensity at mass 488/490.   The doublet of the molecular ions is caused by the existence of two chlorine isotopes
of mass 35 and 37 in a ratio of 3 : 1;   this typical ratio also
allows the detection of all other fragments containing a chlorine
atom.   The main fragment at mass 423/425 was produced by expul-

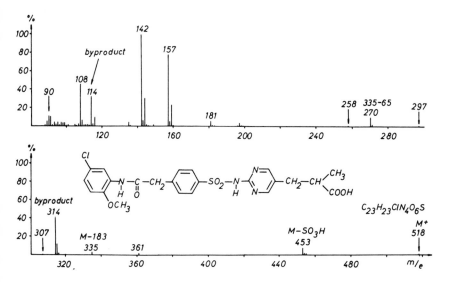

Fig.10b : Mass spectrum of sulfonamid BS 1051 metabolite.

sion of SO_2+H.

A further key fragment of mass 305 ($C_{15}H_{19}N_3O_2S$) corresponded to the loss of the aromatic part as isocyanate. But not only the loss of the aromatic end of the compound was indicated, additional key ions mass 157/159 (C_7H_8ClNO) and 142/144 (C_6H_5ClNO) correspond to positively charged ions containing this aromatic end of the molecule. The exact mass determination of

the fragment of mass 108 $C_5H_6N_3$ allowed us to deduce that it corresponded to the pyrimidine part of the molecule as well as a fragment of mass 151. The spectrum of the metabolite showed

also a very small molecular ion of mass 518/520.   The exact
mass determination allowed the assignment of the molecular formula to be $C_{23}H_{23}ClN_4O_6S$.   This corresponds to the conversion
of a CH_3-group into a COOH-group.

The presence of the unaffected aromatic part of the molecule was corroborated by ions of mass 142/144 and 157/159 and
at M-183.   The peak at mass 108 proved also that the pyrimidine part of the molecule was still present.   The fragment of
mass 151 was shifted to mass 181, indicating that only one of
the methyl groups of the butyl-side chain could be oxidized and
therefore the structure of the metabolite was 7.

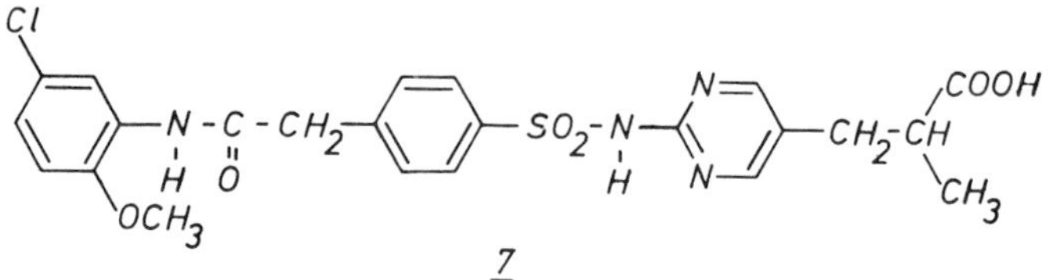

$\underline{7}$

OUTLOOK

Mass spectrometry has just started to become a method of
analysis in all branches of physiological sciences.   The low
amount of sample necessary to run a spectrum combined with low
requirements concerning the purity of the samples makes this
method extremely useful for the solution of problems in forensic medicine, in the investigation of drug metabolites in the
course of the development of new remedies, and in the search
for unknown compounds with biological activity.   It is therefore certain that mass spectrometry will play a dominant role
in all these fields in the near future.

REFERENCES

1. Brown, R.A., Young, W.S. and Nicolaides, N. *Analytic. Chem. 26*: 1653 (1954).

2. Nilsson, M., Ryhage, R. and v. Sydow, E. *Acta Chem. Scand. 11*: 634 (1957).

3. For a detailed discussion see:
   (a) Spiteller, G. Massenspektrometrische Struktur-analyse organischer Verbindungen. Verlag Chemie, Weinheim (1966).
   (b) Budzikiewicz, H., Djerassi, C. and Williams, D.H. Mass Spectrometry of Organic Compounds. Holden Day, San Francisco (1967).

4. Spiteller-Friedmann, M. and Spiteller, G. *Chem. Ber. 100*: 79 (1967).

5. Ryhage, R. *Analytic. Chem. 36*: 759 (1964).

6. Watson, I.T. and Biemann, K. *Analytic. Chem. 36*: 1135 (1964).

7. Kaiser, K., Obermann, H., Remberg, G., Spiteller-Friedmann, K. and Spiteller, G. *Monatsch. Chem. 101*: 240 (1970).

8. Hites, R.A. and Biemann, K. *Analytic. Chem. 40*: 1217 (1968).

9. Heitz, H.S., Hites, R.A. and Biemann, K. *Analytic. Chem. 43*: 681 (1968).

10. For a review see McCloskey, J.A. *In:* Advances in Mass Spectrometry, Vol. IV. In press.

HIGH RESOLUTION

ELECTRON-MICROSCOPIC IMAGING

OF BIOLOGICAL SPECIMEN S*

A.K. KLEINSCHMIDT

Department of Biochemistry,
New York University School of Medicine, New York, U.S.A.

INTRODUCTION

Resolvancy (1) or resolving limit is the distance in Ångström units between two ideal, self-luminous, non-coherent electron-microscopic object points which are separable in their image contrast. Resolvancy is solely dependent on the instrumentation, and modern high resolution electron microscopes have indeed reached atomic distances in resolving limits. Resolution, however, is an object-related distance in Å which is as low as about 20 Å (2nm) at present for complex protein or nucleic acid structures. This discrepancy indicates a still existing gap in the utilization of the actual electron-microscopic resolving power when biological macromolecules are studied.

To evaluate the electron-microscopic image of biological macromolecules and their complexes, it is appropriate to comment

---

* The author's experiments were supported by grants from the
 John A. Hartford Foundation, Inc., New York, New York and from
 National Institutes of Health, United States Public Health
 Service (GM 17377-02).

first on high resolution imaging by conventional transmission
electron microscopy.   The biological object itself is of poor
contrast; contrast-enhancing stains are added to the complex
specimen which often reveals more details of internal structures
after staining.   A few examples of image protein complexes,
protein polysaccharides and nucleoprotein particles will further
illustrate the present limitations in the approach to resolve
structure in highest details of biological macromolecular matter.

In the preparative and analytical procedures of electron
microscopy of biomacromolecules, we are concerned with (1) the
instrumentation capabilities, (2) characteristics of biological
specimens specifically prepared for interacting with the electron beam in a vacuum and (3) high resolution image analysis
based on imaged contrast computation and object reconstruction.
Those details are discussed here which hold promise to clarify
the limits of our present knowledge of high resolution imaging
of biological structure.

THE INSTRUMENTAL CAPABILITIES

How does the electron microscope function to form a two-
dimensional image of a small three-dimensional layered object?
Image formation can be considered from the unifying view of
physical optics (1,2,3).   Images of an electron-microscopic
specimen are formed by elastic and inelastic scattering of electrons in the object.   This scattering process occurs within a
pencil of high energy electrons (40 to 100 ke Volts) emitted
from the tungsten filament of an electron gun.   Magnetic fields
of variable high electron-diffracting power act as electron-
microscopic lenses.   In the illumination system the condenser
lenses collimate the electron beam to a small pencil at the object plane.   In the imaging system the powerful objective lens
acts in forming a diffraction pattern in the back focal plane
caused by the scattering density of the specimen.   This dif-

fracted image is enlarged to the operational magnification by a
set of combined lenses.

    There are two general ways to observe the small three-
dimensional object in the final image: by bright field and by
dark field illumination.   In the most common bright field
image, the image recorder is reached predominantly by those
electrons which are undeflected in the scattering process.   The
electron beam is only limited by microscopic apertures.   Electrons within the beam, however, do not behave individually when
interacting with the object.   Since the electrons are emitted
coherently from a point-like source (pointed tungsten filament),
the illuminating beam passes the object plane as a highly coherent wave front which is wave-modulated by the object (Fig.1).

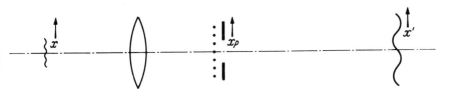

Fig.1 : Scheme of contrast transfer from the scattering object
        (arrow x) to the image (arrow x').   At the back focal
        plane (xp), a diffraction pattern of the object is
        created.   The objective lens has aberrations which in-
        flict the image in the image plane.   Magnification is
        symbolized by the image wave which actually is of the
        same wavelength (according to Hanszen (4)).

The wave modulation is due to a scattering effect of the object
atoms of various scattering densities as they are distributed
in the specimen.   The recorded image contrast at the fluorescent screen or the photographic negative is generated by the
scattering adsorption (area) contrast (due to elastic scattering), and by a variable phase contrast (due to inelastic
scattering) which arises as the spatial interference (phase
shift) of the electron waves at the final image.   When the
image is defocused, most prominent contrast changes can be ob-

served at high magnification. The biological specimens being weak amplitude objects contribute to much higher phase contrast in details, also when electron-dense stain is applied in excess (negatively-stained specimens).

The relations of the scattering density of the object to the contrast distribution in the image are regulated theoretically by contrast transfer functions (4,5). These indicate that contrast in any image detail below 20 $\overset{\circ}{A}$ is increasingly due to phase contrast and can accumulate to a parasitic image of the smallest details. In observing an optimal contrast, it is most important to underfocus the image details with a coherent electron beam.

Electron-microscopic imaging in bright field is then considered as a process of wave propagation of highly coherent electrons in which axially symmetric, magnetic fields act as diffracting media. The electron microscope has a variable amount of aberrations. The most important static aberration is the spherical aberration of the objective lens. It appears to the observer that a hypothetical object point (e.g. a heavy metal atom) can be imaged at highest magnification only as a small disk of uneven contrast. Irregular, astigmatic imaging can be compensated for by stigmators. Statistical aberrations influencing the image are the mechanical and electric instability in various parts of the microscope.

In dark field illumination the image collects preferentially those elastically scattered electrons which pass through the microscopic channel while undeflected electrons and most of the inelastically scattered electrons (travelling within the center beam) are eliminated. Hence, the source of contrast is mainly elastically scattered electrons as if they come from self-luminous object points. An appropriately small object renders elastically scattered electrons in a small proportion of the electron beam into the image. Thus, no or very little phase

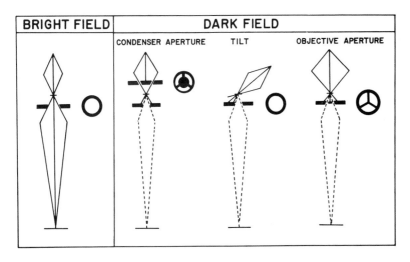

Fig.2 : Schematic instrumental lay-out for bright field and
        dark field electron microscopy.   (a) *Conventional*
        bright field.   The objective aperture (30, 40 or 50 μm
        φ) is assumed to be in or close to the back focal plane
        of the objective lens.   The final image is formed by a
        two-stage (objective-projector) or a three-stage
        (objective-intermediate lens-projector) lens system.
        (b) *Dark field condenser aperture* can be set at various
        positions of the condenser system.   The ring-shaped
        illumination favors minimum damage to the object.
        (c) The *tilt* in the gun-condenser system is produced
        mechanically or by using an electromagnetic beam devi-
        ation system.   The objective aperture is of regular
        size (50 or 30 μm diameter).   (d) *Dark field objective*
        apertures (beam stop) of various forms are positioned
        at the back focal plane or adjustable in the objective
        lens system (6).

contrast is observed in the dark field illumination.

    The various dark field modes are shown in Fig.2.   The most
efficient dark field illumination of biological macromolecules
seems to be the annular condenser ring aperture system (7) which
provides center-symmetric illumination of the object;   the magnetic tilt with unilateral illumination may be of equal performance.   A properly dimensioned beam stop at the position of the
objective aperture (dark field objective aperture) is under a

heavy electron load of the high energy electrons from which
charging by contamination can arise.   This leads to image deterioration.   For all these dark field modes, longer exposure
times than in bright field (up to 10 sec) have to be adapted
for micrography.

Derived from the theory of contrast transfer functions (5),
a few image peculiarities are discussed here which are detectable in bright field at high resolution.   Granularity caused
by the phase contrast in the image of a carbon film and studied
by defocusing (8) can be described better by optical (Fourier)
transforms from an electron micrograph than by having averaged
the contrast variation itself.   The optical Fourier transformation shows a diffraction pattern of the granularity forming a
diffuse ring system.   By stepwise defocusing, the size of rings
and the ring contrast changes.   The amount of defocusing
(Δz in $\overset{\circ}{A}$) is measured from the size and diameter of the dif-
fracted rings.   The method of optical Fourier transformation of
electron micrographs can be used not only for diffused objects
(background) such as carbon supporting films but also for harmonic or repeat structures of a complex biological specimen.
To achieve the analysis of optical transforms, electron micrographs of the biological specimen at high magnification are
selected for "best" optical transforms in which most of the
details under study are resolved.   A diffractometer of proper
design (9) serves best - such as a folded diffractometer (10).

CHARACTERISTICS OF BIOLOGICAL SPECIMENS

Preparative techniques (11) for high resolution electron
micrography of biological macromolecules include extremely thin
supporting films, usually made of carbon or of a carbon-backed
plastic layer (12).   The thinner the carbon supporting film the
less background "noise" (phase contrast) should be expected.   A
limit seems to be reached at about 20 $\overset{\circ}{A}$.   The macromolecular

specimen when mounted should extend minimally in the axial direction of the electron beam.   Either the macromolecules are
dispersed randomly over the supporting film, or macromolecular
aggregates are layered and spread in ordered fashion.   This
prevents the formation of superimposed molecular layers that
eventually give rise to multiple scattering.

Beam damage in high vacuum inflicts regularly the specimen
in respect to loss of material (13) or to molecular translo-
cation of specimen material (14), and rarely to a gross trans-
location (collapse) of a complex structure.   It is difficult
to assess quantitatively the beam damage of a spread biological
specimen since the first few seconds of exposure to the electron beam seem to be crucial for structural change.   The beam
damage is minimized by stabilization of object details during
exposure, such as embedding in electron-dense stains (15) and
short exposure of a virginal region of the object to the bright
field illumination (16).

Many of the newer techniques (11) aim at dispersing the
specimen to the highest degree or utilize droplets from a highly
diluted solution which are then dried onto the supporting film.
Protein added to the test solutions helps to spread large biological macromolecules such as nucleic acid filaments.   These
extremely long filaments, often of many million molecular
weight, are best prepared by protein monolayer techniques (17).
Of course, such preparations finally contain all non-volatile
material mounted dry onto the grid, such as a nucleic acid-
protein layer of unknown structure.   Consequently, staining and
rinsing only remove unwanted materials, such as salts, when they
are soluble after mounting.

HIGH RESOLUTION IMAGE ANALYSIS

Scattering density of a given object causes the image contrast in the electron micrograph of biological macromolecules by

additive amplitude and phase contrast.   Usually, the biological
specimens are positively stained (with minimum of stain deposited
on the macromolecular outline only) or negatively stained (with
stain in excess surrounding the object).   The transmission electron microscope, however, possesses a depth of focus which is
largely compared to the dimensions of the object.   The image is
considered to be a two-dimensional linear contrast transmission
of the total three-dimensional biological specimen and supporting film.   This image includes all artificial (positive and
negative) contrast added by defocusing and aberrations.

Methods of optical Fourier transformation were developed
for analysis of electron micrographs (18,19) using optical dif-
fractometers with monochromatic laser light.   An optical diffraction pattern from a micrograph of a bacteriophage tail and
head is given in Fig.3.   It is mainly the high regularity and
the minimum of background diffraction in the masked micrograph
which provides a suitable diffraction pattern of detailed information.   The reciprocal space in which the diffractogram is recorded provides spatial frequency patterns of repeat structures
which can easily be transferred to lattice spacings in the two-
dimensional image (18).   From densitometric tracing of an electron micrograph, diffraction patterns are also open to computed
lattice spacings (20).   These diffraction patterns are selectively subject to filtering.   Filtering is a process in which
a conjugate proportion of the diffractogram is masked-off allowing those spots to pass through forming a filtered reconstructed
image in real space by Fourier inversion (21).   Numerous electron micrographs of the same object are searched so as to select
proper spatial frequency patterns in reciprocal space.   Helical
structures, as they are presented for example in phage tails,
can be processed from a single image (20), and the same can be
done for contracted tails (22), while spherical virus particles
have to be imaged many times at different specimen tilts (23).

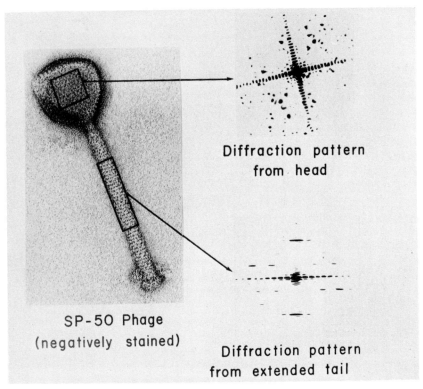

**Diffraction pattern
from head**

**SP-50 Phage
(negatively stained)**

**Diffraction pattern
from extended tail**

Fig.3 : SP50 bacteriophage (magnification x 141,000) and corresponding optical diffractograms of the head and tail.
The rectangles indicate the masked area used in the
diffractometer.

Those electron micrographs provide the various views necessary
to implement the model building process of a three-dimensional
object of sufficient clarity.

EXAMPLES

Bacteriophage structure

An example of the reconstruction of the phage head of DNA
bacteriophage CbK (host, *Caulobacter crescentus*) (24) is given
in Fig.4.   These studies aim at developing a model of this
cylindrical-bipyramidal protein complex (25).   Two types of
electron microscopic images of the phage are used: the negatively stained image actually combines images of the top layer

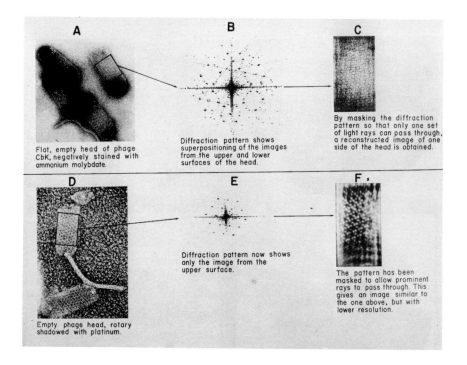

Fig.4 : Electron micrographs, diffraction patterns and filtered
        images of CbK bacteriophage (*Caulobacter crescentus*).
        Magnifications - A x 46,000; D x 33,000.

and the bottom layer of the collapsed phage head.   Frequently,
this image shows regularities forming a kind of Moiré pattern
(Fig.4d).   The diffraction patterns are indexed (Fig.4b,e),
and those spots which contribute either to a second overlaid
pattern (Fig.4b, dashed lines) or to a blurred "contaminated"
pattern (Fig.4e) are masked-off.

     To arrive at the hexagonal structure of the protein sub-
units (Fig.4c,f) of the phage head, more than 80 phage head
images were used.   The filtered images display the order, the
internal details (Fig.4c) or the prominent components (Fig.4e)
of the subunits.   The phage head surface structure (in dimen-

sions of a hexagon of about 120 to 130 Å diameter) is represented in a planar geometric surface lattice of finite size.
By these means one can determine the number of hexagonal sub-
units (hexons) as well as the number and the angle of the helical turns which make up the cylindrical body of the phage. The
pyramid at each end of the phage head consists of a limited
number of five-fold linking proteins (pentons), a few hexons and
a six-fold linking protein at the pyramid tip. Details in the
design of the surface lattice of the phage head may also depend
on findings of the protein composition (25).

Details in the phage head structure can vary. Its planar
geometry which displays all structural subunits is found to
differ only in a few micrographs (25). One searches for poly-
morphic structures which have a fitting surface lattice of
slightly different design; however, they use similar hexagonal
units. Polymorphs will probably command increasing attention
as a geometric expression of genetic mutants (26).

Decorated actin filaments

Optical transforms also can be computed from high resolution
densitometric tracings of electron micrographs. A three-
dimensional image can be reconstructed because, to a suitable
approximation, an electron micrograph represents a projection of
the scattering density distribution onto the image plane. The
general method for reconstructing a three-dimensional image (20)
depends on the assumption that a Fourier transform of a two-
dimensional projection (electron micrograph) is identical with a
congruent proportion of a three-dimensional Fourier transform of
the object. Provided sufficient data are available, the object
is reconstructed, plane by plane, using transforms of different
projected views by Fourier inversion of a resulting three-
dimensional transform. The computation techniques have been
explicitly described for general application (27).

An example of such a three-dimensional image reconstruction
may be seen in the study of filaments by Moore *et al.* (28). The
electron micrographs of characteristic arrow-head structures of
negatively-stained F-actin filaments decorated with heavy meromyosin were improved to the point that sufficiently large
stretches of the myosin-actin complexes could be imaged for analysis.   The reconstructed models gave the shape and distinctive
attachment sites of the heavy meromyosin fragment to the actin
filament at a resolution below 20 Å.   The studies permit the
assignment of a hinge-like interaction between heavy meromyosin
and F-actin filaments underlying the gliding mechanism of muscle
contraction.

Catalase image processing

In another example given by Erickson and Klug (29), electron micrographs of catalase (250,000 daltons) were traced by
densitometry and Fourier-processed in order to compensate for
underfocusing and spherical aberration.   The Fourier processing also compensated for the background noise and allowed to
"purify" the average image of the catalase molecule from the
discrete optical transforms.   While determining the changes in
the optical transform of a layered catalase crystal image,
efforts were made to process the image by relative changes of
amplitude and phase contrast to the formation of scattering
density layers.   The model spacing is first observed in the resolution range (100 to 25 Å) as a function of defocusing. The
compensation for spherical aberration is also computed, thus
providing images and models of higher resolution (to about 10 Å).

Protein polysaccharides

Many biological macromolecular specimens of harmonic structure are not ordered properly to be used for optical diffraction
from micrographs.   These are often filamentous and branched,

non-crystallizable macromolecular complexes. An example for
branched macromolecules are protein polysaccharides. After
purification, they are prepared by the protein monolayer technique (30). By adding a spreadable protein to a $MgCl_2$-extracted
protein polysaccharide from bovine nasal cartilage (ground substance), where the primarily entangled polysaccharide molecules
dispersed into the air/water interphase of a protein film. This
was possible when spread over 0.3 M ammonium acetate, pH 5. The
linear core protein (180,000 daltons) is stretched, and the covalently bonded chondroitin sulfate side chains (30,000 to
50,000 daltons) extended. By extraction and column fractiona-
tion, various sizes of aggregates were found with a constant
number and length of side chains. These latter are the only
stainable components (by their negatively-charged sulfate
groups) while the core protein (or a link protein joining to
higher aggregates) fails to be well-contrasted. Models then
can be formed with spacings of side-chains along the helically
built core protein.

Nucleic acid electron microscopy

The electron microscopy of nucleic acids (13) has seen a
number of modifications in the preparation of DNA or RNA filaments in a protein monolayer. Bright field and dark field
imaging (6) is frequently used. The procedure applicable to
double- and single-stranded DNA and RNA are threefold: the
spreading procedure, the diffusion procedure and the one-step
release procedure (17). All three use an adequate basic protein film to adsorb the nucleic acids; modifications mainly lay
in the spreading solutions and in the subsolutions, their salt,
temperature conditions and substrates added. In general, the
nucleic acid duplexes do not require an *in vitro* pretreatment
and are displayed as semi-rigid filaments measurable in contour
length and gross configuration (linears, circles, catenated

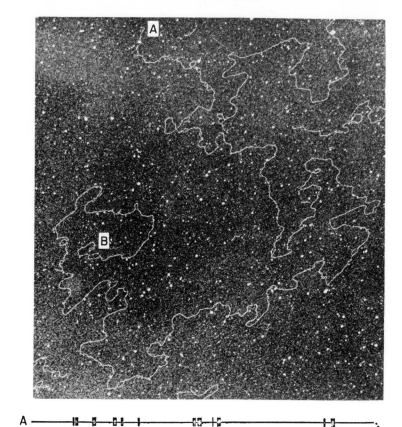

Fig.5 : Electron micrograph of T4 phage DNA (dark field) after
        partial strand separation of the linear duplex molecule
        (from A to B).   Denaturation pattern (total length
        54 μm) displayed from A to B (see (32)).   The purified
        DNA was partially denatured by 10% HCHO in alkali
        (pH 10.6; 10 minutes at room temperature).   After
        chilling, salt and cytochrome c were added and spread
        over 0.3 M ammonium acetate, pH 5.   The contrast for
        dark field micrographs is enhanced by uranyl acetate
        staining.   Magnification, x 5,500.

structures).

    New methods were developed (31) in which *in vitro* partial

strand separation of viral DNA was shown by electron microscopy

to be organized linearly as non-permuted (as in lambda phage

DNA (31) or as circularly permuted (as in T4 phage DNA (31)).
Single-stranded loops of adenine-thymine rich regions appeared
on partial DNA denaturation with formaldehyde and elevated temperatures or with alkaline formaldehyde. When ordered and
normalized to the expected length, one can obtain denaturation
maps of a number of individual molecules. This method of partial denaturation mapping is generally applicable to any DNA
collection in which a repeat pattern is assumed to exist.
While most of these studies are done with various DNA viruses
(lambda phage DNA (33), P2 phage DNA (34), adeno 2 virus (35)),
eukaryotic nucleolar DNA isolated from Triturus eggs recently
showed specific patterns of partial denaturation of regions
known to transcribe the ribosomal RNA (ribosomal DNA (36)).

Other methods of microscopic mapping of some short DNA's
are based upon the ability to select one DNA strand of deletion
mutants which is used to hybridize with the complementary wild-
type strand (37). The linear wild-type DNA is looped-out as a
single strand at unmatched regions; the rest of the DNA is completely base-paired. This method leads to physical mapping of
heteroduplex DNA of lambda phages (38) and of other related
bacteriophages (39).

REFERENCES

1. Zeitler, E.   *Adv. Electronics and Electron Physics 25*:
    277 (1968).

2. (a) Haine, M.E.   The Electron Microscope.   Interscience
    Publishers, New York (1961).
    (b) Meek, G.A.   Practical Electron Microscopy for
    Biologists.   J. Wiley and Sons, New York (1971).

3. (a) Heidenreich, H.D.   Fundamentals of Transmission Elec-
    tron Microscopy.   Interscience Publishers, New York (1964).
    (b) Huxley, H.E. and Klug, A. (Eds.).   New Developments
    in Electron Microscopy.   Royal Society, London, Part 1
    (1971).

4. Hanszen, K.J.   *Naturwissensch. 54*:   125 (1967).

458   *A.K. Kleinschmidt*

5.   Hanszen, K.J.   *Adv. Optical and Electron Microscopy 4*:   1 (1971).

6.   Kleinschmidt, A.K.   *Phil. Trans. Roy. Soc. Lond. B 261*: 143 (1971).

7.   de Harven, E., Leonard, K. and Kleinschmidt, A.K.   Twenty-ninth Annual Electron Microscopy Society of America Meeting, Boston 1971, p.426.

8.   Thon, F.   *Z. Naturforsch. 21b*:   476 (1966).

9.   Taylor, C.A. and Lipson, H.   Optical Transforms, Cornell University Press, Ithaca, 1964, p.25.

10.   Kellenberger, E. and Boy de la Tour, E.   Personal communication.

11.   (a)   Kay, D.H.   Techniques for Electron Microscopy, F.A. Davis, Co., Philadelphia, 1965.
(b)   Schimmel, G.   Elektronenmikroskopische Methodik, Springer-Verlag, Berlin, 1968.
(c)   Hayat, M.A. (Ed.).   Principles and Techniques of Electron Microscopy - Biological Application, Van Nostrand Reinhold, New York, Vol.2, 1971.

12.   Kleinschmidt, A.K.   *Ber. Bunsenges. Physik Chem. 74*:   1190 (1970).

13.   Stenn, K. and Bahr, G.F.   *J. Ultrastructure Res. 31*:   526 (1970).

14.   Thach, R.E. and Thach, S.E.   *Biophysical J. 11*:   204 (1971).

15.   Brenner, S. and Horne, R.W.   *Biochim. Biophys. Acta 34*: 103 (1959).

16.   Williams, R.C. and Fisher, H.W.   *J. Mol. Biol. 52*:   121 (1970).

17.   Kleinschmidt, A.K.   Methods in Enzymology.   L. Grossman and K. Moldave (Eds.).   Vol.12B, Academic Press, New York, 1968, p.361.

18.   Klug, A. and DeRosier, D.J.   *Nature 212*:   29 (1966).

19.   Markham, R.   *Methods in Virology 4*:   503 (1968).

20.   DeRosier, D.J. and Klug, A.   *Nature 217*:   130 (1968).

21.   Marechál, A. and Croce, P.   *C.R. Acad. Sci., Paris 235*: 181 (1971).

22.   Moody, M.F.   *Phil. Trans. Roy. Soc. Lond. B 261*:   181 (1971).

23.   Crowther, R.A., Amos, L.A., Finch, J.F., DeRosier, D.J. and Klug, A.   *Nature 226*:   421 (1970).

24. (a) Agabian-Keshishian, N. and Shapiro, L.   *J. Virology* *5*: 795 (1970).
    (b) Shapiro, L., Agabian-Keshishian, N. and Bendix, I. *Science 173*: 884 (1971).

25. Leonard, K., Kleinschmidt, A.K., Shapiro, L. and Maizel, J. In preparation.

26. Yanagida, M., Boy de la Tour, E., Alff-Steinberger, C. and Kellenberger, E.   *J. Mol. Biol. 50*: 35 (1970).

27. DeRosier, D.J. and Moore, P.B.   *J. Mol. Biol. 52*: 355 (1970).

28. Moore, P.B., Huxley, H.E. and DeRosier, D.J.   *J. Mol. Biol. 50*: 279 (1970).

29. Erickson, H.P. and Klug, A.   *Phil. Trans. Roy. Soc. Lond.* *B 261*: 105 (1971).

30. Rosenberg, L., Hellmann, W. and Kleinschmidt, A.K.   *J. Biol. Chem. 245*: 4123 (1970).

31. Inman, R.B.   *J. Mol. Biol. 18*: 464 (1966); *J. Mol. Biol. 28*: 103 (1967).

32. Mazaitis, A.J. and Kleinschmidt, A.K.   (1971).   In preparation.

33. Inman, R.B. and Schnös, M.   *J. Mol. Biol. 49*: 93 (1970).

34. Inman, R.B. and Bertani, G.   *J. Mol. Biol. 44*: 533 (1969).

35. Doerfler, W. and Kleinschmidt, A.K.   *J. Mol. Biol. 50*: 579 (1970).

36. Wensink, P.C. and Brown, D.D.   *J. Mol. Biol. 60*: 235 (1971).

37. Davis, R.W. and Davidson, N.   *Proc. Natl. Acad. Sci. 60*: 243 (1968).

38. Westmoreland, B.C., Szybalski, W. and Ris, H.   *Science 163*: 1343 (1969).

39. Fiandt, M., Hradecna, Z., Lozeron, H.A. and Szybalski, W. The Bacteriophage Lambda, Cold Spring Harbor Press, New York, 1971.

THE LABORATORY DIGITAL COMPUTER IN BIOMEDICAL RESEARCH APPLICATIONS*

T.T. SANDEL

*Department of Psychology, Washington University,
St. Louis, Missouri, U.S.A.*

Every experimenter has seen on occasion his fallibility displayed in the form of curves which terminate for lack of data just as they become interesting.   He may have been enlightened to find that his results were spurious because of erroneous parametric values of experimental variables.   A list of possible errors of this variety can go on endlessly.   These errors lead to great frustration on the part of the experimenter. However, the emotional situation induced does not represent the real domain of the disaster perpetrated.   The real loss is of irretrievable time.   In a sense, the economy of science is the economy of time.   Time may be thought of as the coin of intellectual or logical transactions.   Because scientific efficiency is measured in the amount of new information attained per unit of time, time should be conserved in optimal experimentation. It has long been recognized that man is a severely limited scientific instrument, particularly with respect to the speed with which he can make decisions and the precision with which he

---

* Partially supported under NIH grant GM 1900.

can control devices in an error-free manner.   As recently as
twenty years ago, we accepted the fact that human constraints of
this variety were an undesirable but necessary ingredient of all
experimentation.

In the recent past, the advent of programmable, logical
devices has radically altered this situation.   In particular,
a properly chosen digital computer has great potential for re-
leasing the experimenter from the tyranny of his own primarily
temporal limitations.

CONCEPTUALIZATION OF THE PROBLEM

It is difficult to conceive of a machine at any foreseeable
time in the future effectively conserving time in this effort.
Indeed, one might question whether there would be any desirability for a mechanical approach to this problem.   This is the
domain in which the experimenter expresses his originality.   The
operations pertaining to such efforts elude the specificity
necessary to program them.

DESIGNING THE EXPERIMENT

There are all degrees of complexity in the design of experiments.   At one extreme in the paraphysical context one sees the
Galilean model of unidimensional variation of a single parameter
while holding all others constant.   At the other extreme, one
finds paradigms of a statistical variety in which the examination
may be of the sources of contingent response variations based on
a plethora of interacting experimental variables.   Strangely
enough, in this undertaking the digital computer properly programmed can play a decisive role.   Here the only burning question is whether the number of experiments to be performed within
the framework of a given general paradigm justifies the relatively large expenditure of time necessary to write the programs
which will make automated experimental design a reality.   It

should be pointed out that the problem of programming and its
necessarily large expenditure of time should be assessed with
as great accuracy as possible by the experimenter in terms of
the long-term gain in time occasioned.   It is conceivable, and
indeed has been shown on numerous occasions, that the programming effort in time has not been justified on the basis of the
final results achieved.   Here an accurate understanding of the
limits of the machine at hand is essential.   A case in point
is the availability of either compilers or conversion routines
that simplify the programming task at hand.   Unfortunately,
the use of compilers is frequently precluded by the realities
of the experiment.   A compiler such as Fortran, while excellent for arithmetic operations, seldom lends itself to time
orders of data acquisition commonly encountered in, say, neuro-
physiology.   This caveat concerning programming applies in all
of the discussion which follows.   It is an easily overlooked
factor, but one which deserves the most searching examination
when a scientist makes the decision to automate his investigations.

EXPERIMENTAL PROCEDURES AND DATA ACQUISITION

This part of the experimental epoch constitutes the heart
of the experiment.   It is also that part of the epoch which
most dramatically justifies the use of a programmable digital
computer.   It should be added parenthetically that it is at
this point in the experiment that the experimenter can, although
not necessarily, intervene in the most disastrous way.   Under
ordinary circumstances, this would be the time when the experimenter would be setting parameters, delivering stimuli, and
measuring and logging data from the experiment.   Every step of
these procedures is a possible source of experimenter, as well
as experimental, error.   Knobs may be incorrectly set;   gauges
may be incorrectly read.   The degree of precision of either

setting or reading is limited by the mechanical or perceptual
skills of the experimenter.   In this situation, a digital computer with appropriate addressable, peripheral devices is an
exceptionally reliable, accurate and fast substitute for the
experimenter with essentially none of the experimenter's fallibility and typically orders of magnitude more accurate in
measurement.   In the temporal domain, where a human decision
under the best circumstances will require 150-500 msec, a computer is capable of making a comparable decision in as little
as 1 μsec or, in the worst case, perhaps 1 msec.   An example
of an experimental situation mediated by a computer can be enlightening.   Assume that the experiment at hand is intended to
ascertain the variability of threshold of neural membrane de-
polarization as a function of the intensity of a depolarizing
potential and of the temporal separation of the depolarizing
potentials.   Assume further that it is desirable to get the
experimenter entirely out of any manipulative role in the experiment.   With these assumptions, it is obvious that not only
will the machine have to log the output of the sensors of the
experiment but will also have to be in control of the spatial
positioning of the sensor as well.   This operation is perhaps
somewhat more subtle than first appears.   If the sensor is a
microelectrode, it will have to be positioned by a device which
is under the control of the machine.   Typically, this would be
a stepping microdrive.   As the electrode is advanced the
machine, via a sampling input, would be continually observing
the potential on the electrode to ascertain whether there had
been a shift in potential commensurate with a membrane penetration.   Should such a potential shift occur, the program of the
machine would dictate that it would cease to advance the electrode and that the mode of experimental variation would be in-
itiated.   This would consist of the delivery of stimuli in
either a preprogrammed or contingent sequence depending upon the

nature of the experiment.   In this sense, "contingent" refers
to the fact that it may be desirable in some situations to modify the stimulus on the basis of the results of prior stimulation within the immediate experimental sequence.   In
threshold measures, this is a particularly powerful procedure.
While the stimulus parameters are being varied, the potentials
from the recording electrode are continuously monitored and
stored for later reductive manipulation and display.   Data acquisition once again can far exceed human capabilities provided
the peripheral devices of the machine are optimized to the problem at hand.   Sampling rates of 100,000 per second are routine
with a magnitude precision of 1 part in 4,096 or .025%.   Here
a comparison with man is enlightening.   Viewing some continuous
time order display of data such as a kymograph record, an experimenter would be hard pressed to log the amplitude of 1 data
point per second and the accuracy with which he could do it
would seldom, if ever, reach 5%.   If we return for a moment to
the question of contingent variation, it is safe to say that,
given normal non-automated modes of data acquisition, it would
be impossible for the experimenter to have the information
necessary to produce a contingent variation within the time
scale of the experiment.   It is likely that he would be aware
of having used the wrong parametric values in his experiment
only after the experiment was terminated and the data had been
examined in detail.   A further example may serve to make this
point clear.   Assume that one wished to block the effects of
inhibitory post-synaptic potentials within a nerve cell which
receives not only inhibitory inputs but also excitatory inputs.
In this case, the experiment would require the recognition of
whether a post-synaptic event was excitatory on the one hand
and inhibitory on the other.   This recognition would have to
occur *before* an action potential could be propagated in order
that the appropriate blocking potential could be induced against

the membrane.   The potential would have to be sampled, categorized, and a decision made as to whether a hyper-polarizing or
de-polarizing potential should be applied against the membrane
in order to block the action potential.   The time scale of this
sequence of events would be from 0.5 to 5 msec, depending upon
the preparation.

DATA REDUCTION AND DISPLAY

Under ordinary circumstances, there is a large hiatus in
time between the termination of an experiment and the final
statement or display of results.   The data are reduced with
varying degrees of labor and painstakingly plotted in various
ways.   In the optimal application of a computer in experimental
usage, not only can the data be reduced and displayed immediately at the termination of the experiment, but frequently it is
possible to reduce and display as the experiment progresses,
providing the experimenter with feedback as to whether what he
is presently doing is appropriate to the question he has asked.
Furthermore, the display can be dynamic in character such that
variations appear as a function of time as opposed to the
simple, static, two-dimensional display of the standard graph.
Reduction and display procedures are the second great time-
saving potentiality of the computer.   Here another caveat is
worth mentioning, and it is a point which has a direct bearing
on the choice of machine on the part of the experimenter.   Data
reduction can be as simple as no reduction at all on the one
hand, or it can consist of the generation of a matrix of eigen-
functions on the other.   In the latter case, each point on the
display may represent literally millions of calculations; it
may be so temporally inefficient to do these calculations within
the constraints of a modest laboratory computer that it may be
preferable to put the data in the form of an adequate input of
some more extensive machine capable of large, fast, arithmetic

operations. This brings up an interesting point seldom discussed with respect to the use of laboratory computers in general. Almost invariably the translation of data into a form adequate for use with a different machine is a time-consuming and costly effort, costly not only in terms of time but also in terms of money. Typically, a laboratory computer is perfectly capable of doing arithmetic operations, albeit relatively slowly, and with no additional cost to the experimenter other than in time. This extra time, however, is frequently balanced by avoiding the necessity of the conversion of data to some other form. Such a conversion might consist of transferring data on magnetic tape in one format to punched cards in another format. Here the experimenter might do well to see how much time a specific data reduction would take outside the context of the experiment. On occasion, he would be agreeably surprised to find that, although some computations are taking hours, he is still saving time in the long run by doing it himself.

In all the foregoing, much reference has been made to peripheral devices. It should be noted that in experimental use the very best computer is only as useful as the access which is possible to the computer. As a consequence, the major responsibility of the experimenter in selecting a computer for his application is to be certain that the devices necessary for his research are either available or readily fabricated and integrated into the structure of the machine he chooses. Unfortunately, the various machines available do not enjoy the same general accessibility. Indeed, one may make the generalization that only highly accessible machines should be considered for scientific applications in the laboratory. Among the capabilities should be the following:

1. Analog-to-digital conversion at high rates.
2. Digital-to-analog conversion at high rates.
3. High speed program loading.

4.   Priority interrupt.

5.   Operational command input and output.

6.   High speed precision display.

7.   Real-time clock.

The above listing is not exhaustive.   However, these functions should cover the needs of most instances.   Their listing
here is predicated on the fact that in most experimental situations, it is possible to transduce signals into electrical form
- a necessity in proper access to the computer.   A word of explanation is in order to justify the desirability of the devices
listed.   Frequently, although not invariably, data might arise
in analogical form.   The time-varying voltages associated with
electrical measures are an example of analog quantities.   In
this instance, depending upon the rate of change of the signal,
a sampling rate in the converter of the order of twice the
highest frequency present in the signal is desirable.   If, for
instance, one were sampling action potentials from a nerve with
a specific interest in the waveform of the potentials, it would
be desirable to have a sampling rate of from 20,000-40,000 times
per second.   There are instances in which it might be desirable
to have a higher rate and other instances in which it might be
adequate to have a lower rate.   These rates are, however, about
the average across applications.   Digital-to-analog conversion
is a desirable feature in machines primarily for driving displays and for controlling external devices which may be of use
in the experiment.   Examples of such applications are driving
a continuous plotter or controlling the output of a voltage control generator for the production of tonal stimuli.   The desirability of high-speed program loading speaks for itself.   In
contingent programming in a machine with a limited memory, it is
frequently desirable to load a new program into core storage,
and this should be done at the fastest rate possible in order to
interrupt the experimental sequence for as short a time as poss-

ible. A priority interrupt is desirable in order that operating modes of the machine can be modified as external exigencies arise which are not under the temporal control of the machine. An example might be a blood pressure sensor which would activate an alarm system should the blood pressure exceed some critical value. The machine should have an operational command structure in order to activate peripheral devices on the one hand and to modify experimental procedures and programs on the other hand on command from external devices. A high-speed display is desirable in order to see the data as it is being generated when this is feasible. The precision of the display will ordinarily be limited by the word length of the machine, but it is desirable that the display matrix be of the general order of 1024×1024 points. The real-time clock is desirable because it precludes the necessity for writing programming loops to time critical functions. It greatly facilitates the minimization of programming time.

As previously noted, nothing in this report is exhaustive. However, an attempt has been made to summarize some of the major aspects of experimentation and machine structures which will be helpful in the majority of experimental situations. The selection of the machine which any given experimenter will use should address itself to questions of the general kind raised here, although the specifics of any given application may deviate in detail from what has been discussed.

THE MEASUREMENT AND COMPARISON OF NEUROLOGICAL AND CARDIOVASCULAR VARIABLES USING A PDP-12 COMPUTER*

C.M. MALPUS

Cardiovascular Unit, Department of Physiology, University of Leeds, Leeds, England

SUMMARY

Methods are discussed by which a laboratory computer installation may be used by research workers or technicians without knowledge of computer programming methods.   Examples of the ways in which the computer is controlled by laboratory personnel are described.   A particular example is given of the analysis of an experiment in which the computer not only speeds the processes of analysis and preparation of graphs and similar plots but also allows extraction of meaningful results from multi-variable situations by elimination of some of the variables.

---

* The author is grateful to Dr. C. Kidd and Dr. P. Penna for their assistance in collecting the data on which Figs 3, 4 and 5 are based.
This work was supported by Medical Research Council, British Heart Foundation and the Wellcome Trust.

INTRODUCTION

The main speaker in this section of the symposium (Professor Sandel) has discussed the main ways in which digital computers are exploited in the laboratory.   Professor Sandel has, of course, had more years of experience in this field than most biomedical workers, and the design of the present-day laboratory computer is an evolution of the Laboratory Instrument Computer (LINC) concept developed in St. Louis, work with which he was closely involved.   The material described in this chapter has been developed on the LINC-8 computer, and its successor, the PDP-12 (Digital Equipment Corporation) which is by way of being the grandson of the classic LINC.

While work in this laboratory has been directed towards the use of the computer to improve the efficiency of present laboratory methods and to allow new techniques to be used, much consideration has also been given to the problem of how a particular method is to be used in practice.   Particular care has been taken to ensure that minimum time is wasted by skilled biomedical workers in the setting up and use of the computer, which must not dominate the laboratory to the detriment of the animal experiments in progress.

Every program which is used in this laboratory has therefore been written in such a way that either it is automatic in operation, or, where user intervention is required, instructions for use are simple and explicit;   it has also been made as difficult as possible to lose valuable data by any means available to an unskilled user.

This chapter describes how these principles are applied in practice, and gives a detailed description of one particular application to demonstrate the power of the laboratory computer.

THE ON LINE CONTROL OF THE COMPUTER

It is a general practical rule of computer use that for any specific task the complexity of the program design is proportional to the ease of use of the program in practice. Good rule-of-thumb guides to the ease of use of a particular program include length of time a new operator takes to master all the operating intricacies, how much the new operator has to know about the program's internal architecture in order to use it properly, and how much valuable data is lost by an operator before and after he masters the system.

The facilities of a LINC-tape computer such as the PDP-12 can be used to provide operating instructions at the time of use to provide fail-safe protection for data and to enable the programmer to provide simple operating controls and procedure. All of these features cost time and money both in the writing and debugging of more complex programs, and in "run-time-overheads", i.e., productive time spent by the computer in processing the added facilities.

All programs developed along these lines are inevitably more complex than the easily developed, jury-rigged simple programs which could do the same job in conjunction with a highly-skilled and experienced full-time operator. The benefits of program complexity, however, are obvious, in reducing additional operation complexity at the time of the experiment.

Most of the instructions given to the operator by the computer are presented in the form of alpha-numeric oscilloscope displays. These have many clear advantages over teleprinter messages, including speed, silence and (for an experienced operator) the ability to ignore them completely. The operator enters his commands or replies to questions via the teleprinter keyboard.

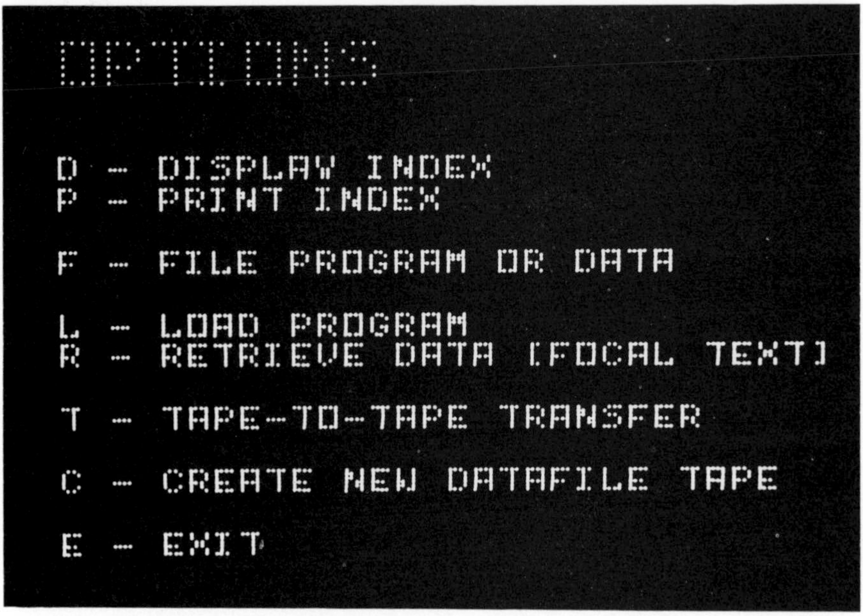

Fig.1 : The basic oscilloscope display of the PDP-12 magnetic
        tape library facility.

A general data-handling system has been developed for general use in all present or envisaged applications of the computer.   A small addition to any program gives it access to a
library facility, which operates from magnetic tape.   Once
called by a program, a display appears on the computer oscilloscope screen (Fig.1) which tells the operator what options are
open to him.   Any of the specified eight options can be called
by typing the corresponding single teletype key;   all other
keys are ignored.

If the first option is selected (type D to display the index) the library index of the contents of a magnetic tape
library are displayed as in Fig.2a.   This shows the title of
the tape (7th July, 1971) and four of the sets of data stored
on that tape, and can be used for easy, quick reference of the
tape contents to find out what data is on it and what program
collected that data.   The results of experiments are stored in

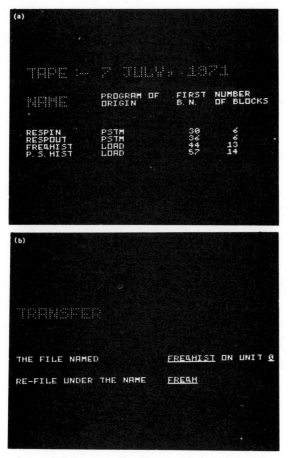

Fig.2 : Examples of library displays.   (a) part of a tape index
        (b) an interactive display sequence.

the library by name for easy and accurate recognition.   This
system can be used both to file data and to retrieve it from
the file, as well as to store and load programs.   When data or
programs must be identified by name by the operator, the name
is displayed as it is typed: for example Fig.2b shows the display frame partially completed for the T (tape-to-tape transfer)
option;   the information already entered at the keyboard is displayed underlined.

All possible error conditions are detected by this system,
For example, displays signal illegal, non-existent or duplicate
names; filed data cannot be accidentally over-written.   In
addition all details of magnetic tape handling, file look-up and
transfer of data between tape and computer memory are taken care
of without the operator intervening or knowing what is happening
in detail inside the computer.   The use of this library can be
taught to researchers, technicians or students in half an hour.

THE ANALYSIS OF THE SPONTANEOUS DISCHARGE OF AN AORTIC ARCH BARORECEPTOR IN THE DOG

Figure 3 illustrates recordings made from an anesthetized
dog which are typical of the recordings used in analysis of
neurological events related to cardiovascular variables.   The
traces illustrated are recordings of (from the top downwards)
intra-tracheal respiratory pressure, electro-cardiogram,
arterial blood pressure and the activity of a single aortic-arch
baroreceptor fibre.

Superficial inspection reveals that the nerve fibre activity
is cyclical, and relates to the arterial pressure in some way;
to investigate the quantitative relationship between the two requires considerable analysis.

The analysis illustrated here uses a modification of the
standard post-stimulus histogram method.   The QRS complex of
the electrocardiogram triggers the computer, which counts nerve
fibre spikes occurring within successive division of time between one QRS and the next; the counting process then starts
again at the first time division and counts are accumulated for
each time division.

The resultant post-QRS histogram appears in Fig. 4a which
is a direct photograph of the complete computer oscilloscope
screen.   The abscissa is calibrated in time in seconds so the
total time-scan of the histogram is 0.4 sec.   The ordinate is

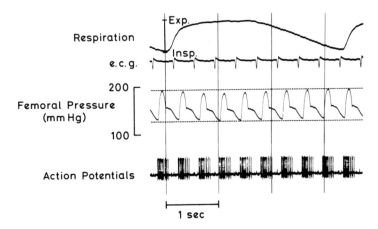

Fig.3 : A typical ultra-violet trace record of an experiment (see text for explanation).

calibrated in percent probability of an event for each particular time division (bin).

It can be seen from this plot that the highest probabilities of the occurrence of a spike potential are 25% at about 0.09 sec and 31% at 0.21 sec after the QRS complex. This can be visually compared with the simultaneous calibrated arterial pressure wave, also collected by the computer; and displayed as in Fig.4b. These records can be stored for measurement and cross-correlation by the computer with other programs using the data filing library described above.

A major problem in this type of experiment is the lack of control of the experimenter over some of the variables. A good example of this can be seen in Fig.3 in which both diastolic and systolic arterial pressures vary according to the stage of the respiratory cycle. This variation in pressure will clearly influence the baroreceptor discharge. Without the power of the computer, the only way to avoid this variation is to stop the respiratory pump for the duration of the recording run, a solution which often creates more problems than it solves.

C.M. Malpus

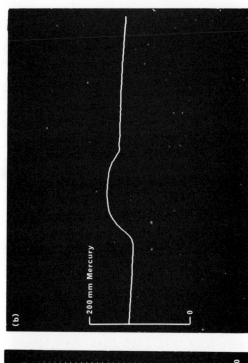

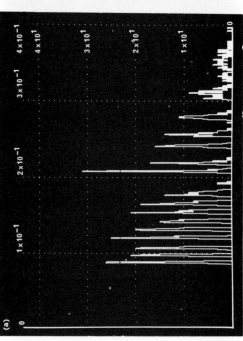

Fig.4 : The oscilloscope displays of collected post-QRS data. (a) the post-QRS histogram. Time (horizontal) in seconds, response percent probability (vertical). (b) the arterial pressure wave on the same time-scale as (a).

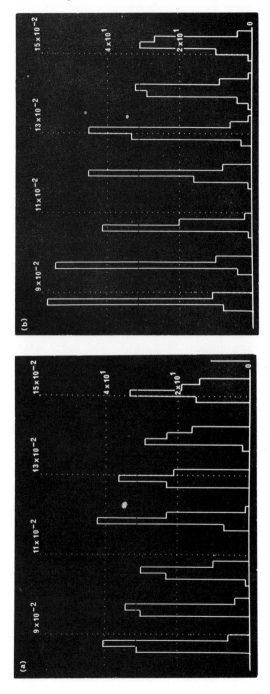

Fig.5 : Portions of two post-QRS histograms. Same calibrations as Fig.4a. (a) was collected without a selective trigger. (b) was selectively triggered at end-expiration only (see text).

The computer, however, can take account of many of these variables. In this example, after the QRS complex of the e.c. g. has triggered the computer, it first checks the respiratory pressure. Only if the ventilatory pressure is at the end-expiration level does processing proceed. In this way post-QRS sampling sweeps occur at a fixed stage in the ventilatory cycle only.

The results of this analysis are seen in Fig.5. These histograms are again direct photographs of the computer oscillo-scope face, and are portions of the same kind of histogram as that in Fig.4a with the same calibration. The expansion is carried out by the computer, under the control of the operator.

The data of Fig.5a was collected without differentiating between parts of the ventilatory cycle; that of Fig.5b consists only of cardiac cycles occurring within end-expiration. It can be seen that a much narrower distribution of responses occurs, with a higher centre probability, when the ventilatory pressure selection method is used.

CONCLUSION

The experience accumulated in this laboratory demonstrates that computers are usually first introduced to the laboratory to reduce the time and labor involved in data analysis, and to supply the results of experiments quickly enough for them to in-fluence the development of the experiment. These roles very rapidly expand until the computer carries out tasks without which experiments would not be possible.

It is also evident that careful planning and programming can make the computer available to be used both by relatively unskilled technicians and also by research workers who do not wish to, or have the time to, understand the details of how a particular problem is solved by the computer.

THE USE OF ANALOG

AND HYBRID COMPUTERS

FOR SIMULATION AND MODELLING

IN PHYSIOLOGICAL RESEARCH

V.C. RIDEOUT

*University of Wisconsin, Madison, Wisconsin, U.S.A. and
Institute of Medical Physics TNO, Utrecht, Holland*

INTRODUCTION

The hybrid computer is a combination of one or more analog computers with one or more digital computers. The analog computer (1) is sometimes called an electronic differential analyzer because its most important component, the integrator, makes possible its use for the solving or simulating of systems of differential equations. It works with continuous variables, and is a parallel machine, i.e. all of its operations proceed simultaneously. Because of this last feature, and because of the bandwidth possible with today's operational amplifiers, the analog machine is fast, and because its parameter settings are easy to change and its output easy to view, it is also known for the very effective man-machine rapport which it can provide. Typically, in analog computer use, the operator is also the person whose problem is being solved, or whose system is being simulated, and he can interact with the machine and problem set-up as the solution proceeds.

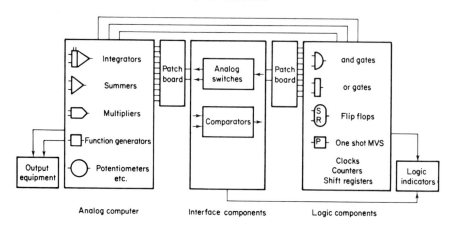

Fig.1 : Block diagram showing the main components of an iterative
       analog computer, which includes, in addition to the ordi-
       nary analog components, a set of patchable logic com-
       ponents which can be linked to the analog components
       through logic-operated analog switches and analog com-
       parators with logic output.   The entire machine can pro-
       vide iterative or equation systems as required in opti-
       mization studies.

The digital computer works with discrete variables, usually
in binary form, and is a serial machine which performs its arithmetic, logic and storage operations sequentially, but with great
rapidity.   It is capable of being programmed to perform a wide
variety of tasks, from difficult mathematical operations to
language translation to business data processing.   It may be
programmed to compete with the analog computer in the simulation
of differential equations, where its lack of speed (because of
its serial type of operation) and poorer man-machine communication are somewhat offset by its talents for logical operations
and for program and data storage in its various memory devices.
   Some attempts have been made to improve the analog machine
by adding logic components and analog memory devices.   These
were not wholly successful, but gave rise to an improved machine,
the iterative analog computer (Fig.1).   Attempts were also made

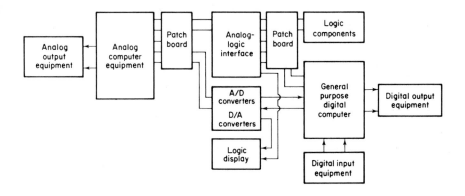

Fig.2 : Block diagram of a hybrid computer which consists of an
        iterative analog computer, a digital computer, A/D and
        D/A converters and logic communication lines, plus peri-
        pheral equipment for input and display purposes.

to design parallel digital machines, including a parallel digital differential analyzer (DDA), but these efforts have not
been very successful as yet.

     In an effort to get the best of both worlds, increasing use
has been made, during the past ten years, of hybrid computers
(2,3,4) which conjoin analog and digital computers by means of
a suitable interface (Fig.2).   Such machines combined the speed
of solution and easy man-machine communication of the analog
computer with the logic manipulation ability and memory capability of the digital computer.   The power of this combination
is not without its costs, however;   the principal one being the
difficulty of programming, which is such that general purpose
hybrid computers are in general only useful if they are rather
large machines used for the simulation of large systems.
Smaller hybrid computers, usually constructed for a single
special purpose, may, however, be used for certain data-
processing uses, as instruments to deal with data from the real,
continuous world by a combination of the analog and digital
approaches.

The systems of biology, which are exceedingly complex, demand the utmost in computing power as well as in man-machine rapport.   If we are concerned with the dynamics of biological systems, they can best be described by means of differential equations, and, for effective description, by rather large sets of such equations.   In many cases, therefore, the hybrid computer, if available, may be preferred for the simulation study of biological systems.

STRUCTURE AND SPECIFICATIONS OF COMPUTER EQUIPMENT FOR HYBRID USE

The analog computer is composed of operational amplifiers, biased-diode devices and various switches, interconnection equipment and power supplies.   A list of analog equipment classified as to function includes the following:

1. Integrators (capable of being individually placed in operate, hold or initial-condition mode).
2. Summer-inverters.
3. Potentiometers.
4. Function generators for generating powers, logs, sine functions, etc. of input signals.
5. Arbitrary function generators.
6. Interconnection patch-boards.

An analog computer to be used as part of a hybrid should have some patchable logic, including:

1. Gates and inverters.
2. Flip flops.
3. Pulsers or one-shots.
4. Counters.
5. Shift registers.
6. Interconnection patch-board.

Interface equipment is required to link the analog and logic equipment.   This must include:

1.   Logic-controlled analog switches.

2.   Analog signal comparators with logic output.

3.   Digitally-set potentiometers.

Display equipment is important and should include a multi-channel strip recorder, X-Y recorder and multi-channel oscilloscopes, of both the regular and memory types.

The above equipment, combined, gives what is termed an iterative analog computer (Fig.1), because it can be used to perform iterative as well as simple repetitive solutions.

A digital computer, for use in a hybrid computer (Fig.2), is usually a medium sized machine, but may have a back-up connection to a larger digital computer for occasional use of its larger memory and speed.   Hardware arithmetic is essential and hardware floating-point is a desirable feature.   Word-length should be at least 16 bits, so that fixed-point single-precision operations will readily match up with the precision capability of the analog machine.

Core memory should be at least 8000 words, and a disk should be available for mass-storage.   Input and output equipment may be somewhat simpler than would normally be desirable in a machine of this size, because the most demanding input and output requirements concern the information flow from and to the analog machine.

The analog-to-digital (A/D) and digital-to-analog (D/A) converters should have a word-length of at least 14 bits to match up with the precision possibilities of the analog equipment, and should be able to operate at a speed such that the interface does not become a bottleneck.   This usually requires D/A output and A/D multiplexer output of no less than 30,000 words per second.   The number of channels is also important; for many biological simulations 16 channels each way is a minimum requirement.

There should also be logic lines connecting the digital
computer to the patchable logic of the analog machine, as well
as interrupt lines to the interrupt structure of the digital
computer.

A most important interface item is the digitally-set poten-
tiometer. This device, constructed as a tree of resistors and
fast switches, may be used to change analog amplifier input
gains rapidly while a solution is in progress. They greatly
decrease the setup time of the analog program over what is required with servo-set potentiometers (by a factor of about
1000), and are essential in parameter estimation studies.

The hybrid software provided with the digital computer is
of great importance. The hybrid executive program should permit easy control of and communication with the analog computer,
and should be usable with either a modern FORTRAN package, or,
where greater speed is needed, an assembler language.

The desirable size of the analog computer has not been so
far discussed; this may best be specified in terms of the number of integrators needed (50 to 100) and the number of multipliers (30 to 50). A more or less standard mix of other components will accompany the selected number of integrators and
multipliers. Note that the minimum number of components is
related to the minimum size of the hybrid computer needed to
handle the problems which are large enough to demand such a
machine; the maximum number is related to the present knowledge
of physiological systems and the skill of today's hybrid users,
and may soon need to be increased.

The user-oriented systems design of hardware and software
of a hybrid installation is a difficult but important task.
Laboratories contemplating the acquisition of a hybrid computer
will be well-advised to give system responsibility to a single
vendor.

In any discussion of computer equipment, the importance of good maintenance should be emphasized. A hybrid computer installation should have at least one, but preferably two, full-time expert maintenance men to provide first-class preventative maintenance as well as to repair equipment. Also, on-line use by the man with the problem (open-shop operation) is essential, with consequent requirements on laboratory supervision as well as good maintenance.

It should be noted that all-digital solution of differential equations, with programming made easy by the use of simulation languages (5,6) is possible, and that such simulation languages have been written to simulate hybrid computers (7). Recent efforts to design convenient on-line operator features (8) help to make such systems competitive with at least the smaller analog computers for solutions of systems of differential equations. For larger problems the hybrid computer remains the best choice, as evidenced by the fact that although many large real-time simulations of dynamic biological systems have been made using hybrid systems (see following section), none have been reported, to the author's knowledge, using digital simulation languages. However, a language of this kind may be a valuable addition to the hybrid computer laboratory. If it is to be used on a medium-sized computer, however, it should be a language which is itself written in assembler language for efficiency and speed.

SIMULATION AND MODELLING WITH ANALOG AND HYBRID COMPUTERS

Mathematical models of physiological systems have long been of interest, but their usefulness has been hampered by the great complexity of these systems and by the mathematical intractability of all but the simplest models of such systems. This problem has recently been eased by the development of such tools as the hybrid computer. Thus in the case of the arterial system, the single-section or single *Windkessel* simulation of Frank (9)

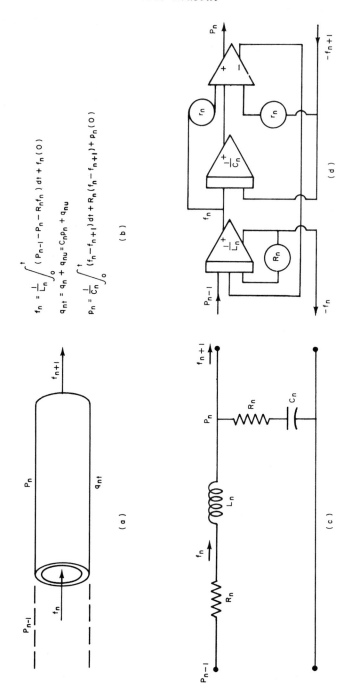

$$f_n = \frac{1}{L_n} \int_0^t (P_{n-1} - P_n - R_n f_n)\, dt + f_n(0)$$

$$q_{nt} = q_n + q_{nu} = C_n P_n + q_{nu}$$

$$P_n = \frac{1}{C_n} \int_0^t (f_n - f_{n+1})\, dt + R_n(f_n - f_{n+1}) + p_n(0)$$

(b)

(a)

(c)

(d)

Fig.3 : Basic description of pressure-flow dynamics in a segment
of a non-collapsing blood vessel, as used in computer
simulation (12). (a) Vessel segment with pressure p_n,
total volume q_{nt}, input and output flows f_n, f_{n+1},
shown as system variables. (b) Equations inter-
relating variables and segment parameters. These latter include inertance L_n, blood viscosity resistance R_n,
vessel wall compliance C_n and damping coefficient r_n.
(c) Electrical circuit representation of equations in
(b). (d) Analog computer setup for simulation of
equations in (b).

---

re-appeared in multiple-section form in the electrical circuit
analog studies of Noordergraaf (10) and De Pater (11), followed
by analog computer simulations (12) (see Fig.3) such as that of
Snyder *et al.* (13). Cardiovascular modelling has been extended
to encompass the entire circulatory system by Beneken (14),
using the analog computer.

Other simulations of the entire cardiovascular system have
been made, using a hybrid computer, by Dick (15) in a study of
major transients in the canine circulation, and by Snyder (16)
in a study emphasizing venous system dynamics (Fig.4). In
these simulations, the hybrid computer was used mainly because
of the size and complexity of the equation set, and in each case
analog simulation of pressure-flow equations was used, with
digital simulation of the control aspects of the system. The
speed advantage of hybrid computer simulation is particularly
important in model studies such as Snyder's venous system simulation (16,17,18) in which output transients were observed over
many heart-beats (Fig.5).

It should be noted that pressure-flow equations are particularly difficult to simulate by all-digital means (19), and
that the control equations, as normally expressed for cardiovascular studies, are slower and more non-linear, so that a
natural division of labor between the analog and digital computers results. This is still more true in multiple modelling

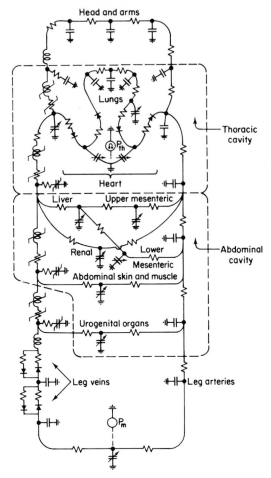

Fig.4 : Pressure-flow model (developed by Snyder (16,17)) em-
        phasizing venous system dynamics, shown in electrical
        circuit form.   This part of the model was set up on the
        analog part of a hybrid computer, while ventricle com-
        pliance waveforms, heart rate and contraction control,
        and venous tone control loops were set up in the digital
        computer.

(20,21,22) used for the simultaneous study of momentum (or
pressure-flow) transport and mass transport (Fig.6).   In the
study of dye transport in the cardiovascular system, for example,
with pulsatility effects included, the mass transport model re-

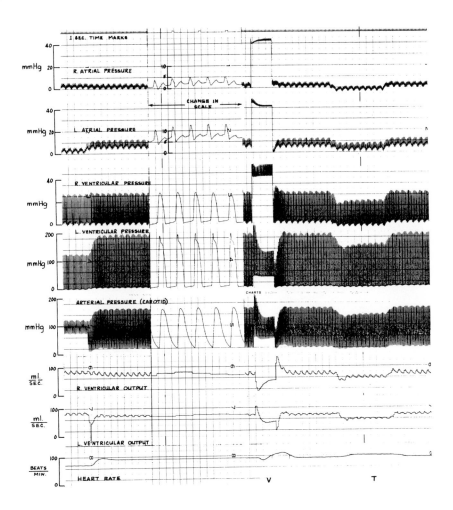

Fig.5 : Traces of several variables recorded using the model
shown, in part, in Fig.4, after aortic valve insuffic-
iency was introduced (arrow) followed by a Valsalva
maneuver (V) and a tiltable test (T). About 3.5 minutes
were needed to record the traces shown, with the computer
simulating the system in real time.

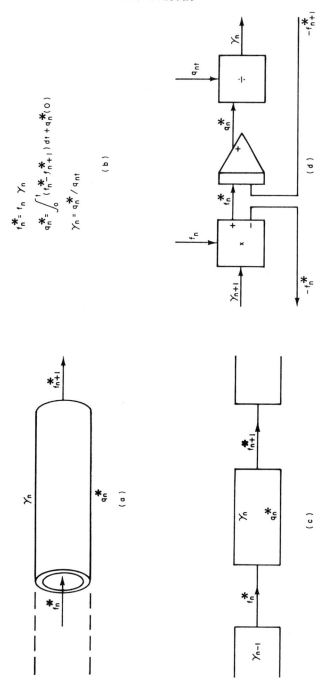

$$f_n^* = f_n \, \gamma_n$$

$$q_n^* = \int_0^t (f_n^* - f_{n+1}^*) \, dt + q_n^*(0)$$

$$\gamma_n = q_n^* / q_{nt}$$

(b)

Fig.6 : Basic description of mass transport (20) (assuming perfect mixing in compartments) in a segment of a blood
vessel, as used in computer simulation.
(a) Vessel segment with concentration γ_n, volume or
mass q_n^*, and input and output flows f_n^* and f_{n+1}^* of a
substance carried by the blood, shown as system variables. (b) Equations inter-relating the variables
described in (a) and blood flow f_n, and segmental blood
volume q_{nt} (see Fig.3), assuming no flow reversals.
(c) Compartment representation of segment.
(d) Analog computer set up for simulation of mass transport in segment. Note "driving terms" f_n, q_{nt} from
pressure-flow simulation. This simulation may be performed digitally (21), with suitable sampling f_n, q_{nt}.

quires about as many equations as the pressure-flow model which
drives it (20). These equations involve more non-linear operations, but relate to slower transients, and for these and
other reasons (21) are often best represented on the digital
part of a hybrid with the pressure-flow equations set up on the
analog part, as before (see Fig.7b). More than one kind of
substance may be transported in the bloodstream; as an example
CO_2 and O_2 are transported through the lungs and bloodstream,
and this might be modelled as a pair of mass transport circuits
driven by the pressure-flow simulations of the cardiovascular
and of the respiratory system. The same division of labor as
before might be used here (Fig.7c).

Another important field of application of the hybrid computer is in random process problems (24), particularly those in
which random waves must be generated and used as part of the
problem. In such studies many runs must be made and averaged,
often with great demands on bandwidth. Such methods, often
referred to as Monte Carlo methods, may also be used for the
study of certain boundary value problems, and hybrid methods
have been suggested for the study of such problems, arising in
studies of oxygen transport in the microcirculation (25).

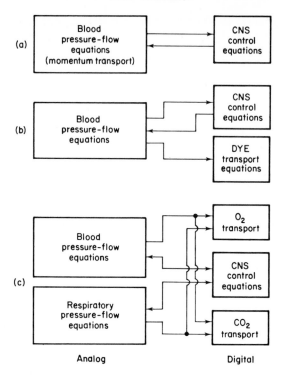

Fig.7 : (a) Pressure flow equations may be set up in the analog part of a hybrid computer, with control equations
involving the central nervous system (CNS) set up in
the digital computer.   (b) If a multiple model (20)
is to be set up to study both pressure-flow dynamics
(Fig.3) and mass transport of an indicator dye (Fig.6),
the digital computer may be used for the mass transport
of the simulation as well as for the control equations.
(c) In a more complex hybrid computer model used to
study the interaction of the cardiovascular and the respiratory system, pressure-flow equations of both systems
may be set up in the analog computer, while O_2 and CO_2
mass transport as well as CNS control is set up in the
digital computer.

As more complex modelling has become possible, the determination of parameter values has assumed greater importance.
Some parameters may be measured directly, but others cannot. In
any case, because of biological variability, it appears desirable to try to devise models in which parameters may be automati-

cally adjusted to correspond to those in an individual, by the use of parameter estimation methods. A hybrid computer method has been used by Katz *et al.* (26) for automatic determination of 13 venous system parameters from pressure and flow records; in a similar study, Sims (27) also determined 13 system parameters, in this case in the canine arterial system. In the methods used by these investigators, a model was set up on the analog computer, and the digital computer was used until the model pressure and flow curves gave, in some sense, a best agreement with corresponding recorded curves from the biological system. Estimated values for such parameters as fluid resistance, inertance and compliance for various segments of the system could then be found from the model (see Fig.8). Here the hybrid computer, with fast digitally-set potentiometers is a most useful tool for all but the simpler systems.

The examples above relate to the modelling of the cardio-vascular system, which is most familiar to the author. Many references can be cited (28-34) to show that hybrid machines

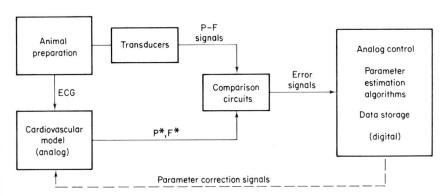

Fig.8 : The hybrid computer is well-suited to parameter estimation studies (26,27). A block diagram shows the division of labor ordinarily used in this case, for a cardiovascular pressure-flow system example. Here the model and comparison circuits are analog, while the control of the model and parameter estimation procedures are handled by the digital computer.

have been used or could be used for studies of many other systems
or sub-systems, such as the respiratory, renal, thermal control,
fluid balance system, etc.   Examples may also be found where it
was advantageous to use an analog (35,36) or hybrid computer for
relatively small but highly non-linear problems.

FUTURE PROSPECTS FOR ANALOG AND HYBRID COMPUTER MODELLING IN PHYSIOLOGICAL RESEARCH

The general-purpose analog computer is not often used for
the study of large models of biological systems.   Hybrid computers, as pointed out in the previous section, are better
suited for large dynamic models.   Digital computers, using
special languages, may be used for dynamic simulations but, because they are serial machines, may be rather slow for large
simulations.   It may be expected that large systems of equations will continue to be of interest.   New developments (37)
may be expected to change the present picture in some of the
following ways:

(1)   Special-purpose analog computers may be designed to improve
       the all-analog or hybrid computer simulation of, for
       example, the arterial system (38), by making use of the
       cheap and compact integrated-circuit operational amplifiers
       which are now available.

(2)   New types of digital computers which permit the parallel
       processing of digital integrators and summers (39) may make
       it possible to combine speed of solution with at least some
       of the other advantages of the ordinary digital computer,
       so that human-on-line digital simulation becomes feasible
       for a greater range of problems.   Combination of such
       parallel digital processors with serial digital machines
       would, in effect, give a new and useful kind of hybrid com-
       puter.

(3)   The present hybrid computer may be improved by various

means, including the use of more than the one digital computer used in most installations.   The slow and difficult part of programming involving the placing of patch-cords in the patch-boards may be eliminated in the future by the introduction of switching systems which will make automatic patching possible by means of digital programs (40, 41).

These various possibilities, which in some elementary form already exist, should make it possible for a modelling team of physiologists and system engineers to proceed with the study of models of considerable complexity and nonlinearity, unhampered by the limitations of the computer used.   They can then give their attention more completely to the important problems of system identification (or determination of model topology) and parameter estimation as the more essential steps in hypothesis verification.

REFERENCES

1.   Randall, J.E.   The Analog Computer in the Biomedical Laboratory.   Chap. IV *in* Computers and Biomedical Research. Stacy and Waxman (Eds.), Academic Press, N.Y. (1965).

2.   Siler, William.   Hybrid Computers in Bioscience.   Chap.V *in* Computers in Biomedical Research.   Stacy and Waxman (Eds.), Academic Press, N.Y. (1965).

3.   Korn, G.A. and Korn, T.M.   Electronic Analog and Hybrid Computers.   McGraw-Hill (1964).

4.   Beckey, G.A. and Karplus, W.J.   Hybrid Computation.   Wiley, New York (1968).

5.   Clancy, J.J. and Fineberg, M.S.   Digital Simulation Languages: A Critique and a Guide.   *Proc. F.J.C.C.* (1965).

6.   Brennan, R.D. and Linebarger, R.N.   A Survey of Digital Simulation: Digital Analog Simulator Programs. *Simulation 3*:   No.6, 22-36 (1964).

7.   Hurley, J.R., Janoski, R.M., Rideout, V.C., Skiles, J.J. and Vebber, W.O.   Simulation of a Hybrid Computer on a Digital Computer.   Proc. 4th International Analogue Computation Meeting, 1964.   Presses Académiques, Bruxelles.

8. Korn, G.A.   Project DARE: differential analyzer replacement by on-line digital simulation. *Proc. F.J.C.C.* (1969).

9. Frank, O.   Die Grundform des arteriellen Pulses; 1e Abhandlung: Mathematische Analyse. *Zeitschrift Biologie 37*: 483-526 (1899).

10. Noordergraaf, A.   Development of an Analog Computer for the Human Systemic Circulatory System. *In:* Circulatory Analog Computers.   A. Noordergraaf, G.N. Jager and N. Westerhof (Eds.).   North Holland Publishing Co., Amsterdam (1963), p.29.

11. De Pater, L.   An Electrical Analogue of the Human Circulatory System. Ph.D. Thesis, University of Groningen, Groningen, Netherlands (1966).

12. Rideout, V.C. and Dick, D.E.   Difference-Differential Equations for Fluid Flow in Distensible Tubes. *IEEE Trans. on Biomedical Engg BME-14*: No.3, 171-177 (1967).

13. Snyder, M.F., Rideout, V.C. and Hillestad, R.J.   Computer Modelling of the Human Systemic Arterial Tree. *J. Biomech. 4*: No.2, 341-353 (1968).

14. Beneken, J.E.W. and de Wit, B.   A Physical Approach to the Hemodynamic Properties of the Human Cardiovascular System. *In:* Physical Bases of Circulatory Transport, Regulation and Exchange.   E.B. Reeve and A.C. Guyton (Eds.). W.B. Saunders Co., Philadelphia (1967).

15. Dick, D.E.   A Hybrid Computer Study of Major Transients in the Canine Cardiovascular System.   Ph.D. Thesis, University of Wisconsin (1968).

16. Snyder, M.F. and Rideout, V.C.   Computer simulation Studies of the Venous Circulation. *IEEE Trans. on Biomedical Engg BME-16*: 325-334 (1969).

17. Snyder, M.F.   A Study of the Human Venous System Using Hybrid Computer Modelling.   Ph.D. Thesis, University of Wisconsin (1969).

18. Snyder, M.F.   Modelling of Venous System Dynamics Using Hybrid Computation.   Proc. 197 Summer Computer Simulation Conf., Denver, Colo., June 1970.

19. Rideout, V.C. and Dick, D.E.   Letter to the Editor, *Simulation 10*: No.4 (1968).

20. Beneken, J.E.W. and Rideout, V.C.   The Use of Multiple Models in Cardiovascular Systems Studies: Transport and Perturbation Methods. *IEEE Trans. on Biomedical Engg BME-15*: No.4, 281-289 (1968).

21. Rideout, V.C. and Schaefer, R.L.   Hybrid Computer Simulation of the Transport of Chemicals in the Circulation. Proc. AICA Conf. on Hybrid Computation, Munich, Aug.-Sept. 1970.   Presses Académiques, Bruxelles, pp.348-354.

22. de Waal, B.M.J.   An Analogue Model of Transport in the Blood Circulation.   Institute of Medical Physics-TNO, Utrecht (1969).

23. Beneken, J.E.W.   Some Computer Models in Cardiovascular Research. *In:* Cardiovascular Fluid Dynamics.   D.H. Bergel (Ed.) (To be published).

24. Korn, G.A.   Random-Process Simulation and Measurements. McGraw-Hill, New York (1966).

25. Halberg, M.R., Bruley D.F. and Knisely, M.H.   "Simulating Oxygen Transport in the Microcirculation by Monte Carlo Methods. *Simulation 15:* No.5, 206-212 (1970).

26. Katz, A.I., Fromm, N.C. and Moreno, A.H.   Parameter Optimization for a Model of the Cardiovascular System.   Proc. Summer Computer Simulation Conf., 1969.   Assoc. Comp. Mach., New York, pp.889-898.

27. Sims, J.B.   A Hybrid Computer Aided Study of Parameter Estimation in the Systemic Arterial System.   Ph.D. Thesis, Univ. of Wisconsin   (1970).

28. Yamamoto, W.S. and Raub, W.F.   Models of the Regulation of External Respiration in Mammals: Problems and Promises. *Computers and Biomedical Research 1:* 65-104 (1967).

29. Harmon, L.D. and Lewis, E.R.   Neural Modelling. *In:* Advances in Biomed. Engg and Med. Physics, Vol.1. Interscience Publishers, Inc. (1968).

30. Zwislocki, J.   Electrical Model of the Middle Ear. *Jl. Acoust. Soc. Am. 31:* 841 (1959).

31. Winton, J. and Linebarger, R.N.   Computer Simulation of Human Temperature Control.   *Simulation 15:* No.5, 213-221 (1970).

32. Gianunzio, J.G.   Multiple Model Study of Flow and Transport Phenomena in the Cardiovascular and Renal System. Ph.D. Thesis, Univ. of Wisconsin (1971).

33. Houk, J.C.   A Mathematical Model of the Stretch Reflex in Human Muscle Systems.   Ph.D. Thesis, M.I.T. (1963).

34. Sugita, M.   A Hybrid Computing Model of the Cellular System.   Proc. 7th Intl. Conf. on Med. and Biol. Engg, Stockholm, 1967, p.479.

35. Apter, J.T.   Computer Analysis of Ocular Tonograms.   Proc.

7th Intl. Conf. on Med. and Biol. Engg, Stockholm, 1967, p.482.

36. Donders, J.J.H. and Beneken, J.E.W.   Computer Model of Cardiac Muscle Mechanics.   *Cardiovascular Research, Supplement 1* (1971).

37. Rideout, V.C.   Future Prospects for Computer Modelling of the Cardiovascular and Other Physiological Systems. Proc. 8th I.C.M.B.E., Chicago, July 1969.

38. Grewe, L.A., Neustedter, P.F., Rideout, V.C. and Rukavina, D.M.   Cardiovascular modelling using integrated-circuit op-amps.   *Simulation 13*:   No.2, 103-106 (1969).

39. Kett, B.L.A. and Rae, W.G.   A New Development in Operational Computing Systems.   Proc. AICA Conf. on Hybrid Computation, Munich, Aug.-Sept. 1970.   Presses Académiques, Bruxelles.

40. Hannauer, G.   Automatic Patching for Analog and Hybrid Computers.   *Simulation 12*:   No.5, 219-232 (1969).

41. Howe, R.M., Moran, R.A. and Beige, F.D.   Time Sharing of Hybrid Computers Using Electronic Patching.   Applied Dynamics, Ann Arbor, Mich. (1971).

THE USE OF

HYBRID COMPUTER TECHNOLOGY

IN PHYSIOLOGICAL RESEARCH*

E.O. ATTINGER and A. ANNÉ

*Division of Biomedical Engineering, University of Virginia,
Charlottesville, Virginia, U.S.A.*

INTRODUCTION

The identification and analysis of living systems require computer facilities quite different from those in most other areas of scientific endeavours. Especially in physiology, simultaneous measurements of many variables either on a continuous or semicontinuous basis for time periods of varying lengths (which are not *a priori* predictable) are usually necessary, particularly if one attempts to establish and characterize performance criteria of physiological systems. Furthermore, the data obtained from most measurement systems are relatively soft because of measurement errors inherent in the instrumentation and because of the difficulties in coupling the latter to the living system. For most biological measurements the total error cannot be reduced to below 5% of the "normal" value despite claims to the contrary by the instrumentation industry. In addition, the signal-to-noise ratio is frequently quite low and the signal may therefore require considerable processing before

* Supported in part by NIH grants HE 11747 and GM 01919.

it becomes suitable for analysis.   On the other hand, the vast
amount of information contained in the record of a single experiment cannot yet be fully retrieved even by automated data
processing techniques.

For reasons to be discussed more fully later, the use of
hybrid computers is in most situations more advantageous for
the experimental physiologist than either one of its two components the digital or the analog computer alone.   Smaller
machines, suitable for most experimental purposes are now available in price ranges that make their use as devoted computers
not only economically feasible but advantageous.

The application of hybrid computers to physiological research can be grouped into four categories:

    a.   Data acquisition and storage;

    b.   Data analysis;

    c.   Control of experiments;

    d.   Simulation.

In this paper, we will first discuss the structure and properties of hybrid computers, and then explore in some detail
each of the four application areas and their potential in general.   Finally, these applications will be illustrated by means
of specific examples drawn from our own work that deals with a
performance analysis of the O_2 transport system.

STRUCTURE AND PROPERTIES OF HYBRID COMPUTERS

The primary assets of hybrid computers lie in the use of
both analog and digital components for the achievement of high-
speed computation and extensive memory combined with logic capabilities (1-6).   The analog computer provides high-speed simulation capability and the digital computer the arithmetic and
logic capability as well as the memory capacity.   A typical
hybrid computer, as shown in Fig.1, is composed of three major
parts:

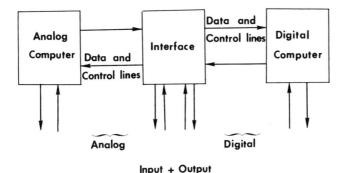

Fig.1 : Schematic diagram of a hybrid computer.

a.   A general-purpose analog computer;
b.   A general-purpose digital computer;
c.   An interface which provides for exchange of data and control information between the two computers.

If these characteristic features are properly combined, both the overall capability and the flexibility of a hybrid computer are superior to those of either an analog or a digital system many times its size, and hence considerably less costly.   In practice the choice of combinations extends over a very wide spectrum, ranging from analog computers with some limited logic capability to the simulation of analog computers on giant digital machines.   However, even in the latter case, recourse to analog components must usually be made for the solution of complex systems of equations in real time.

The three components of a hybrid computer can briefly be described as follows:

The analog computer

The basic computing elements of the analog computer are electrical networks designed to perform mathematical operations such as addition, multiplication, integration, comparison and the generation of various functions.   Depending upon the need

of the investigator these components are connected together
through a program, which results in a sequence of mathematical
operations, usually in real time.

The computer is provided with electronic mode control of
all time-dependent components (particularly the integrators).
All three modes, RESET, HOLD and OPERATE, are controllable
either by manual push-button control or by commands from the
digital computer. By means of special circuits, the integrators may be operated at high speed, cycling up to 10,000 times
a second between the three modes. When under control of digital logic, the mode selection of the integrators is carried out
by logic signals.

A special kind of hybrid computer component is the track-
store circuit which is not found in a conventional analog computer. The track-store circuit is essentially an integrator
under electronic mode control with input solely through the initial condition input terminal. When the integrator is in the
RESET mode it "tracks" the input, whereas in the OPERATE mode
it "stores".

The main advantage of the analog computer is the fact that
complicated differential equations or systems thereof can be
solved either in real or in scaled time (i.e., at higher or
lower speeds). Computation accuracy of 0.1 to 1% can be
achieved with reliable modern circuit devices. The disadvantages of the analog computer are its poor performance for differentiation (because of electrical noise) and the absence of a
memory except for limited track-store circuits.

The digital computer

A typical digital computer in a hybrid system has the following characteristics:

(i) Memory size of 4000-8000 words of 12 to 18 bits;

(ii) Ability to interface with a number of external devices;

(iii) Multilevel priority interrupt system both internal and external;*

(iv) Hardware capability permitting both arithmetic and logical operations;

(v) Secondary memories such as discs or magnetic tapes;

(vi) Input/output devices such as paper tape reader and punch, plotter or scope.

Several inexpensive "mini" digital computers are available with the characteristics needed for fast hybrid operations. Computers that fall into this category are the PDP 12, PDP 11, Raytheon 704 and CDC 1700. A most desirable property is high speed, providing the ability to perform operations such as adding two data words or shifting a word in about 2 μsec. This implies memory cycle times of the order of 1 μsec and an instruction structure (hard wired) which has few memory references for any command. More complex operations such as multiplication which are not as critical for most hybrid operations may have execution times of about 10 μsec or less.

In hybrid operations, it is convenient to have a binary machine for direct manipulation of the individual bits. The computer should have an instruction structure which makes direct use of machine language easy and straightforward, because for many hybrid operations, the programs must be written in machine language. This necessity can be made practical through the use of a simple language.

---

* An interrupt facility permits the computer program to maintain control of the computer until an infrequent event (such as reading an external device) requires its attention. Otherwise, the program would have to check periodically for the event, a time-consuming activity. In a multilevel priority interrupt scheme, each interrupt is assigned a certain priority so that the computer can attend to the most urgent need first when more than one interrupt occurs simultaneously.

The input-output characteristics of the digital computer are also of critical importance. It should be possible to perform several input or output operations at high speed, while continuing computations. This generally means that the machine should have several independent, buffered input-output data channels. The control of input and output operations should lend itself to easy and flexible connection of equipment. The methods of selecting and controlling inputs and outputs should be simple, logically straightforward and independent of critical timing problems.

In most medical and biological applications small computers with data words of 12 bits are satisfactory and allow a precision well beyond that needed for most biological work. The primary requirements for such small computers are speed, convenient access to inputs and outputs and ease of programming (including multiple precision arithmetic) rather than advanced arithmetic facilities or long words. On the other hand, recent developments in digital computer technology permit the time-sharing of digital computers for a variety of uses. Thus, it may be possible to time-share a larger digital computer with a hybrid installation, provided that the hybrid computer users are granted a "foreground" (high) priority, while all other users during hybrid computation can be handled on a "background" (low) priority. This means that the digital computer makes available *all the time needed at the right moment* to the hybrid installation; all other users must be satisfied with whatever time is left over from hybrid computation. The main advantage of the time-shared hybrid computer installation would appear to be the availability of a larger digital computer at less cost. However, given the uncertainties associated with the time requirements of physiological experiments, it is questionable whether such priority requirements can always be satisfied. On the other hand, the uses of devoted computers in individual labora-

tories can often be justified on economic grounds alone because
of a marked increase in yield from each experiment, *provided*
the basic criteria for sound experimental design and for quality
of experimental methodology are met.

The interface

The interface provides the link between the two computer
systems for purposes of control and information exchange.
Since the two computers work with physically different signals
(the analog computer uses analog voltages, the digital computer
two discrete voltage levels), the primary function of the interface is to provide a number of high speed analog-to-digital
(A-D) and digital-to-analog (D-A) channels through which intercomputer information flow takes place.   The data to be handled
will determine the speed, accuracy, and type of A-D and D-A
converters to be used.   Conversion accuracies of 8 to 12 bits
(including sign bit) are ample for most physiological problems.
Table I shows the relative conversion accuracy obtainable as a
function of word length in terms of percentile figures.

In addition to conversion accuracy, conversion time (time
required to perform the conversion) is often of great importance.   Typically, an A-D converter can handle 20,000 to 50,000
conversions per second of words ranging in length from 8 to 12
bits.   D-A converters operate at speeds of a few (about 4)
microseconds.   A-D converters usually require sample-and-hold
devices for maintaining the accuracy of the conversion process.

If the input signal is changing with respect to time, it
is very important to know at exactly what time the analog signal
has been converted to a digital output.   The uncertainty in
this time measure is termed the aperture time (sometimes also
called window time).   The size of the aperture and the time
when the aperture occurs vary depending on the conversion method
used.   For a successive approximation converter (Fig.2a), the

E.O. Attinger and A. Anné

TABLE I

CONVERSION ACCURACY FOR A-D CONVERTERS

Bits (word length)	Value (base 10)	Percentage Accuracy
1	2	50
2	4	25
3	8	12.5
4	16	6.2
5	32	3.1
6	64	1.6
7	128	0.8
8	256	0.4
9	512	0.2
10	1024	0.1
11	2048	0.05
12	4096	0.02

normal range { (8 through 12)

digital output may correspond to any one of the values of the
analog input during the conversion.   Thus, the aperture is
equal to the total conversion time.   Aperture time of a successive approximation converter can be reduced by using a
sample-and-hold circuit, thus increasing the precision with
which an analog input is reconstructed (Fig.2b).

A multiplexer is often used to carry out the conversion of
several analog signals by time-sharing a single A-D converter.
The multiplexer selects the channel whose input is to be converted.   Because of the sequential operation of the multiplexer, the consecutively generated samples correspond to different moments of time, when more than one analog signal is converted (Fig.3, left).   This time difference, called time-

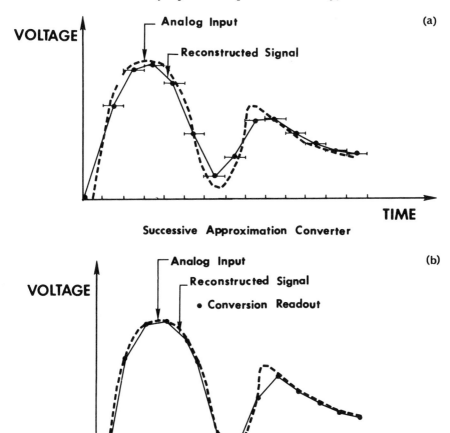

Fig.2 : A-D conversion with and without sample-and-hold circuits.
The horizontal lines in (a) indicate digital readout and
aperture, which are reduced to a single point in time by
means of a S and H circuit (b), thus improving conversion
accuracy.

skewing, could be of significance in a given situation.   The
problem can be minimized if an A-D converter is selected with a
conversion speed that is higher than the speed required for a

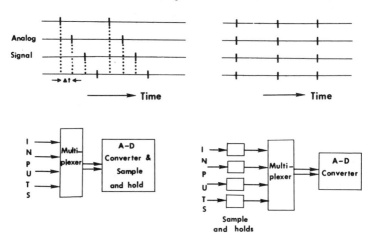

Fig.3 : Schematic diagram of time skewing (on left) and its correction by sample-and-hold circuits (on right).

single channel, or it can be completely eliminated if sample-and-hold circuits are provided for each channel (or for a selected number of channels) and synchronized. The sampling time then becomes identical for all channels although the conversion still occurs sequentially. The sample-and-hold circuit precedes the multiplexer (Fig.3, right).

The interface also provides for intercomputer control for purposes of timing, interrupt, input and output, and logic decisions during the execution of a computer program. The operation of the interface can be under control of the analog or the digital computer. Digital control is usually preferred because of greater flexibility and ease of programming. More specific interface requirements can, however, only be discussed in connection with specific computers and experimental requirements.

APPLICATIONS

Data acquisition and storage

It is clear that the reliability of data processing by computers depends primarily on the quality of the experimental
data.   The accuracy of computer processing is usually at least
an order of magnitude better than the accuracy of physiological
measurements.   It is thus crucial that primary attention and
extensive care be given to the reliability of the experimental
methodology as well as to the accuracy, calibration and stability
of the instrumentation.   It is particularly important that the
calibration procedure includes the entire system (both with respect to magnitude and to phase) since it is not uncommon that
one or more amplifiers in individual system components are subject to occasional changes in gain. The data obtained may either
be acquired and stored (on or off line) or simultaneously pre-
processed (averaging or filtering, linearization, derivation of
secondary variables, etc.).   Data channels, experimental procedures and events as well as calibration signals have to be
appropriately coded during the experiment if full use of the
potential of automated data processing is to be made.   Because
of the limited memory capacity of small hybrid systems, the
computer memory can only be used for temporary storage of small
data blocks.   Permanent storage requires either disks or tape.
Magnetic tape storage is usually considerably slower, but somewhat more versatile and cheaper than disk storage.

Particular attention must be given to the rate at which
data are sampled in order to provide an adequate but not
excessive data base for further analysis.   There can be no
justification for sampling and storing data which will never be
processed.   Through the use of digital filtering techniques
even large experimental records can be stored continuously in
terms of means and confidence limits over selected time periods

in a minimum of space.   The fundamental sampling theorem states
that an analog signal without spectral (frequency) components
beyond B Hz can be reconstructed exactly from 2 B periodic
samples per second.   In actual practice at least five times
this minimum sampling rate is generally used.   Unambiguous data
reconstruction requires, furthermore, careful removal of any
signal and/or noise components beyond B Hz by low pass filters
or averaging circuits preceding the sampler (A-D converter).
Sample inputs beyond B Hz can result in spurious low frequency
components due to aliasing.   The bandwidth of biological data
is usually less than 1 kHz.   This implies that hybrid methods
have in general sufficient speed to process the flow of data in
real time because of their capability of making the necessary
calculations and transformations at a rate which is fast compared to the rate of data flow.   When dealing with periodic
signals, careful attention must also be given to the period
over which data are sampled.   Erroneous period measurements
result in spurious frequency spectra (7).

Secondary variables such as heart period, respiratory
period, flow resistance, etc. can be measured on line and
stored on analog tape along with the primary variables such as
pressures and flows.   For example the heart period can be
determined by measuring the R-R time interval from the electrocardiogram, using analog elements for signal conditioning
(amplification, baseline shifting, filtering, etc.) and the
digital computer for measuring the time interval between two
consecutive R waves (Fig.4).

Data analysis

Data analysis may be carried out on or off line.   While
most techniques of data analysis rely primarily on digital computer processing, the hybrid computer has unique capabilities
for time-series analysis, such as auto- and cross-correlations,

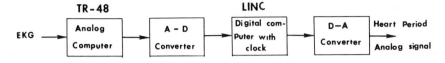

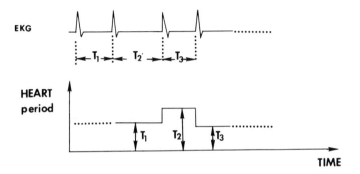

Fig.4 : Schematic representation of a method for continuous
determination of the heart period, using a hybrid computer, triggered from the R-wave of the ECG.   Note
that the analog tracings for heart period and ECG are
shifted by one cardiac cycle with respect to each other.

or for the analysis of transport processes through multicompart-
mental systems.

Auto- and cross-correlations can be performed easily by
using the analog computer for scaling, multiplication and integration, while the digital computer memory is used to provide
the necessary delays.   With variables having bandwidths of less
than 200 Hz, 20 or more points of auto-correlation function
can be computed for each pass of the analog tape depending on the
size of the digital computer memory.   The evaluation of convo-
lution integrals represents a similar example.

Flow determinations by dilution techniques are simple
examples of analyses of transport processes.   For instance,
cardiac output can be determined from dye or thermodilution curves

using the analog computer for averaging and smoothing operations
to remove most of the pulsations (1).   Simultaneously the curve
can be transformed by an analog element to a logarithmic function and transferred to the digital computer for the detection
of the true maximum and the generation of a straight line of
best fit to the descending portion of the curve.   The fitted
straight line is extrapolated to the baseline, the area computed
and converted to cardiac output by means of previously established calibration factors.

Control of experiments

Many physiological experiments, particularly those involving the analysis of biological control systems, depend heavily
on artificial inputs as driving functions.   Such inputs are
easily generated in any desired shape by either the analog or
the digital computer and can be delayed at will through the
latter.

In addition, in electronic mode control the hybrid computer
acts as an excellent high speed control device using signals
derived from biological sources.   In such a function, it may
perform complex calculations on biological variables and then
process the results for direct control of the course of the experiment, playing the part of a large and versatile feedback
system or error-correcting device.

As an example consider the experimental setup (Fig.5) used
in our laboratory to study the open- and closed-loop carotid
sinus baroreceptor response in anesthetized dogs (8).   The experiments were designed to evaluate quantitatively various
assumptions underlying a number of previously proposed linear
and nonlinear models of this prototype biological control system.   The carotid sinus was isolated bilaterally and connected
to an external servo-controlled infusion-withdrawal perfusion
system.   Carotid sinus pressure was measured through the lingual

OVERALL BLOCK DIAGRAM

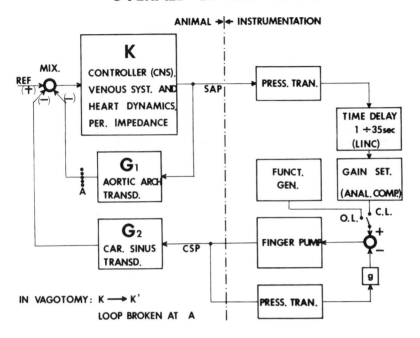

Fig.5(a) : Schematic diagram of our experimental method for the study of open-loop (OL) and closed-loop (CL) transfer function G_1 and G_2 of the baroreceptor control system. SAP = systemic arterial pressure, CSP = carotid sinus pressure, g = gain of pressure transducer. Time delays, ranging from 1 to 35 seconds are generated by the LINC, driving functions and gains by the analog computer.

artery and controlled by the difference between actual carotid sinus pressure and the desired reference signal. Open-loop input forcing functions, generated by the analog computer, consisted of impulse functions, square waves and sinusoidal variations of carotid sinus pressure. The feedback loop was closed by sampling systemic arterial pressure, storing the digital signal for variable time intervals (0-35 sec), reconstructing

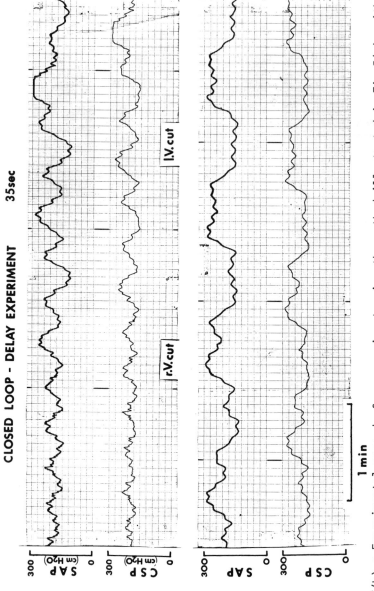

Fig.5(b) : Experimental record of an experiment using the method illustrated in Fig.5(a), with an artificial delay of 35 seconds. R.V. and L.V. stand for right and left vagus. Other symbols as in Fig.5(a).

the systemic arterial pressure signal with different gains and using it to actuate the servo-controlled perfusion system. The heart period was calculated from the R-R interval of the electrocardiogram as indicated in Fig.4. The physiological data were digitally filtered to remove high frequency components and analyzed for harmonic content on the digital component (LINC) of the hybrid computer. As a result of these experiments, we were able by means of simulation techniques to separate the total gain of this particular control system into frequency- and amplitude-dependent components and to explain the differences between open- and closed-loop performance by nonlinear superposition of these components.

The development of a perfusion system for generating time-varying concentrations of blood gases at constant flows and pressures may be cited as another example for the use of hybrid computers in the control of physiological experiments (9). In this particular study, we attempted to identify the dynamic properties of the carotid sinus chemoreceptor reflex (Fig.6). Two servo-controlled perfusion pumps, fed from blood reservoirs with different O_2 concentrations, were coupled in such a way that total blood flow to the carotid sinus remained constant while its blood oxygen concentration was defined by the relationships:

$$\left[O_2\right]_t = \frac{F_t}{\dot{Q}_t} = \frac{1}{2}\left[(O_2)_1 + (O_2)_2\right] + \frac{1}{2}\left[(O_2)_1 - (O_2)_2\right]\frac{V_{(t)}}{V_0}$$

where

F_t = Total Oxygen flow generated by both pumps

$\dot{Q}_t$ = Total blood flow generated by both pumps

$(O_2)_1$; $(O_2)_2$ = Oxygen concentration of blood in reservoirs 1 and 2

$V(t)$ = Time variation of flow generated by the function generator

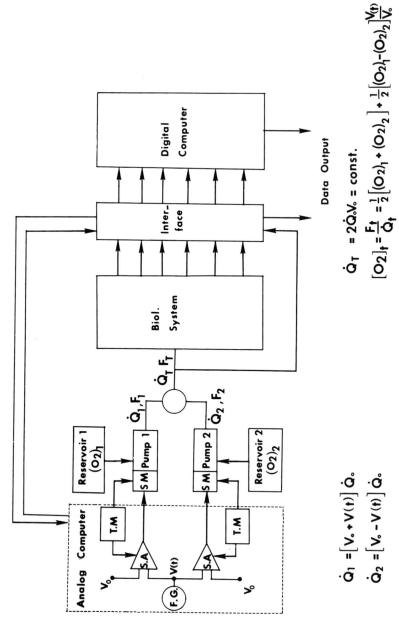

$$\dot{Q}_1 = [V_\circ + V(t)]\,\dot{Q}_\circ$$

$$\dot{Q}_2 = [V_\circ - V(t)]\,\dot{Q}_\circ$$

$$\dot{Q}_T = 2\dot{Q}_\circ V_\circ = \text{const.}$$

$$[O_2]_{lt} = \frac{F_t}{\dot{Q}_t} = \frac{1}{2}\big[(O_2)_1 + (O_2)_2\big] + \frac{1}{2}\big[(O_2)_1 - (O_2)_2\big]\frac{V_{(t)}}{V_\circ}$$

Fig.6 : Method for the generation of time-varying O_2 concentrations in blood perfusing the chemo-receptors. The analog computer generates the driving functions. S.A. = servoamplifier, T.M. = tachometer, S.M. = servomotor, F.G. = function generator. For details see text.

V_o = Mean flow upon which time variations are superimposed.

It is apparent that such a system is applicable to a wide variety of studies dealing with the analysis of control systems
that are sensitive to changes in concentration of a particular
substance.

Simulation in real time

Simulation is one of the major and most promising tolls for
the exploration of complex systems. Hybrid systems are particularly well suited for such tasks where the analog computer
serves to solve system equations in real time while the digital
computer executes the control equations through the generation
of mathematical functions.

As an example of the practical use of such a model consider
the estimation of cardiac output based on respiratory measurements and a mathematical model of the respiratory system only
(10). The method takes advantage of the fact that cardiac output is a parameter in the respiratory system (Fig.7). A mathematical model of this system consisting of the material balance
equations for CO_2 in the lungs and in peripheral tissue is simulated on the computer and forced on line with a signal proportional to the subject's total ventilation (for example a pneumo-
tachogram). The output of the model, namely the end-tidal P_{CO_2}
is continuously compared with the measured P_{CO_2} obtained from a
CO_2 analyzer. A criterion function is formed by squaring the
difference between these two signals at the end of each breath,
and the difference is minimized by adjusting the model cardiac
output using a discrete steepest-descent procedure. Animal experiments indicate that the computer-generated estimates agree
with direct measurements of cardiac output by the dye dilution
technique to within 10% with over half of the estimates agreeing
within 4%.

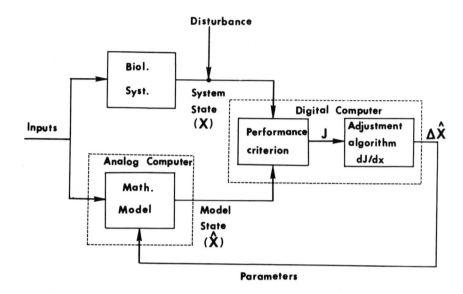

Fig.7 : Schematic diagram of a parameter-identification method
for the indirect determination of cardiac output. (For
details see text.)

There are three major problems which have to be kept in
mind in the application of such parameter-identification techniques to biological systems.   The first deals with the sensitivity of the selected state variables to the parameter in
question.   In the above example, this problem is reduced to
the question of how sensitively end-tidal CO_2 responds to a
change in cardiac output.   The second problem involves the
cross-coupling of parameters.   For example, if in the quoted
respiratory model, cardiac output depends as strongly on other
parameters, such as functional residual capacity or metabolic
rate, as on end-tidal CO_2, the validity of the model is severely
limited.   Finally, the accuracy of the estimate and the range
of conditions over which this accuracy is obtainable must be
determined.   If these problems can be satisfactorily solved,

parameter identification techniques may well present a unique opportunity for the development of noninvasive measurement techniques.

A SPECIFIC EXAMPLE: THE HYBRID COMPUTER IN BIOMEDICAL ENGINEERING AT THE UNIVERSITY OF VIRGINIA

In the following paragraphs, we re-emphasize some of the salient points brought up in previous sections, using our main research program, namely a performance analysis of the oxygen transport system (11), for which our computer facility was designed, as an example. The needs of our program include all four application categories for hybrid computers listed in the introduction to this paper: data acquisition and storage, data analysis, control of experiments, and simulation.

The system

The system has been developed over the past eight years around one of the original LINC computers (5), which already possessed some hybrid characteristics. At the beginning of this period, we were primarily interested in the dynamic pressure-flow relationships in both the circulatory and respiratory system and made therefore extensive use of frequency-domain methods in the analysis of our experimental data. It soon became apparent, however, that the original eight-bit (including sign bit) A-D conversion capability of the LINC was inadequate for our needs, particularly with respect to the estimation of the magnitude and phase relationships of the harmonic components of vascular impedance. Since we were also somewhat hampered by the limited direct memory capacity of the LINC (originally one thousand words), we acquired after two years a TR 48 analog computer as well as an interface consisting of a 64-channel multiplexed 12-bit A-D converter system with 14 sample-and-hold circuits, three high speed D-A converters, an external clock,

and a multilevel priority-interrupt system. This system satisfied most of our needs until recently, although at the cost of
a rather slow processing time. During the past few years,
however, the size of the department as well as the scope and
magnitude of its research activities grew rapidly and it became
apparent that a more powerful digital machine would be required.
A comparison of the salient characteristics of the new and old
system is given in Table II. It may be noted that because of
the developments in the computer industry, the total price of
the new system is less than that of the old system despite the
fact that its capacity is larger by at least an order of magnitude.

The research program

In our research program, we have adapted as the basic
hypothesis that in mammals the O_2 transport is optimized under
conditions of stress in terms of minimizing transport costs
while optimizing the supply to those organs which bear the
brunt of the stress load. The physical aspects of the transport system involve three types of sub-systems: pumps, systems
of conduits, and interfaces which permit the transfer between
blood and air or tissues. The plant can thus be represented
schematically as in Fig.8, and the evaluation of its performance
involves eight sets of prime variables:

1.   the concentration of inspired oxygen ($F_{I_{O_2}}$)

2.   the ventilation (l/min) and the energy requirements of the
      respiratory pump (cal/min)

3.   the cardiac output (l/min) and the energy requirements of
      the heart (cal/min)

4.   the local ventilation-perfusion ratios in different parts
      of the lungs $\left\{(\dot{V}_{A_j}/\dot{Q}_j)\right\}$

5.   the oxygen-carrying capacity of the blood

TABLE II

CHARACTERISTICS OF HYBRID COMPUTER SYSTEM
IN BIOMEDICAL ENGINEERING AT UNIVERSITY OF VIRGINIA

	Initial System	New System
Digital computer	LINC	PDP 11/20
memory size	2k, 12 bit word	8k, 16 bit word
memory cycle time	8 μsec	1.2 μsec
Analog computer	TR 48	TR 48
Interface	64 A-D channels, multiplexed with 14 S & H, 12 bits	
	3 D-A channels	10 D-A channels
	external clock	same
	1 multilevel priority inter-rupt with additional digital I/O channels	4 line multilevel dynamically adjustable priority interrupt.
Input-output device	LINC-Tapes	DEC-Tapes
	Teletype	Teletype
	Keyboard	Paper punch/reader
	Oscilloscope	same
	Analog and digital plotters	same
	No disk capability	Magnetic disk capability (256 k words)

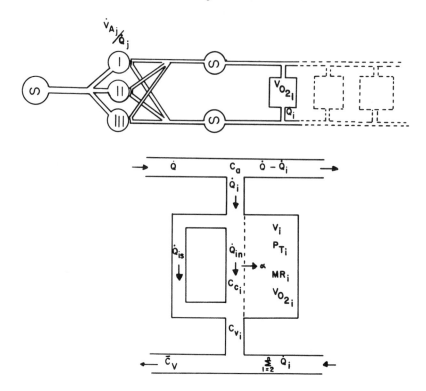

Fig.8 : Model of the O_2 transport system.   The upper diagram
        indicates the three pumps (respiratory, left and right
        heart, identified by $\sim$), three lung regions with dif-
        ferent ventilation-perfusion ratios, and a number of
        peripheral vascular beds arranged in parallel.   The
        lower diagram illustrates the model of one peripheral
        vascular bed in detail. $\dot{Q}$ = blood flow, $\dot{Q}_i$ = blood
        flow into the bed, $\dot{Q}_{is}$ and $\dot{Q}_{in}$ = shunt flow and nutri-
        tive flow respectively, $C = O_2$ concentration in blood
        (a = arterial, c = capillary, v = venous), V = tissue
        volume, P_T = tissue O_2 tension, MR = metabolic rate,
        V_{O_2} = O_2 consumption.   (For details see text.)

6.   the local blood flow through a peripheral bed ($\dot{Q}_i$) and the
     local extraction ratio for oxygen (α_i)

7.   the local metabolic rate (MR_i)

8.   the temperature

In economic terms, an increase in peripheral oxygen requirements can be met at considerably smaller metabolic costs to the transport system through a redistribution of blood flow and a change in local extraction rates as compared to an increase in ventilation or in circulation. The kybernetic aspects thus involve the organization of the system and its hierarchical control, which provide it with the capability to retain an efficient operation over a wide range of metabolic requirements. Our experimental evidence, supported by results from computer simulations, indicates that local control represents the preferred choice under conditions where O_2 supply to a given bed can be increased without seriously interfering with the function of other organs. Central control, on the other hand, is called into play if the needs of various organs have to be met under conditions of limited supply and excessive local demands.

This concept implies at least five different hierarchy levels. The lowest, and the least expensive in terms of transport costs, consists in local redistribution of blood flow. At the second level, individual beds are borrowing blood from one or more other vascular beds with which they are arranged in parallel. The third, fourth and fifth levels imply changes in ventilation-perfusion ratios, in minute ventilation, and in cardiac output. One of the essential aspects of our hypothesis is the fact that hierarchy levels may change as a function of either stress or the condition of the individual.

Data sampling, storage and analysis

Testing our hypothesis involves experiments in which a large number of variables are simultaneously measured, recorded over extended periods of time, and analyzed both in the frequency and in the time domain. Figure 9 illustrates the record of an experiment in which the time course of the change in blood flow and oxygen consumption of the hind leg and the biceps

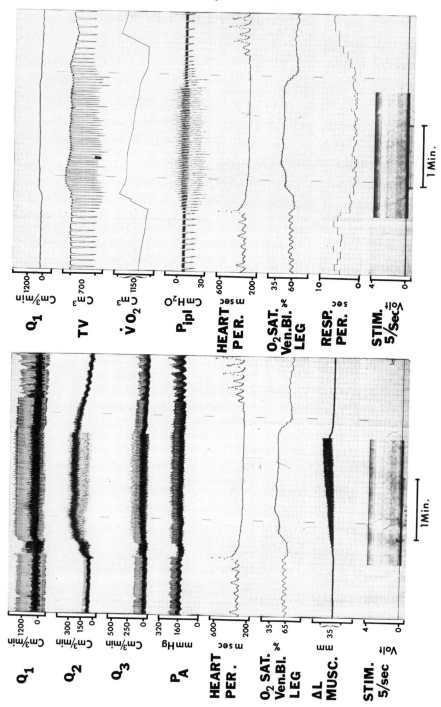

Fig.9 : Change in blood flow and O_2 consumption in the hind leg
and the biceps femoris of a greyhound during stimulation
of the sciatic nerve. The tracings represent from top
to bottom, on the left: iliac flow, upper inflow to
the biceps, lower inflow to the biceps, arterial pressure, heart period, O_2 saturation of the venous outflow
of the leg, change in muscle length, output from the
electrical stimulator; on the right: mean iliac flow,
tidal volume, total O_2 consumption, intrapleural pressure, heart period, venous O_2 saturation, respiratory
period, and stimulator output. For details see text.

---

femoris muscle of a greyhound was examined during stimulation of
the sciatic nerve (12).   On the left the tracings are, from top
to bottom, iliac flow, upper biceps arterial supply, lower biceps arterial supply, arterial pressure, heart period (or the
inverse of heart rate), O_2 saturation of the venous outflow from
the leg, change in length of the biceps muscle and the output of
the Grass Stimulator. On the right hand side the tracings represent, from top to bottom, mean flow in the iliac artery,
tidal volume, total O_2 consumption, intrapleural pressure, heart
period, venous O_2 saturation, respiratory period, and stimulator
output.   Visual inspection of the record and manual analysis
of the data reveal that upon stimulation there is an instantaneous but temporary bradycardia, and a decrease in pulsatility
of flows and pressures.   These transients are associated in
this particular experiment with the stimulus and *not* with muscular contraction and are followed by an increase in blood flow,
venous O_2 desaturation and ventilation.   The respiratory response starts with the first breath following stimulation and
increases to a maximum after about five breaths.   Of particular
interest is the fact that the respiratory modulation of heart
rate and venous oxygen saturation present during the control
state disappears completely during stimulation and reappears
shortly after stimulation has ceased.   Computer analysis of the
data increases the yield in information by an order of magnitude.

Time factors, the magnitude of changes as well as changes in
total flow can be evaluated by statistical methods at any desired confidence level.   Both frequency- and time-domain techniques can be employed for the evaluation of the biophysical
properties and the control characteristics of the system.

For data analysis in the frequency domain, we make extensive use of Fourier analysis techniques, using either cardiac
cycles, respiratory cycles or the cycle length of artificially
generated forcing functions (Figs 5 and 6) as the basic period.
Since the frequency response of our instrumentation is only
reliable up to 25 Hz, we are able to choose sampling frequencies
which are relatively low in accordance with the fundamental
sampling theorem.   For example, for a heart rate of 60 beats
per minute a sampling rate of 100 per second is more than adequate.   Higher sampling rates simply increase data processing
time without any improvement in accuracy.   As already mentioned,
particular care must be given to the accurate determination of
the period over which the data are sampled.

For analysis in the time domain, the data are usually
digitized at a sampling rate of 250 samples per second, averaged
over one or more cardiac cycles and stored on digital tape (Fig.
10).   Because of the limited memory capacity of the LINC sampling of a record of the type shown in Fig.9, one pass per four
channels is required for each three minutes of experimental procedures.   With the PDP 11, fourteen samples can be sampled
simultaneously at the same rate for experimental periods of up
to 30 minutes.   The initial analysis is based on pattern-
recognition techniques that measure delay times between stimulus
and response, the initial slope of the response, the magnitude
and timing of all the maxima and minima of the response, and the
characteristics of the steady state response.   High frequency
components of the signals are removed by analog and digital filtering and the lower limits for the recognition are set with

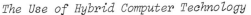

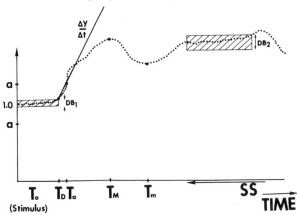

Fig.10 : Sampling techniques in the time domain.   The dotted
line represents the analog signal, having a normalized
value of 1 during the control period.   Following the
initiation of the stimulus (T_O) the delay time (T_D) is
measured and the initial slope of the response dy/dt
is determined using an appropriate choice for the
value of a.   The magnitude and timing of maxima (T_M)
and minima (T_m) as well as the characteristics of the
steady-state response (SS) are automatically established.   The rectangular boxes indicate the width of
the dead band which eliminates noise based on the
statistics of the control run analysis.   (See text.)

reference to the statistics obtained from the analysis of the
control run immediately preceding each experimental stress
period.   Further data analysis involves time-series analysis,
phase-space plots and cluster-analysis techniques.

Simulation

The interpretation of the analysis as well as the design of
new experiments is based largely on simulation and modelling
techniques.   An example of such a simulation using the model of
Fig.8, is shown in Fig.11 (13).   The physical system was represented on the analog part of the hybrid computer whereas the
digital part served as the controller.   Each peripheral vascular bed was divided into two flow channels representing nutrient
flow, where oxygen fully equilibrates between blood and tissue,

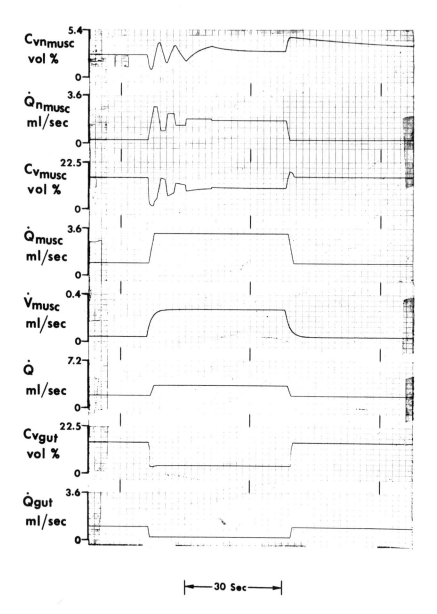

Fig.11 : Recording from a simulation experiment using the model
of Fig.8 with two parallel vascular beds (muscle and
gut).   Tracings are from top to bottom: O_2 concentration and blood flow in the nutritive outflow channels

of the muscular bed, O_2 concentration and blood flow in the total outflow channels of the muscular bed, O_2 consumption of the muscle, total flow to muscle and gut, O_2 concentration and blood flow of the outflow channels in the gut. A stepwise increase in the O_2 consumption of the muscle was used as the driving function. Note the hunting for a new steady-state during stress in the three uppermost tracings and the asymmetry in the on and off transients.

---

and shunt flow, where no gas exchange occurs. This is, of course, a very simplified model of the real system, because it represents only the two limits of a continuous spectrum. The amount of oxygen available within the tissue at any time is a function of the volume of the organ and its partial O_2 pressure; the consumption is a function of aerobic metabolic rate. Using the volumes of O_2 and CO_2 in the tissue as the state variables, the mass-balance equations can be written as follows:

Mass-balance equations for a single bed

For O_2:

$$\frac{dx_1(t)}{dt} = a_{11}(t) \cdot x_1(t) + n_1(t) \, u_1(t) - \dot{V}_1(t)$$

$$y_1(t) = h_1(t) \, x_1(t) + s_1(t) \cdot u_1(t)$$

For CO_2:

$$\frac{dx_2(t)}{dt} = a_{22}(t) \cdot x_2(t) + n_2(t) \, u_2(t) + \dot{V}_2(t)$$

$$y_2(t) = h_2(t) \, x_2(t) + s_2(t) \, u_2(t)$$

The state equations of the system in matrix form are:

$$\dot{X}(t) = A(t) \, X(t) + N(t) \, U(t) + W\dot{V}(t)$$

$$Y(t) = H(t) \, X(t) + S(t) \, U(t)$$

The first four equations describe the changes in tissue oxygen
and CO_2 as well as in venous outflow in a single bed as functions of metabolic rate ($\dot{V}$), total blood-gas supply (u), shunt
flow (s) and nutrient flow (n), the tissue volume and the dissociation characteristics of the blood.   The last two equations
formulate these relationships for the entire model in the form
of vector-matrix differential equations where

$\dot{X}(t)$ = State vector (volumes of O_2 and CO_2 in tissue)
$\dot{V}(t)$ = Metabolic rate vector
$U(t)$ = Input vector (blood-gas supply)
$Y(t)$ = Output vector (venous outflow of the gases)
$A(t)$ = Systems matrix (n × n) (determines the dynamic behav-
          ior of the system)
  W   = Consumption-production matrix (n × n)
$N(t)$ = Nutrient matrix (n × 2)
$S(t)$ = Shunt matrix (2 × 2)
$H(t)$ = Output matrix

The elements of these vectors and matrices are time-
dependent.   Coupling between either the characteristics of the
individual beds or the state variables can easily be introduced
by including appropriate functions in the systems matrix, A.
The function of the controller then lies in adjusting the vector
and matrix elements, and hence the system parameters, in terms
of optimization criteria as a function of the metabolic rate.
Because this controller acts not only on individual beds and
pumps, but also on their interactions, it must have a hierarchical organization.

The results of such a simulation for the vascular beds of
the muscle and the gut during stepwise increases in exercise are
illustrated in Fig.11.   Because the hepatic-splanchnic circulation receives approximately 25% of the resting cardiac output
and extracts only 10 to 20% of the available oxygen, this region

is ideally suited for redistribution of blood flow to other regions.   Three levels of control were included: intraorgan blood flow distribution, interorgan blood flow distribution and control of cardiac output.   The tracings represent, from top to bottom:

Venous oxygen content of the nutrient flow channel in the muscle.

Nutrient flow in the muscle.

Venous oxygen content of the total outflow from the muscle.

Total flow into the muscle.

Oxygen consumption in the muscle.

Fractional cardiac output supplying muscle and gut.

Venous oxygen content of the total flow from the gut.

Total flow into the gut.

As the oxygen consumption is increased in a stepwise fashion in the muscle, the O_2 content of the nutrient flow channel begins to fall rapidly, reaching a lower limit, at which point the blood flow distribution within the bed changes;   nutrient flow increases, thus reducing shunt flow.   If this compensation is insufficient, the O_2 content of the venous outflow decreases, eventually reaching a lower limit where muscle blood flow begins to increase at the expense of the total blood flow into the gut. At higher levels of exercise this compensation becomes insufficient, and cardiac output begins to rise when the total blood flow into the gut cannot be reduced further.   The similarity of the time course in these simulations to the time course of the variables measured in our animal experiments (Fig.9) is striking, particularly with respect to the hunting for a new steady state during the stress period.

SUMMARY

The special characteristics of hybrid computers, namely the combination of the high-speed simulation capability of analog

computers with the arithmetic and logic capability as well as
the memory capacity of digital computers are particularly suited
for use in physiological research. Their application can be
grouped into four areas:

Data acquisition and storage

     It is clear that the reliability of data processing by
computers depends primarily on the quality of the experimental
data.   The accuracy of computer processing is usually at least
an order of magnitude better than the accuracy of physiological
measurements.   It is thus crucial that primary attention and
extensive care be given to a critical evaluation of experimental
methodology and instrumentation.   Data may be acquired and
stored either on or off line as well as simultaneously prepro-
cessed (filtering, linearization, derivation of secondary variables, etc.).   Particular attention has to be given to the
rate and the periods over which data are sampled in order to
provide an adequate but not excessive data base for further analysis.   Data channels, experimental procedures and events as
well as calibration signals have to be appropriately coded
during the experiment if full use is to be made of automated
data processing.

Data analysis

     Data analysis may be carried out on or off line.   While
most techniques of data analysis rely primarily on digital computer processing, the hybrid computer has unique capabilities
for time-series analysis, such as auto- and cross-correlations,
where the analog computer provides the mechanics for integration and the digital computer the required time delays.

Control of experiments

Many physiological experiments, particularly those involving the analysis of biological control systems, depend heavily
on the generation of artificial inputs as driving functions.
Such inputs are easily provided in any desired shape by the
analog part of the computer and can be delayed at will through
digital components.

Simulation in real time

Simulation is one of the major and most promising tools for
the exploration of complex systems, and can be made most efficient by the use of hybrid systems where the analog computer
serves to solve the system equations in real time while the
digital computer executes the control equations through the
generation of mathematical functions.

Problems and potentials within these application areas have
been illustrated by means of specific examples, drawn from our
experience with the hybrid computer, which was specifically
designed for the analysis of biological systems and their simulation, in the Division of Biomedical Engineering at the University of Virginia.

REFERENCES

1. Macy, J. Jr.   Hybrid computer techniques.   *Ann. N.Y.
    Ac. Sci. 115*:   568-590 (1964).

2. Korn, G.A. and Korn, T.M.   Electronic Analog and Hybrid
    Computers.   McGraw-Hill, New York (1964).

3. Cadzow, J.A. and Martens, H.R.   Discrete-Time and Com-
    puter Control Systems.   Prentice-Hall, New York (1970).

4. Stephenson, B.W.   Analog-Digital Conversion Handbook.
    Digital Equipment Corporation, Maynard, Mass. (1964).

5. Sandel, T.T.   The Laboratory Digital Computer in Biomedical
    Research Applications.   This volume.

6. Rideout, V.C.   The Use of Analog and Hybrid Computers for Simulation and Modelling in Physiological Research.   This volume.

7. Attinger, E.O., Anné, A. and McDonald, D.A.   Use of Fourier series for the analysis of biological systems.   *Biophys.J.* *6*: 291-304 (1966).

8. Kenner, T., Stegemann, J., Allison, J., Anné, A. and Attinger, E.O.   Systems analysis of the open- and closed-loop carotid sinus baroreceptor response.   *Fed. Proc. 29*: 1506 (1970).

9. Bartschi, A., Allison, J. and Levine, J.   A perfusion system for generating time varying concentrations in blood gases.   *Fed. Proc. 29*: 951 (1970).

10. Bekey, G.A. and Maloney, J.C., Jr.   On-line estimation of cardiac output from respiratory data.   *Fed. Proc. 30*: 697 (1971).

11. Attinger, E.O.   Mechanical behavior of biological systems. *In:* Future Goals of Engineering in Biology and Medicine. J.F. Dickson and J.H.U. Brown (Eds.).   Academic Press, New York (1968).

12. Attinger, F.M., Attinger, E.O., Ono, K., Megerman, J., Mudd, C. and Rubenstein, H.J.   Control hierarchies in limb blood flow.   *Fed. Proc. 30*: 211 (1971).

13. Anné, A., Attinger, E.O. and Rubenstein, H.J.   Time domain analysis of biological systems.   *Fed. Proc. 29*: 951 (1970).

DATE DUE
